METHODS IN MOLECULAR BIOLOGY

Series Editor
John M. Walker
School of Life and Medical Sciences
University of Hertfordshire
Hatfield, Hertfordshire, UK

For further volumes:
http://www.springer.com/series/7651

For over 35 years, biological scientists have come to rely on the research protocols and methodologies in the critically acclaimed *Methods in Molecular Biology* series. The series was the first to introduce the step-by-step protocols approach that has become the standard in all biomedical protocol publishing. Each protocol is provided in readily-reproducible step-by-step fashion, opening with an introductory overview, a list of the materials and reagents needed to complete the experiment, and followed by a detailed procedure that is supported with a helpful notes section offering tips and tricks of the trade as well as troubleshooting advice. These hallmark features were introduced by series editor Dr. John Walker and constitute the key ingredient in each and every volume of the *Methods in Molecular Biology* series. Tested and trusted, comprehensive and reliable, all protocols from the series are indexed in Pub Med.

Stem Cell Mobilization

Methods and Protocols

Edited by

Gerd Klein

Department of Internal Medicine II, Center for Medical Research, University of Tübingen, Tübingen, Germany

Patrick Wuchter

Institute of Transfusion Medicine and Immunology, Medical Faculty Mannheim, Heidelberg University, Mannheim, Germany

German Red Cross Blood Service, Baden-Württemberg — Hessen, Mannheim, Germany

Editors
Gerd Klein
Department of Internal Medicine II
Center for Medical Research
University of Tübingen
Tübingen, Germany

Patrick Wuchter
Institute of Transfusion Medicine and Immunology
Medical Faculty Mannheim
Heidelberg University
Mannheim, Germany

German Red Cross Blood Service
Baden-Württemberg — Hessen
Mannheim, Germany

ISSN 1064-3745 ISSN 1940-6029 (electronic)
Methods in Molecular Biology
ISBN 978-1-4939-9576-9 ISBN 978-1-4939-9574-5 (eBook)
https://doi.org/10.1007/978-1-4939-9574-5

This Humana imprint is published by the registered company Springer Science+Business Media, LLC part of Springer Nature.
The registered company address is: 233 Spring Street, New York, NY 10013, U.S.A.

Preface

For over 50 years, patients with malignant hematological diseases have been treated using hematopoietic stem cells for transplantation as a curative therapy. Successful engraftment requires the collection of an adequate number of hematopoietic stem cells in sufficient quality. Currently, the primary source for harvesting hematopoietic stem cells is the peripheral blood, following mobilization of the cells from the bone marrow.

Although this technique is performed over ten thousand times per year in the clinic, stem cell mobilization is still an empirical method because the underlying molecular mechanisms are not yet fully elucidated. Our lack of knowledge of these exact mechanisms is also reflected by the fact that a considerable number of allogeneic donors (~5%) and autologous patients (up to 20%) mobilize inadequately low numbers of hematopoietic stem cells. Therefore, there is still a need to improve stem cell mobilization protocols in order to enhance the efficiency and quality of stem cell harvesting.

The main purpose of this book is to provide detailed state-of-the-art protocols for analyzing individual aspects of stem cell mobilization at the molecular and cellular level. As an introduction, the first chapter is a very personal review written by a pioneer in the field, Anthony D. Ho. The chapter details the first peripheral blood stem cell transplantation and the evolution of the technique since that time. The majority of the following chapters encompass new technologies which were not included in the 2012 edition of *Stem Cell Mobilization* in the *Methods of Molecular Biology* series. Specifically, this current book covers cell adhesion and migration assays in three-dimensional models, imaging techniques for the determination of blood flow or hypoxic status, innovative biomechanical studies of stem cells, and assays to determine the involvement of proteases or complement factors. One chapter also describes optimal stem cell freezing which is a clinically important issue and is a prerequisite for short- or long-term storage of hematopoietic stem cells. Since animal models can be very helpful in elucidating cellular and molecular interactions, two protocols on aging hematopoietic stem cells in mice and on human blood cells in zebra fish are also included. Finally, an important mathematical model is provided for analyzing the dynamics of human hematopoietic stem cells.

We are grateful to all our contributing authors for sharing their time, their efforts, and especially their protocols and application notes. It is our hope that this compilation of methods will be helpful to all laboratories in their endeavors to decipher the mechanisms of stem cell mobilization. Progress in this field will ultimately help all stem cell donors and patients.

Tübingen, Germany — *Gerd Klein*
Mannheim, Germany — *Patrick Wuchter*

Contents

Contributors

MARIA GABRIELE BIXEL • *Tissue Morphogenesis, Max Planck Institute for Molecular Biomedicine, Münster, Germany*
NINA CABEZAS-WALLSCHEID • *Max Planck Institute of Immunobiology and Epigenetics, Freiburg, Germany*
WALTER DE BACK • *Carl Gustav Carus Faculty of Medicine, Institute for Medical Informatics and Biometry, TU Dresden, Dresden, Saxony, Germany; Center for Information Services and High Performance Computing, TU Dresden, Dresden, Saxony, Germany*
STEFAN GISELBRECHT • *Department of Complex Tissue Regeneration, MERLN Institute for Technology-Inspired Regenerative Medicine, Maastricht University, Maastricht, The Netherlands*
ERIC GOTTWALD • *Institute of Functional Interfaces, Karlsruhe Institute of Technology, Eggenstein-Leopoldshafen, Germany*
ANNA I. GRABOWSKA • *Department of Regenerative Medicine, Center for Preclinical Studies and Technology, Warsaw Medical University, Warsaw, Poland*
JOCHEN GUCK • *Biotechnology Center, Center for Molecular and Cellular Bioengineering, Technische Universität Dresden, Dresden, Germany*
JAKUB M. HAWRYLUK • *Department of Regenerative Medicine, Center for Preclinical Studies and Technology, Warsaw Medical University, Warsaw, Poland*
REINHARD HENSCHLER • *Institute of Transfusion Medicine, University Hospital Leipzig, Leipzig, Germany*
MAIK HERBIG • *Biotechnology Center, Center for Molecular and Cellular Bioengineering, Technische Universität Dresden, Dresden, Germany*
ANTHONY D. HO • *Department of Medicine V (Hematology, Oncology, Rheumatology), Heidelberg University, Heidelberg, Germany*
JULIA HÜMMER • *Institute of Functional Interfaces, Karlsruhe Institute of Technology (KIT), Eggenstein-Leopoldshafen, Germany; Institute of Cell Biology and Biophysics, Leibniz University Hannover, Hannover, Germany*
ANGELA JACOBI • *Biotechnology Center, Center for Molecular and Cellular Bioengineering, Technische Universität Dresden, Dresden, Germany*
GERD KLEIN • *Department of Internal Medicine II, Center for Medical Research, University of Tübingen, Tübingen, Germany*
JULIAN KOC • *Analytical Chemistry-Biointerfaces, Ruhr University Bochum, Bochum, Germany*
MARTINA KONANTZ • *Department of Biomedicine, University of Basel and University Hospital Basel, Basel, Switzerland*
MARTIN KRÄTER • *Biotechnology Center, Center for Molecular and Cellular Bioengineering, Technische Universität Dresden, Dresden, Germany*
KATHARINA KRIEGSMANN • *Department of Medicine V (Hematology, Oncology, Rheumatology), Heidelberg University, Heidelberg, Germany*
ROMY KRONSTEIN-WIEDEMANN • *Institute for Transfusion Medicine, German Red Cross Blood Donation Service North-East, Medical Faculty Carl Gustav Carus, Technical University Dresden, Dresden, Germany*

SASCHA LAIER • *Stem Cell Laboratory, Institute of Clinical Transfusion Medicine and Cell Therapy Heidelberg GmbH, Heidelberg, Germany*
CORNELIA LEE-THEDIECK • *Institute of Functional Interfaces, Karlsruhe Institute of Technology (KIT), Eggenstein-Leopoldshafen, Germany; Institute of Cell Biology and Biophysics, Leibniz University Hannover, Hannover, Germany*
CLAUDIA LENGERKE • *Department of Biomedicine, University of Basel and University Hospital Basel, Basel, Switzerland*
ANDREAS MAURER • *Department of Preclinical Imaging and Radiopharmacy, Werner Siemens Imaging Center, University of Tübingen, Tübingen, Germany*
JOËLLE S. MÜLLER • *Department of Biomedicine, University of Basel and University Hospital Basel, Basel, Switzerland*
CORDULA NIES • *Institute of Functional Interfaces, Karlsruhe Institute of Technology, Eggenstein-Leopoldshafen, Germany*
CÉSAR NOMBELA-ARRIETA • *Department of Medical Oncology and Hematology, University of Zurich, Zurich, Switzerland*
PETRA PAVEL • *Stem Cell Laboratory, Institute of Clinical Transfusion Medicine and Cell Therapy Heidelberg GmbH, Heidelberg, Germany*
ANNAMARIJA RAIC • *Institute of Functional Interfaces, Karlsruhe Institute of Technology (KIT), Eggenstein-Leopoldshafen, Germany; Institute of Cell Biology and Biophysics, Leibniz University Hannover, Hannover, Germany*
RUDOLF RICHTER • *Department of Internal Medicine, Clinic of Immunology, Hannover Medical School, Hannover, Germany; Institute of Transfusion Medicine, Blood Donation Service, Bavarian Red Cross, Regensburg, Germany*
INGO ROEDER • *Carl Gustav Carus Faculty of Medicine, Institute for Medical Informatics and Biometry, TU Dresden, Dresden, Saxony, Germany; National Center for Tumor Diseases (NCT), Partner Site Dresden, TU Dresden, Dresden, Saxony, Germany*
PHILIPP ROSENDAHL • *Biotechnology Center, Center for Molecular and Cellular Bioengineering, Technische Universität Dresden, Dresden, Germany*
AXEL ROSENHAHN • *Analytical Chemistry-Biointerfaces, Ruhr University Bochum, Bochum, Germany*
RAINER SAFFRICH • *Institute for Transfusion Medicine and Immunology, Medical Faculty Mannheim, Heidelberg University; German Red Cross Blood Service Baden-Württemberg, Heidelberg, Germany; Department of Medicine V (Hematology, Oncology, Rheumatology), Heidelberg University, Mannheim, Germany*
NICOLE D. STAUDT • *Pharmaceutical Biology, University of Tübingen, Tübingen, Germany*
JONAS STEWEN • *Tissue Morphogenesis, Max Planck Institute for Molecular Biomedicine, Münster, Germany*
UTE SUESSBIER • *Department of Medical Oncology and Hematology, University of Zurich, Zurich, Switzerland*
MOTOMU TANAKA • *Physical Chemistry of Biosystems, Institute of Physical Chemistry, Heidelberg University, Heidelberg, Germany; Center for Integrative Medicine and Physics, Institute for Advanced Study, Kyoto University, Kyoto, Japan*
TORSTEN TONN • *Institute for Transfusion Medicine, German Red Cross Blood Donation Service North-East, Medical Faculty Carl Gustav Carus, Technical University Dresden, Dresden, Germany*
ROMAN TRUCKENMÜLLER • *Department of Complex Tissue Regeneration, MERLN Institute for Technology-Inspired Regenerative Medicine, Maastricht University, Maastricht, The Netherlands*

MARTA URBANSKA • *Biotechnology Center, Center for Molecular and Cellular Bioengineering, Technische Universität Dresden, Dresden, Germany*
PATRICK WUCHTER • *Institute of Transfusion Medicine and Immunology, Medical Faculty Mannheim, Heidelberg University, Mannheim, Germany; German Red Cross Blood Service Baden-Württemberg — Hessen, Mannheim, Germany*
THOMAS ZERJATKE • *Carl Gustav Carus Faculty of Medicine, Institute for Medical Informatics and Biometry, TU Dresden, Dresden, Saxony, Germany*
YU WEI ZHANG • *Max Planck Institute of Immunobiology and Epigenetics, Freiburg, Germany; International Max Planck Research School for Molecular and Cellular Biology (IMPRS-MCB), Freiburg, Germany*
SABRINA ZIPPEL • *Institute of Functional Interfaces, Karlsruhe Institute of Technology (KIT), Eggenstein-Leopoldshafen, Germany; Institute of Cell Biology and Biophysics, Leibniz University Hannover, Hannover, Germany*

Chapter 1

Evolution of Peripheral Blood Stem Cell Transplantation

Anthony D. Ho

Abstract

Blood-derived progenitors have become the predominant source of hematopoietic stem cells for clinical transplantation. The main advantages compared to the bone marrow are as follows: harvesting blood stem cells is less painful for the donor, utilizes much less resources such as operating theater time and general anesthesia, and, above all, is associated with *significantly accelerated reconstitution*. The latter has ultimately improved patient safety as a consequence of significantly shortened aplastic phase and hence reduced morbidity and mortality after transplantation. Basic and translational research efforts in the 1960s to the mid-1980s have made the first blood stem cell transplantation in Heidelberg in 1985 possible. Diverse groups around the world have contributed to incremental knowledge that culminated in the first successful attempts in blood stem cell transplantation. These efforts have spawned modern research into stem cell biology and the immune modulatory effects of allogeneic transplantations.

Key words HSC mobilization, HSC transplantation, Engraftment

1 Introduction

The concept of stem cell was first introduced by Alexander Maximow in 1909 as the common ancestor of different cellular elements of blood [1]. It took, however, almost another 60 years, i.e., in 1963, before McCullough and his coworkers provided evidence for the existence of hematopoietic stem cells (HSC) in the bone marrow [2, 3]. In a murine model, their series of experiments demonstrated that, first of all, cells from the bone marrow could reconstitute hematopoiesis and hence rescue lethally irradiated recipient animals. Secondly, by serial transplantations, they have established the self-renewal ability of these cells. When cells from the spleen colonies in the recipients were harvested and re-transplanted into other animals that received a lethal dose of irradiation, colonies of white and red blood corpuscles were again found in the secondary recipients. Based on these experiments, HSC were defined as cells with *the abilities of self-renewal as well as multilineage differentiation*. This discovery marked the beginning of modern stem cell research.

Gerd Klein and Patrick Wuchter (eds.), *Stem Cell Mobilization: Methods and Protocols*, Methods in Molecular Biology, vol. 2017, https://doi.org/10.1007/978-1-4939-9574-5_1,

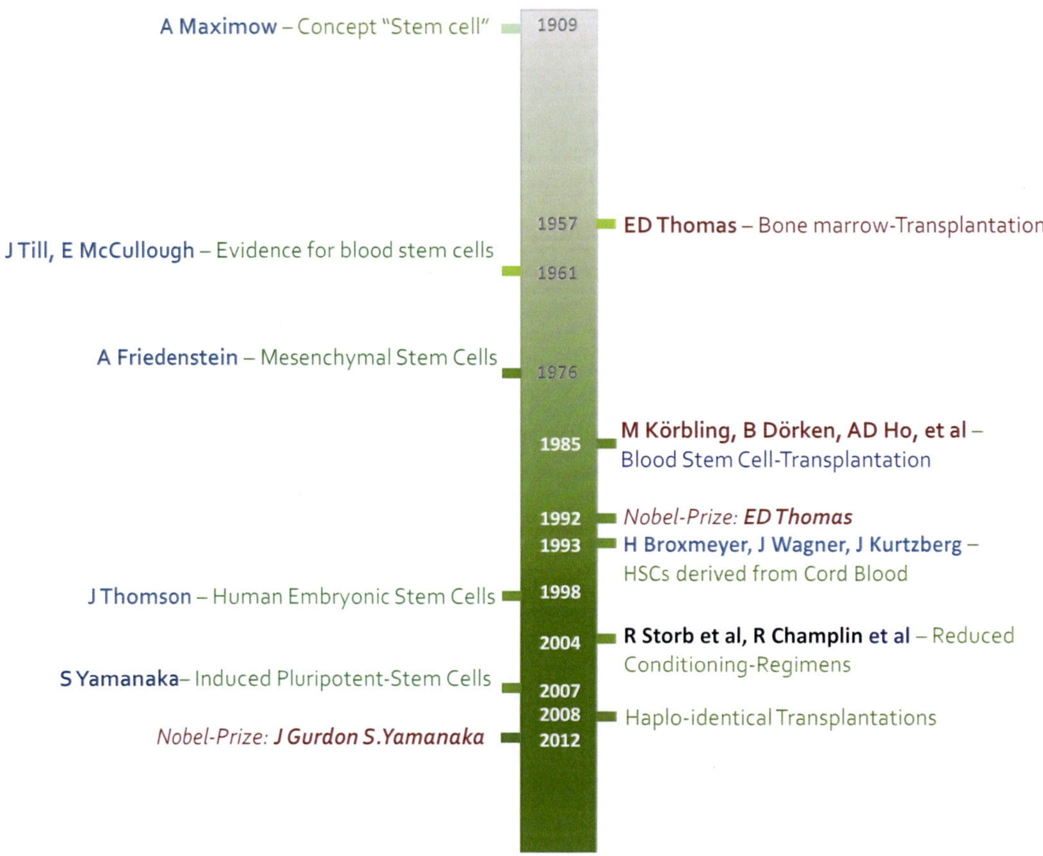

Fig. 1 Milestones of stem cell therapy. The key preclinical concepts and discoveries are depicted on the left side whereas the clinical developments and achievements on the right side of the time axis in the middle

Albeit the present enthusiasm for stem cell research being spurred by the establishment of human embryonic stem cell lines in 1998 [4] and by the generation of induced pluripotent stem cells [5, 6], hematologists have been applying stem cell technology successfully in clinical practice since the 1960s. Some of the milestones in the development of stem cell research and stem cell transplantation are depicted in Fig. 1.

2 Development of Bone Marrow Transplantation

The first *successful* attempts using bone marrow transplantation as a treatment strategy for patients with hereditary immunodeficiency or acute leukemias were performed in the 1960s [7–10]. The original idea was to replace the diseased bone marrow with a healthy one after myeloablation. Without the benefits of knowledge of immunology and supportive care at the time, morbidity and mortality rates associated with this treatment procedure were high [10]. Nevertheless, the results were considered encouraging as compared to those obtained with conventional treatment options.

Bone marrow transplantation has in the meantime been proven to be the only chance of cure for some patients with leukemia and some hereditary immunodeficiencies [11]. As demonstrated years later by McCullough's group, the success of bone marrow transplantation was due to the presence of HSC in the marrow graft, which were able to reconstitute the blood and immune systems after myeloablation [2, 3].

Until the mid-1980s, BM was the only source of HSC for allogeneic or autologous transplantations. In the second half of the 1980s, several groups have shown that stem cells could be harvested from the peripheral blood after chemotherapy. Myelosuppressive drugs induced a significant increase in the number of peripheral blood-derived stem cells (PBSC) during the recovery phase [12–16]. The yield was, however, unpredictable, and most patients had to undergo multiple of up to six leukapheresis procedures before an adequate amount was collected.

3 Trafficking of HSC in Peripheral Blood

Numerous researchers around the globe have made relevant contributions and provided incremental knowledge, leading to the first clinical attempts in 1985. In a mouse model, Goodman and Hodgson provided evidence for HSC in the peripheral blood that were able to restore hematopoiesis in radiation-induced marrow aplasia [17]. Theodor Fliedner at the University of Ulm, Germany, was one of the pioneers who discovered dividing, non-leukemic DNA--synthesizing cells in peripheral blood as evidence for the existence of progenitor cells in circulating blood [18]. In the 1970s, Martin Körbling established various procedures to separate mononuclear leukocytes by density-gradient centrifugation and defined reagents that might be able to mobilize stem cells from the bone marrow into circulating blood in a dog model [18–21].

Initially identified only in the bone marrow, HSC could be found in the peripheral blood upon stimulation such as during the recovery phase after myelosuppressive therapy [18]. Simultaneously, Juttner and coworkers in Adelaide, Australia, showed that peripheral blood stem cells collected in early remission from acute non-lymphoblastic leukemias could represent a realistic alternative to bone marrow for transplantation [22].

Thus, a number of scientists around the globe laid the foundations for resolving the myriad of challenges such as stimulation and mobilization of stem cells from the marrow into the circulation, optimizing the procedures for collection of stem cells by leukapheresis, and establishing protocols for the cryopreservation of stem and progenitor cells and quality control of the stem cell preparations [18–22]. Provided with all these results from basic and clinical research, the first transplantation in October 1985 in Heidelberg represented the culmination of decades of arduous and meticulous research work by many institutions across continents.

4 The First Blood Stem Cell Transplantation: Need Is the Mother of Invention

The patient, 38-year-old male, presented with a fist-size tumor on the right neck in 1985 [12]. Biopsy showed B-lymphoblastic non--Hodgkin's lymphoma of Burkitt type. Computerized axial tomography (CAT) revealed abdominal tumor manifestations. Induction regimen consisted of two cycles of cyclophosphamide 1.0 g/m^2, vincristine 1.4 mg/m^2, methotrexate 12.5 mg/m^2, and prednisone 1.0 g/m^2 all IV (COMP). Simultaneously two intrathecal administrations of methotrexate (12.5 mg/m^2) were administered. Autologous bone marrow transplantation was considered as the appropriate intensive consolidation treatment. The patient was found to suffer from hereditary hyperostosis such that a regular marrow harvest from both pelvic areas was not possible. The decision was made to use blood-derived stem cells instead of bone marrow for autologous engraftment. Approval for this experimental therapy was granted by the Institutional Research Ethics Board of the University of Heidelberg, and the appropriate informed consent form was signed. At the recovery phase 16 days after the second cycle of chemotherapy, blood stem cell collections were initiated. A total of seven consecutive leukaphereses were performed. The pretransplant regimen consisted of total body irradiation fractionated over 4 days at 120 rad single doses up to a total of 1320 rad, followed by cyclophosphamide 50 mg/kg administered on each of four consecutive days. To our great surprise, blood cell recovery occurred rapidly and WBC reached 1000/μL on day 9 and 500 polymorphonuclear cells/μL and 50,000 platelets/μL on day 10. The kinetics of hematologic recovery after autologous blood stem cell transplantation in this patient is shown in Fig. 2 (reproduced from [12]) and the kinetics of recovery of CFU-GM and CFU-GEMM in Fig. 3.

This very patient has remained in complete remission ever since then, and his blood count has been in the normal range for more than 33 years. He is now 71 years old and his performance is excellent at the time of writing.

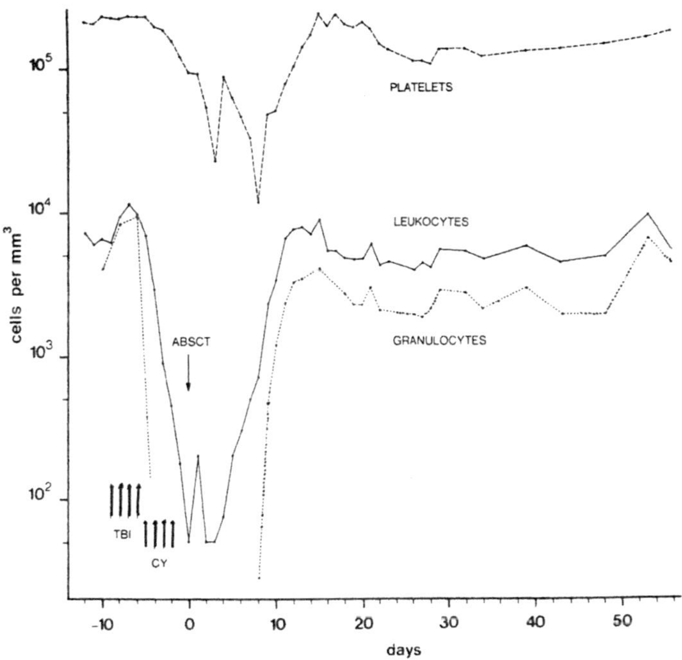

Fig. 2 Kinetics of recovery of leukocytes, granulocytes, and platelets in peripheral blood after PBSCT in the first patient (reproduced from [12])

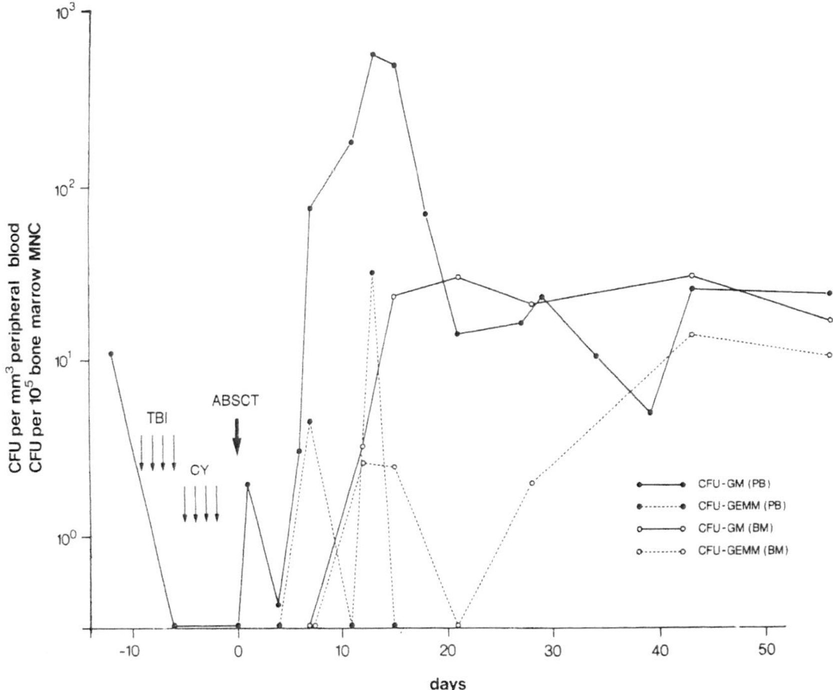

Fig. 3 Kinetics of recovery of CFU-GM and CFU-GEMM in peripheral blood after PBSCT in the first patient (reproduced from [12])

5 Bone Marrow Versus Peripheral Blood as the Source of Stem Cells

The use of peripheral blood-derived stem cells (PBSC) offers several advantages such as harvest of cells without general anesthesia and elimination of pain resulting from multiple aspirations from the BM and is above all associated with more rapid engraftment [23–25]. The latter came initially as a pleasant surprise. The median time to recovery of peripheral white blood count of over 1000/μL after bone marrow transplant is 21 days versus 10 days for PBSCT. The main challenge of PBSC is that they exist in the circulation in very small numbers. Hematopoietic progenitor and stem cells (HSC) reside in the BM, and they have to be mobilized into the circulation prior to being collected by apheresis. The number of apheresis procedures needed and the success of transplantation are determined by the efficiency of stem cell mobilization [26–28] (reviewed in reference [29]).

Stem cells adhere to their BM niche by interactions between SDF1α, which is produced by BM stromal cells, and CXCR4, which is expressed on $CD34^{+}$ cells [30, 31]. G-CSF, the standard and most widely used agent for this purpose over the past 20 years, mobilizes stem cells from the marrow niche by secretion of neutrophil-associated extracellular proteases, such as MMP-9, which subsequently releases HSC from their niche [32]. In the past decade, an inhibitor of the SDF1α-CXCR4 axis, plerixafor, has been shown to be efficient and approved for mobilization of HSC [33].

6 Challenges in Mobilization of PBSC

In the early days of PBSCT, stem cell mobilization was achieved with chemotherapeutic drugs as chemotherapy induces a significant increase in the number of HSC in circulating blood at the time of recovery [12–16]. The yield was unpredictable, and many patients had to undergo four to six leukaphereses before an adequate number of cells were collected. Many patients also failed to mobilize sufficient PBSC for transplantation in response to chemotherapy.

The availability of GM-CSF and G-CSF has greatly revolutionized the mobilization of PBSC for autologous transplantations (for a detailed review, see reference [29]). In the 1980s, a surrogate marker for HSC activity, CD34, has been identified [34], and GM-CSF and G-CSF were made available [35–38]. G-CSF and GM-CSF were approved for use as HSC mobilizing agents, but G-CSF (in combination with chemotherapy or alone) has become the standard. Unfortunately, some patients, especially those who have been heavily pretreated with chemotherapy or irradiation, still fail to mobilize sufficient numbers of PBSC for transplantation in response to G-CSF with or without chemotherapy.

7 "Poor Mobilizers"

There were numerous reports on patients or healthy donors who failed to mobilize sufficient numbers of PBSC for transplantation [39–41]. We have previously shown that there is a highly significant correlation between the $CD34^+$ concentration in peripheral blood and the potential to collect an adequate amount of $CD34^+$ cells within one or up to three leukapheresis procedures [42]. Based on previous reports, and based on a retrospective analysis of 840 patients at the Heidelberg center who were mobilized with chemotherapy and growth factors with the intent of autologous transplantation, we have defined the criteria for poor-mobilizers and have confirmed that pre-apheresis $CD34^+$ count would predict reliably the quality of collection [43].

Another controversial issue is whether higher concentrations of $CD34^+$ cells transplanted would produce better long-term outcome. Present evidence indicated that once an adequate amount has been transplanted, $CD34^+$ cell numbers above 2.0×10^6 cells per kg body weight (bw) of $CD34^+$ cells would not necessarily confer more advantage in terms of engraftment of leukocytes and platelets. Our previous observation showed that although higher doses of $CD34^+$ cells (i.e., $>6.5 \times 10^6$/kg bw) might marginally but significantly shorten the time to platelet recovery, stable engraftment was achieved with transplantation of 2.0×10^6 $CD34^+$ cells/kg bw. Thus, there is no solid evidence that for autologous transplant, levels of >5.0 to 6.0×10^6/kg bw would improve long-term clinical outcome [43].

8 Future Perspectives of Hematopoietic Stem Cell Transplantation

The concept of stem cell was introduced by a hematologist more than a century ago and the first attempts in bone marrow transplantation performed in 1957 [44]. The use of bone marrow transplantation for cure of acute leukemias gained momentum in the late 1970s to early 1980s as knowledge in transplantation immunology, human leukocyte antigen system, transfusion medicine, and control of opportunistic infections evolved. In the 1980s discoveries of novel cytokines, blood progenitor subpopulations, and of their roles in proliferation, differentiation, and regulation of stem cell fate spawned another wave of innovations. Coupled at the same time with advances in enrichment and large-scale separation of HSC in the early 1980s, many research groups around the globe have contributed to the scientific foundation for the success of the first peripheral blood-derived HSC transplantations in the mid-1980s. In the past years, we have witnessed the exponential increase in the number of transplant centers and, correspondingly, the number of patients transplanted worldwide. Some of the then novel technologies and innovative indications for transplantation have become clinical routine, while others have been discarded. Since the 1990s the immunological sequelae of allogeneic transplantation have been recognized as the major mechanism for cure of leukemias. Transplantation strategies have been developed to exploit the selective immune modulatory effects of allogeneic transplantation without myeloablation in the past 15 years. Conditioning regimens have also been established that permit haplo-identical transplantations such that every patient practically might have a potential donor. Advances in management of opportunistic infections and of graft-versus-host disease have simultaneously contributed significantly to the reduction in transplant-related morbidity and mortality. Finally, decades of research into cellular and genetic manipulations of T lymphocytes and tumor-infiltrating lymphocytes have led to the development of adoptive immune therapy and subsequent chimeric antigen receptor T lymphocytes (CAR T-cell) as a therapeutic principle. The latter treatment strategy has generated encouraging results in the past 5 years such that specific protocols have been approved by the Food and Drug Administration of the USA.

References

1. Maximow A (1909) Der Lymphozyt als gemeinsame Stammzelle der verschiedenen Blutelemente in der embryonalen Entwicklung und im postfetalen Leber der Säugetiere. Folia Haematol (Leipzig) 8:125–141
2. Siminovitch L, McCulloch EA, Till JE (1963) The distribution of colony-forming cells among spleen colonies. J Cell Comp Physiol 62:327–336
3. Becker A, McCulloch EA, Till JE (1963) Cytological demonstration of the clonal nature of spleen colonies derived from transplanted mouse marrow cells. Nature 197:452–454
4. Thomson JA, Itskovitz-Eldor J, Shapiro SS, Waknitz MA, Swiergiel JJ, Marshall VS, Jones JM (1998) Embryonic stem cell lines derived from human blastocysts. Science 282:1145–1147
5. Takahashi K, Tanabe K, Ohnuki M, Narita M, Ichisaka T, Tomoda K, Yamanaka S (2007) Induction of pluripotent stem cells from adult human fibroblasts by defined factors. Cell 131:861–872
6. Yu J, Vodyanik MA, Smuga-Otto K, Antosiewicz-Bourget J, Frane JL, Tian S, Nie J, Jonsdottir GA, Ruotti V, Stewart R, Slukvin II, Thomson JA (2007) Induced pluripotent stem cell lines derived from human somatic cells. Science 318:1917–1920
7. Bach FH, Albertini RJ, Joo P, Anderson JL, Bortin MM (1968) Bone-marrow transplantation in a patient with the Wiskott-Aldrich syndrome. Lancet 2:1364–1366
8. Gatti RA, Meuwissen HJ, Allen HD, Hong R, Good RA (1968) Immunological reconstitution of sex-linked lymphopenic immunological deficiency. Lancet 2:1366–1369
9. de Koning J, van Bekkum DW, Dicke KA, Dooren LJ, Radl J, van Rood JJ (1969) Transplantation of bone-marrow cells and fetal thymus in an infant with lymphopenic immunological deficiency. Lancet 1:1223–1227
10. Thomas ED, Bryant JI, Buckner CD, Clift RA, Fefer A, Fialkow PJ, Funk DD, Neiman PE, Rudolph RH, Slichter SJ, Storb R (1971) Allogeneic marrow grafting using HL-A matched donor-recipient sibling pairs. Trans Assoc Am Phys 84:248–261
11. Thomas ED, Flournoy N, Buckner CD, Clift RA, Fefer A, Neimen PE, Storb R (1977) Cure of leukaemia by marrow transplantation. Leukemia Res 1:67–70
12. Körbling M, Dörken B, Ho AD, Pezzutto A, Hunstein W, Fliedner TM (1986) Autologous transplantation of blood-derived hemopoietic stem cells after myeloablative therapy in a patient with Burkitt's lymphoma. Blood 67:529–532

13. Reiffers J, Bernard P, David B, Vezon G, Sarrat A, Marit G, Moulinier J, Broustet A (1986) Successful autologous transplantation with peripheral blood hemopoietic cells in a patient with acute leukemia. Exp Hematol 14:312–315
14. To LB, Dyson PG, Branford AL, Russell JA, Haylock DN, Ho JQ, Kimber RJ, Juttner CA (1987) Peripheral blood stem cells collected in very early remission produce rapid and sustained autologous haemopoietic reconstitution in acute non-lymphoblastic leukaemia. Bone Marrow Transplant 2:103–108
15. Bell AJ, Figes A, Oscier DG, Hamblin TJ (1987) Peripheral blood stem cell autografts in the treatment of lymphoid malignancies: initial experience in three patients. Br J Haematol 66:63–68
16. Kessinger A, Armitage JO, Landmark JD, Smith DM, Weisenburger DD (1988) Autologous peripheral hematopoietic stem cell transplantation restores hematopoietic function following marrow ablative therapy. Blood 71:723–727
17. Goodman JW, Hodgson GS (1962) Evidence for stem cells in the peripheral blood of mice. Blood 19:702–714
18. Fliedner TM (1995) Blood stem cell transplantation: from preclinical to clinical models. Stem Cells 13(Suppl 3):1–12
19. Körbling M, Fliedner TM, Calvo W, Nothdurft W, Ross WM (1977) In vitro and in vivo properties of canine blood mononuclear leukocytes separated by discontinuous density gradient centrifugation. Biomedicine 26:275–283
20. Körbling M, Ross W, Pflieger H, Arnold R, Fliedner TM (1977) Procurement of human blood stem cells by continuous flow centrifugation. Blood 50:747–754
21. Korbling M, Fliedner TM, Pflieger H (1980) Collection of large quantities of granulocyte/macrophage progenitor cells (CFUc) in man by continuous flow leukapheresis. Scand J Haematol 24:22–28
22. Juttner CA, To LB, Haylock DN, Branford A, Kimber RJ (1985) Circulating autologous stem cells collected in very early remission from acute non-lymphoblastic leukemia produce prompt but incomplete haemopoietic reconstitution after high dose melphalan or supralethal chemoradiotherapy. Br J Haematol 61:739–745
23. Beyer J, Schwella N, Zingsem J, Strohscheer I, Schwaner I, Oettle H, Serke S, Huhn D, Stieger W (1995) Hematopoietic rescue after high-dose chemotherapy using autologous peripheral-blood progenitor cells or bone marrow: a randomized comparison. J Clin Oncol 13:1328–1335
24. Schmitz N, Linch DC, Dreger P, Goldstone AH, Boogaerts MA, Ferrant A, Demuynck HM, Link H, Zander A, Barge A (1996) Randomised trial of filgrastim-mobilised peripheral blood progenitor cell transplantation versus autologous bone-marrow transplantation in lymphoma patients. Lancet 347:353–357
25. Hartmann O, Le Corroller AG, Blaise D, Michon J, Philip I, Norol F, Janvier M, Pico JL, Baranzelli MC, Rubie H, Coze C, Pinna A, Méresse V, Benhamou E (1997) Peripheral blood stem cells and bone marrow transplantation for solid tumors and lymphomas: hematologic recovery and costs. A randomized, controlled trial. Ann Intern Med 126:600–607
26. Fruehauf S, Haas R, Conradt C, Murea S, Witt B, Möhle R, Hunstein W (1995) Peripheral blood progenitor cells (PBPC) counts during steady-state hemopoiesis allow to estimate the yield of mobilized PBPC after filgrastim (R-metHuG-CSF)-supported cytotoxic chemotherapy. Blood 85:2619–2626
27. Weaver CH, Hazelton B, Birch R, Palmer P, Allen C, Schwartzberg L, West W (1995) An analysis of engraftment kinetics as a function of the CD34 content of peripheral blood progenitor cell collections in 692 patients after the administration of myeloablative chemotherapy. Blood 86:3961–3969
28. Bensinger W, Appelbaum F, Rowley S, Storb R, Sanders J, Lilleby K, Gooley T, Demirer T, Schiffman K, Weaver C (1995) Factors that influence collection and engraftment of autologous peripheral-blood stem cells. J Clin Oncol 13:2547–2555
29. To LB, Haylock DN, Simmons PJ, Juttner CA (1997) The biology and clinical uses of blood stem cells. Blood 89:2233–2258
30. Möhle R, Bautz F, Rafii S, Moore MA, Brugger W, Kanz L (1998) The chemokine receptor CXCR-4 is expressed on CD34+ hematopoietic progenitors and leukemic cells and mediates transendothelial migration induced by stromal cell derived factor-1. Blood 91:4523–4530
31. Lapidot T, Petit I (2002) Current understanding of stem cell mobilization: the roles of chemokines, proteolytic enzymes, adhesion molecules, cytokines, and stromal cells. Exp Hematol 30:973–981
32. Petit I, Szyper-Kravitz M, Nagler A, Lahav M, Peled A, Habler L, Ponomaryov T, Taichman RS, Arenzana-Seisdedos F, Fujii N, Sandbank J, Zipori D, Lapidot T (2002) G-CSF induces stem cell mobilization by

decreasing bone marrow SDF-1 and up-regulating CXCR4. Nat Immunol 3:687–694

33. Broxmeyer HE, Orschell CM, Clapp DW, Hangoc G, Cooper S, Plett PA, Liles WC, Li X, Graham-Evans B, Campbell TB, Calandra G, Bridger G, Dale DC, Srour EF (2005) Rapid mobilization of murine and human hematopoietic stem and progenitor cells with AMD3100, a CXCR4 antagonist. J Exp Med 201:1307–1318
34. Civin CI, Strauss LC, Brovall C, Fackler MJ, Schwartz JF, Shaper JH (1984) Antigenic analysis of hematopoiesis. III. A hematopoietic progenitor cell surface antigen defined by a monoclonal antibody raised against KG-1a cells. J Immunol 133:157–165
35. Gianni AM, Siena S, Bregni M, Tarella C, Stern AC, Pileri A, Bonadonna G (1989) Granulocyte-macrophage colony-stimulating factor to harvest circulating haemopoietic stem cells for autotransplantation. Lancet 2:580–585
36. Haas R, Ho AD, Bredthauer U, Cayeux S, Egerer G, Knauf W, Hunstein W (1990) Successful autologous transplantation of blood stem cells mobilized with recombinant human granulocyte-macrophage colony-stimulating factor. Exp Hematol 18:94–98
37. Elias AD, Ayash L, Anderson KC, Hunt M, Wheeler C, Schwartz G, Tepler I, Mazanet R, Lynch C, Pap S (1992) Mobilization of peripheral blood progenitor cells by chemotherapy and granulocyte-macrophage colony-stimulating factor for hematologic support after high-dose intensification for breast cancer. Blood 79:3036–3044
38. Bensinger W, Singer J, Appelbaum F, Lilleby K, Longin K, Rowley S, Clarke E, Clift R, Hansen J, Shields T (1993) Autologous transplantation with peripheral blood mononuclear cells collected after administration of recombinant granulocyte stimulating factor. Blood 81:3158–3163
39. Gordan LN, Sugrue MW, Lynch JW, Williams KD, Khan SA, Wingard JR, Moreb JS (2003) Poor mobilization of peripheral blood stem cells is a risk factor for worse outcome in lymphoma patients undergoing autologous stem cell transplantation. Leuk Lymphoma 44:815–820
40. Kuittinen T, Nousiainen T, Halonen P, Mahlamaki E, Jantunen E (2004) Prediction of mobilisation failure in patients with non--Hodgkin's lymphoma. Bone Marrow Transplant 33:907–912
41. Pavone V, Gaudio F, Console G et al (2006) Poor mobilization is an independent prognostic factor in patients with malignant lymphomas treated by peripheral blood stem cell transplantation. Bone Marrow Transplant 37:719–724
42. Fruehauf S, Haas R, Conradt C, Murea S, Witt B, Möhle R, Hunstein W (1995) Peripheral blood progenitor cell (PBPC) counts during steady-state hematopoiesis allow to estimate the yield of mobilized PBPC after filgrastim (R-metHuG-CSF)-supported cytotoxic chemotherapy. Blood 85:2619–2626
43. Wuchter P, Ran D, Bruckner T, Schmitt T, Witzens-Harig M, Neben K, Goldschmidt H, Ho AD (2010) Poor mobilization of hematopoietic stem cells – definitions, incidence, risk factors and outcome of autologous transplantation. Bio Blood Marrow Transplant 16:490–499
44. Thomas ED, Lochte HL Jr, Lu WC, Ferrebee JW (1957) Intravenous infusion of bone marrow in patients receiving radiation and chemotherapy. N Engl J Med 257:491–496

Chapter 2

In Vitro Dynamic Phenotyping for Testing Novel Mobilizing Agents

Motomu Tanaka

Abstract

A new method to quantify the influence of mobilization agents on the dynamics of human hematopoietic stem and progenitor cells (HSPC) is introduced. Different from the microscopy-based high-content screening relying on multiple staining, machine learning, and molecular-level perturbation, the proposed method sheds light on the "dynamics" of HSPC in the presence of extrinsic factors, including SDF1α and mobilization agents. A well-defined model of the bone marrow niche is fabricated by the deposition of planar lipid membranes on glass slides (called supported membranes) displaying ligand molecules at precisely controlled surface densities. The dynamics of human HSPC, $CD34^{+}$ cells from umbilical cord blood or peripheral blood, are monitored by time-lapse, live cell imaging with a standard phase-contrast microscopy or a specially designed microinterferometry in the absence or presence of mobilization agents. After extracting the contour of each cell, one can analyze the dynamics of cell "shapes" step-by-step, yielding various levels of information ranging from the principal mode of deformation, the persistence of deformation patterns, and the energy consumption by HSPC in the absence and presence of mobilization agents. Moreover, by tracking the migration trajectories of HSPC, one can gain insight how mobilization agents influence the "motion" of HSPC. As these readouts can be connected to a theoretical model, this strategy enables one to classify the influence of not only mobilization agents but also target-specific inhibitors or other treatments in quantitative indices.

Key words Supported membrane, Cell adhesion, Cell migration, Theoretical model

1 Introduction

The dormancy of the most primitive hematopoietic stem and progenitor cells (HSPC) is maintained by the bone marrow niche via several ligand-receptor interactions. Mounting evidence suggests that mesenchymal stem or stromal cells (MSC) play key roles in sustaining the niche functions [1, 2]. One of the important molecular axes is the homophilic interaction between N-cadherin molecules expressed on both HSPC and MSC, which supports the long-term maintenance of the primitive HSPC pool in the bone marrow niche [3–5]. On the other hand, CXCR4 expressed on HSPC specifically recognizes stromal cell-derived factor 1α (SDF1α or

Gerd Klein and Patrick Wuchter (eds.), *Stem Cell Mobilization: Methods and Protocols*, Methods in Molecular Biology, vol. 2017, https://doi.org/10.1007/978-1-4939-9574-5_2,

CXCL12) that coexists on MSC surface and in the marrow liquid. SDF1α is identified as the chemoattractant that regulates homing and migration of HSPC in the marrow niche [6–9].

Peripheral HSPC have largely replaced bone marrow-derived cells as the major source for autologous and allogeneic transplantation [10–12]. Needless to say, efficient mobilization of HSPC is a prerequisite for the successful stem cell collection and consecutive transplantation. So far, G-CSF, which leads to secretion of neutrophil-associated extracellular proteases, has been most widely used as the mobilization agent over the past 25 years [13, 14]. However, among the patients intended for autologous transplantation, about 10–15% of them have difficulties in mobilizing an adequate amount of HSPC for transplantation [15].

To improve the efficacy of therapeutic treatments, new and highly effective mobilizing reagents are needed. Plerixafor (trade name: Mozobil, produced by Sanofi) is an effective mobilizer of $CD34^{+}$ cells for autologous transplantation, which was initially regarded as an antagonist to CXCR4 [16, 17], but recent studies suggested a revised view, even claiming that plerixafor is a partial agonist [18, 19]. On the other hand, NOX-A12 (NOXXON Pharma), which is an L-enantiomeric RNA oligonucleotide, was designed to bind and neutralize SDF1α. This compound showed a half-maximal inhibitory concentration value of 300 pM (4.3 ng/mL) in a migration assay using Jurkat cell line [20]. - However, in general, very little is understood how these mobilization agents function.

One major trend is to increase the phenotypic readouts by a microscopy-based high-content screening using the combination of multiple staining, machine learning, and perturbation (RNA interference, mutation, etc.) [21]. A completely different but complementary strategy to the static and qualitative phenotyping of fixed objects is to extract quantitative information from the "dynamic phenotypes" of living cells, such as active shape deformation and motion. Recent developments in interface science and soft matter physics enable to design well-defined surrogate surfaces that mimic the bone marrow niche for human HSPC from donors and patients. The combination of label-free microscopy and analytical platforms based on non-equilibrium statistical physics helps us quantify how mobilization agents influence the mechanical strength of HSPC-niche interactions, the cellular activity (energy consumption), and the mode of migration, which cannot be accessed otherwise.

2 Materials

2.1 Functionalization of Supported Membranes

- Use analytical grade reagents and double deionized water ($R > 18$ MΩcm).
- Acetone p.a.
- Ethanol p.a.
- Methanol p.a.
- H_2O_2 (30%)/NH_4OH (25%)/H_2O, 1:1:5 (v/v/v).
- Use synthetic lipids with defined phase behaviors. The use of natural lipid mixtures, such as soybean lecithin or egg yolk lecithin, is not recommended (*see* **Note 1**).
- Chloroform ($CHCl_3$).
- N_2 stream (technical grade).
- Buffer for generating small unilamellar vesicles (SUVs): 150 mM NaCl, 10 mM Hepes, pH 7.4.
- Use microscope-grade slides (thickness = 0.13–0.16 mm). For structural characterization with reflectivity-based techniques, use silicon wafers. For neutron reflectivity, silicon substrates with native oxide with the thickness of 1–2 nm are recommended, whereas silicon wafers with thermal oxide with the thickness of 140–150 nm are recommended (*see* **Note 2**).
- Any ligand that can be harnessed to anchor lipids can be coupled (Fig. 1). Widely used anchor lipids are functionalized with biotin or nitrilotriacetic acid (NTA) head groups (Avanti Polar Lipids, Alabama, USA) [22]. More recently, the synthesis of lipid with SNAP-tag ligands was reported [23].
- Ni^{2+}-loaded buffer: 1 mM $NiCl_2 \times 6\ H_2O$, 150 mM NaCl, 10 mM Hepes, pH 7.4.
- Normal buffer: 150 mM NaCl, 10 mM Hepes, pH 7.4.
- Histidine-tagged proteins (10 μg/mL).
- Neutravidin solution (40 μg/mL).
- Biotinylated ligand solution (10 μg/mL).
- If ligands with biotin-tags, histidine-tags, or SNAP-tags are not commercially available, synthesize recombinant proteins by your own, or use commercial kits and couple these tags to the ligands [24, 25].

2.2 Isolation of Human HSPC

- $CD34^+$ cells, isolated by magnetic activated cell sorting.
- Iscove's Modified Dulbecco's Medium (IMDM), pre-warmed.
- Long-term bone marrow culture (LTBMC) medium: 75% IMDM supplemented with 12.5% FCS, 12.5% horse serum, 2 mM L-glutamine, 100 U/mL penicillin/streptomycin, and 0.05% hydrocortisone 100.

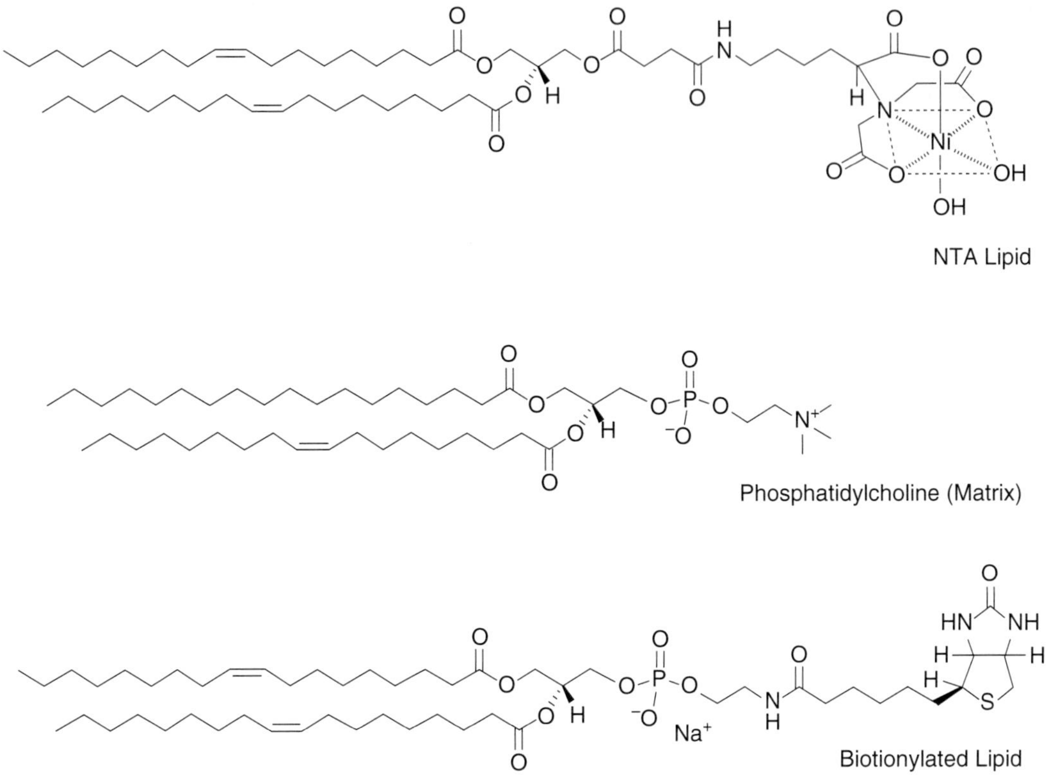

NTA Lipid

Phosphatidylcholine (Matrix)

Biotionylated Lipid

Fig. 1 Chemical structures of matrix lipids and anchor lipids

- Propidium iodide (2 μg/mL, dissolved in 150 mM NaCl, 10 mM Hepes, pH 7.4).

2.3 Quantifying Significance of Cell Adhesion

- Reflection interference contrast microscope (RICM) consisting of an inverted microscope with a polarizer, an analyzer, and an oil-immersion objective (63×, N.A. 1.25) with a built-in λ/4 plate.
- Laser pulse generator (Nd: YAG laser ($\lambda = 1064$ nm)).
- Inverted microscope objective (10×, N.A. 0.40).
- Needle-type pressure sensor (e.g., from Müller Instruments, Oberursel, Germany).

3 Methods

3.1 Cleaning of Glass Slides

- The cleanliness of glass slides is one of the most crucial factors. Prior to the cleaning, sonicate substrates in acetone, ethanol, methanol, and water (3 min for each solvent) to remove crude impurities (fats, dusts, etc.).

- Clean the substrates following the modified RCA methods, which are widely used in semiconductor industry [26]. In brief, sonicate the substrates in H_2O_2 (30%)/NH_4OH (25%)/H_2O (1:1:5) at room temperature for 3 min, and then soak them at 60 °C for another 30 min in the same solution. Afterward, rinse the substrates intensively with ultrapure water, dry them at 70 °C, and store them in a vacuum chamber at room temperature.
- The longer storage is not recommended, as the surface hydrophilicity, which is very important for the membrane deposition, decays over time.

3.2 Supported Membrane Preparation

The ultraclean substrates are then covered with a lipid layer (Fig. 2a), generating supported membranes [27, 28].

- Mix the stock solutions of lipids in $CHCl_3$ (1–5 mg/mL) to obtain the desired molar fraction of anchor lipids, such as lipids with NTA or biotin head groups.

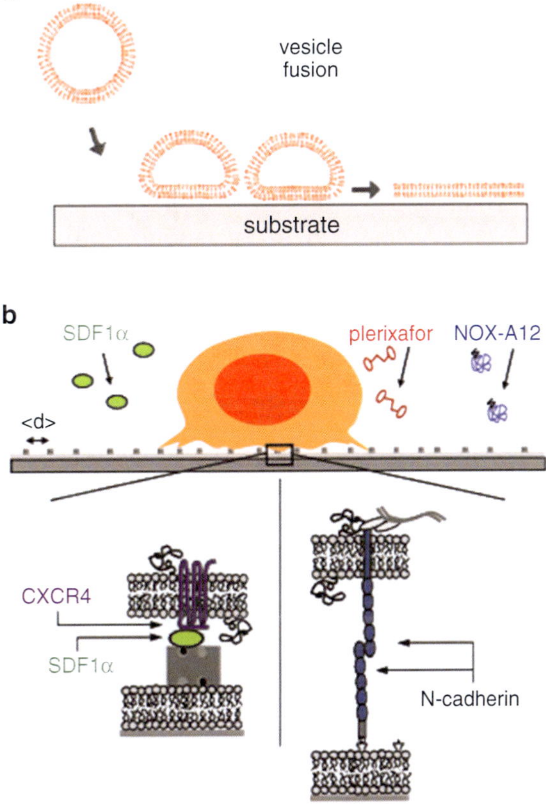

Fig. 2 Schematic illustrations of (**a**) supported membrane preparation and (**b**) functionalization with ligand molecules biotin-tags (left) and histidine-tags (right)

- Evaporate $CHCl_3$ under N_2 stream and store under vacuum overnight.
- Suspend lipids in buffer (150 mM NaCl, 10 mM Hepes, pH 7.4), and sonicate the suspension with a tip sonicator for 30 min to obtain small unilamellar vesicles (SUVs).
- Centrifugate vesicle suspensions for 10 min (13,400 × g) to remove residual titanium particles.
- After the preparation, SUV suspensions (0.5 mg/mL) were stored at +4 °C.
- Dilute the suspension to a final concentration of 0.5 mg/mL before use.
- Incubate the SUV suspensions with the substrates for 30 min at +40 °C.
- Rinse the sample intensively with the buffer to remove excess SUVs.

3.3 Functionalization with Ligands

The supported membranes can be functionalized with various ligand molecules (Fig. 2b).

- Functionalization with histidine-tagged proteins [29, 30]:
  - Incubate the membrane with Ni^{2+}-loaded buffer to saturate NTA head groups with Ni^{2+} (45 min).
  - Exchange the buffer to the normal buffer without Ni^{2+}.
  - Incubate with histidine-tagged proteins (10 μg/mL) for 12 h at room temperature (*see* **Note 3**).
- Functionalization with biotin-tagged ligands [24, 25]:
  - Incubate the membrane with the neutravidin solution (40 μg/mL) at room temperature for 2 h.
  - Remove the unbound neutravidin by intensive rinsing.
  - Exchange the solution to the biotinylated ligand solution (10 μg/mL).
  - After 12 h incubation, remove unbound proteins by intensively rinsing the sample.
- Functionalization with histidine- and biotin-tagged ligands:
  - To avoid any potential damage caused by the Ni^{2+}-loaded buffer, functionalize the membrane first with histidine-tagged ligands.
  - Incubate the membrane with neutravidin and biotinylated ligands, as described above.

3.4 Isolation of Human HSPC

- All samples of primary cells must be collected from voluntary donors after obtaining informed consent according to the guidelines approved by the Ethics Committee of each institution.

- Collect human HSPC, defined as $CD34^+$ cells (*see* **Note 4**), from umbilical cord blood.
- Alternatively use HSPC from healthy allogeneic stem cell donors who received a mobilization regimen with G-CSF (10 μg/kg body weight per day) for 5 days. Take peripheral blood (60 mL) prior to leukapheresis.
- Isolate HSPC following the previous accounts [5, 31]:
  - Isolate mononuclear cells by density-gradient centrifugation.
  - Enrich $CD34^+$ cells by using magnetic beads.
  - Sort the cells by 2× elution through an affinity column.
  - Allow cells to rest at least for 2 h at 37 °C and 5% CO_2.
- Store the cells in long-term bone marrow culture (LTBMC) medium [32].
- Stain nonviable cells with propidium iodide, and remove them by flow cytometry, which should result in a purity of $CD34^+$ cells of >95%.

3.5 Incubation with Mobilizing Agents

- Select carefully the concentrations of each mobilization agent.
- For example, the concentration of [SDF1α] = 5 ng/mL was used in our previous report [30], as it corresponds to the physiological level in human bone marrow.
- The concentration of plerixafor, 50 ng/mL (corresponding to ≈ 0.1 μM), was selected, as previous in vitro studies used 500 ng/mL [19].
- [NOX-A12] = 50 ng/mL (≈3.5 nM) was used in our recent study [33], as this lies between the IC50 level found in in vitro chemotaxis study on chronic lymphatic leukemia cells and lymphoid cell lines (≈0.3 nM) [20] and the plasma level at which effective mobilization of leukocytes in human was observed (~1 μM) [34].
- Incubate HSPC with the agent for 2 h in suspension, and allow them to adhere onto the supported membrane surfaces for 1 h.

3.6 Quantifying Significance of Cell Adhesion

- Separate HSPC in LTBMC into different portions, and preincubate them at 37 °C and 5% CO_2 for 2 h.
- Exchange the medium to the pre-warmed IMDM.
- As stated above, incubate HSPC with mobilization agent for 2 h prior to the seeding.
- Seed HSPC at a density of 1×10^5 cells/cm^2 and incubate them for 1 h.

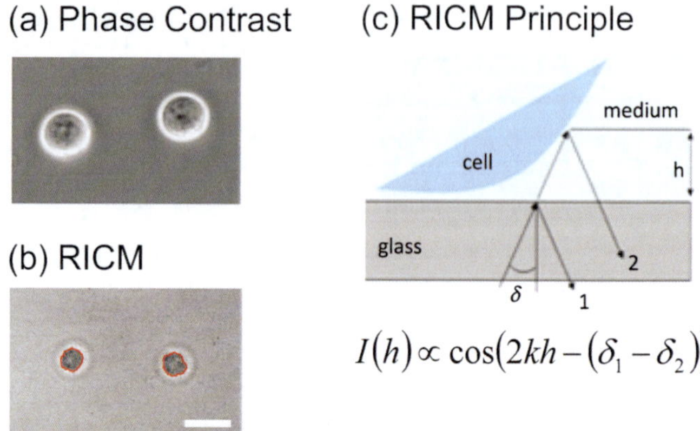

Fig. 3 (**a**) Phase-contrast and (**b**) reflection interference contrast microscopy (RICM) images of HSPC seeded on supported membrane displaying SDF1α. The average intermolecular distance between SDF1α $\langle d \rangle \approx 18$ nm. (**c**) Principle of RICM, detecting the cell-substrate distance from multiple reflection of polarized light

- For compact and less deformable cells like HSPC, the adhesion area per cell is distinctly different from the projected area obtained by the phase-contrast microscopy (Fig. 3a).
- To visualize the region of tight cell-substrate contact, a micro-interferometry, called reflection interference contrast microscopy (RICM) [24, 35, 36], is very useful (Fig. 3b).
  - RICM (Fig. 3c) requires three additional components inserted into the optical path of a standard inverted microscope: a polarizer, a λ/4-plate, and an analyzer.
  - Need to use an antiflex objective with an integrated λ/4-plate, which converts the polarization of the illumination from linear to circular.
  - The rays reflected at interfaces with different refractive indices were converted again into the linear polarization.
  - Taking the intensity of the light reflected on the substrate surface I_1 and that reflected on the cell surface I_2, the total intensity detected by a CCD camera is given as:

$$I = I_1 + I_2 + 2\sqrt{I_1 I_2}\cos\left(2kh(x, y) + \phi\right), \tag{1}$$

    where $k = 2\pi/\lambda$, $h(x,y)$ is the cell-substrate distance at the position (x,y) and ϕ the phase shift.
  - This enables one to gain the cell-substrate distance at each pixel.
  - As the contact zone appears black due to the destructive interference, the area of adhesion for each cell can be

calculated from a simple intensity thresholding using an ImageJ plugin, for instance.

- To measure the significance of cell adhesion, commonly used techniques include (1) "scratching" or "pulling" cells by atomic force microscopy (AFM) [37, 38] and (2) "washing off" cells by shear stresses in a microfluidic chamber [25, 39].
  - AFM scratching records the sum of bond rupture forces over time. The deflection of a cantilever δ is proportional to the drag force following the Hook's law: $F_{\mathrm{drag}} \propto \delta$. In case the tension σ and the cell-substrate contact area A are constant throughout the measurements, $F_{\mathrm{drag}} = \sigma A$. However, in reality, neither σ nor A takes a constant value during the scratching process. One technical issue of this technique is the limited throughput of the measurements, as the cantilever must be replaced after each measurement. In reality, 3–4 cells per day are the normal throughput, which is not sufficient to follow the time evolution of adhesion with reliable statistics.
  - Wash-off experiments rely on the Newton's second law, where the shear stress γ exerted on the cells can be given by [40]:

$$\gamma = \frac{12\eta\Phi}{wh^2}, \tag{2}$$

where η is the viscosity of the medium [$\mathrm{Nm}^{-2}\mathrm{s}$], Φ the volume flux [$\mathrm{m}^3\ \mathrm{s}^{-1}$], w the channel width, and h [m] the height of the channel [m]. Parallel plate flow chambers are convenient tools, which can also be used to monitor the dynamic adhesion of cells, too. Different from AFM, this technique can offer sufficiently high statistics under physiological shear conditions.
  - Both techniques, however, suffer from several fundamental drawbacks. First, the bond rupture force is extremely rate dependent. As shown by Merkel et al., the single force spectroscopy data suggest that the bond rupture force increases exponentially with the increase in the pulling rate [41]. Second, it is often observed that the cells "react" to the scratching force and shear stress, i.e., cells change their shapes and remodel cytoskeletons, and focal adhesions during the measurements.
- As an alternative technique, we developed a noninvasive, high-throughput technique to quantify the critical force required for the cell detachment (Fig. 4a) [42].
  - Focus a single picosecond (ps) laser pulse (Nd: YAG laser ($\lambda = 1064$ nm), pulse length $\tau_L = 60$ ps) into the culture medium through an inverted microscope objective (10×, N.A. 0.40). The focal point is typically 2 mm apart from the target cells and at 100 μm above the surface.

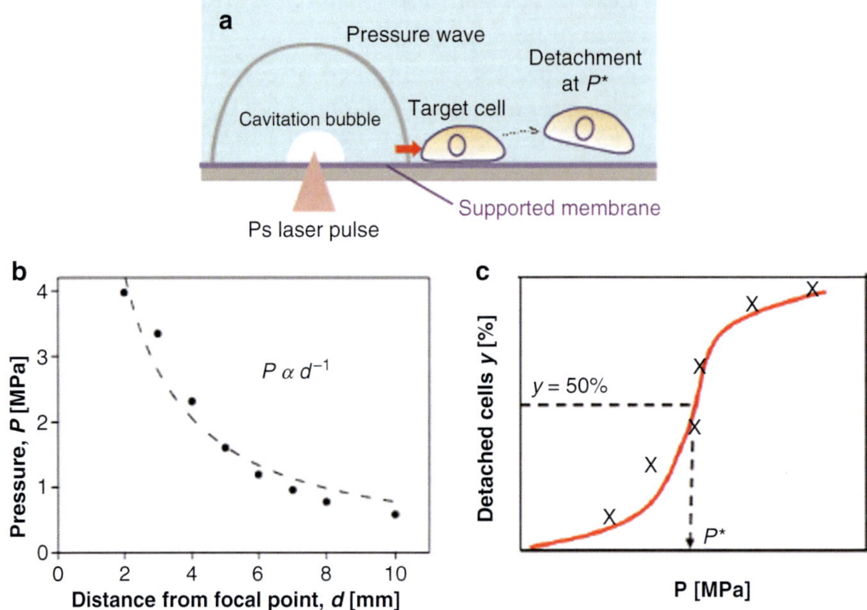

Fig. 4 (**a**) Schematic illustration of the setup quantifying adhesion strength by using pressure waves. A 60 ps laser pulse was focused into an incubation chamber, inducing an ultrasound pressure wave. (**b**) Pressure P can be calibrated with a distance from the focal point d. (**c**) With the aid of the calibration curve, the fraction of detached cells determined from the image $y(d)$ can be converted to $y(P)$. The error function fit (Eq. 3, red line) yields the critical pressure for detachment $P*$, corresponding to $y = 50\%$

- A multiphoton absorption locally boils water in the close proximity of the focal point, generating a cavitation bubble. The bubble immediately collapses, which induces a very powerful pressure wave that travels beyond the sound velocity. This wave, called shock wave, is not accompanied by the mass transfer.
- Measure the pressure P by a needle-type pressure sensor. As P depends on the pulse energy E and the distance from the focal point d, one can gain a calibration curve P vs. d by fixing the pulse energy E constant (Fig. 4b).
- Allow cells to adhere on the substrates prior to the measurements (typically, 1–2 h). The positions of individual cells before and after the exposure to laser pulse are recorded by comparing the bright field images. Utilizing the calibration curve (P vs. d), the fraction of detached cells y gained as a function of d was re-plotted as a function of P (Fig. 4c). If one assumes that the probability distribution function of detachment pressure P follows the normal distribution with the average $P*$ and standard deviation P_{SD}, the percentage of the detached cells $y(P)$ can be fitted with:

$$y(P) = \frac{1}{2}\left(1 + \mathrm{erf}\left(\frac{P - P^*}{\sqrt{2}P_{\mathrm{SD}}}\right)\right), \tag{3}$$

where $\mathrm{erf}(x) = \frac{2}{\sqrt{\pi}} \int_0^x e^{-t^2} dt$ is the error function [43].

- Note that this new technique offers several unique advantages over the commonly used techniques. As the full width at half maximum of the pressure wave is shorter than 100 ns, the cell is not able to remodel focal adhesions or cytoskeletons. This makes the readouts $P*$ independent from the loading rate. Moreover, this technique does not rely on any probe, such as AFM, and the pressure wave does not kill the cell. After detachment, cells swim by inertia and then adhere to the new place. Last but not least, this assay can achieve an extremely high throughput, as there is no need to measure the pressure exerted on each cell. The throughput of the first generation was about 30–50 per hour [25, 30, 42], but the recent version with a motorized XY-stage could reach >500 cells within 30 min [43].

3.7 Extraction of Dynamic Cellular Phenotypes (Shape)

- Characteristic spatiotemporal patterns of cell dynamics can be extracted from various length scales. Since "shape" and "motion" are the key readouts, no labeling is necessary. This is in fact a major advantage in dealing with primary HSPC from donors or patients.
- Extract the cell periphery either from phase-contrast image or RICM image. Phase contrast is more recommended for HSPC, as the adhesion area is small (Fig. 3) [30, 42]. For cells flattened on the substrates, such as cancer cells, RICM is more recommended [24].
- Plot the radial distance from the center of mass in a polar coordinate $r(\theta)$ (Fig. 5a), and plot them over time (Fig. 5b). Note that the features are very noisy.
- Calculate the autocorrelation function map (Fig. 5d):

$$\Gamma_{rr}(\theta, t) = \frac{\langle r(\theta + \Delta\theta, t + \Delta t) \cdot r(\theta, t) \rangle}{\left\langle [r(\theta, t)]^2 \right\rangle}. \tag{4}$$

Autocorrelation functions yield the information about the number of rotational symmetry axes and the persistence of characteristic deformation patterns. Comparable to highly deformable cancer cells [24], the autocorrelation maps of HSPC are much more featureless as the cytoplasmic spaces are mostly occupied by nuclei [30, 31].

- Calculate the power spectrum $\hat{\Gamma}_m$ (Fig. 5e) to extract more information: $\hat{\Gamma}_m = c_m c_{m-}$, where c_m is the Fourier transform of the deformation $r(\theta,t)$ at mth mode:

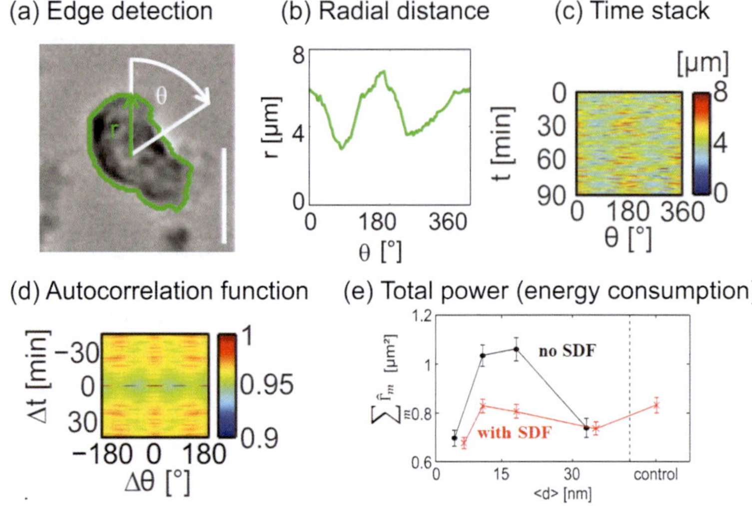

Fig. 5 Spatiotemporal analysis of dynamic cell shape. (**a**) Edge detection from phase-contrast image. (**b**) Plot of radial distance $r(\theta)$ in a polar coordinate. (**c**) Recording the radial distance over time $r(\theta, t)$. (**d**) Autocorrelation function map following Eq. 4. (**e**) Total power in the absence (black) and presence (red) of 5 ng/mL SDF1α, plotted as a function of average distance $<d>$ between SDF1α

$$c_m = \frac{1}{2}\int_0^{2\pi} d\theta r(\theta, t) e^{im\theta}. \quad (5)$$

The definition of each mode is schematically presented in Fig. 5c: $m = 0$ is an isotropic expansion/contraction, $m = 1$ a translational displacement, $m = 2$ an elliptic deformation, and $m = 3$ a triangular deformation. $m = 0$ and $m = 1$ are not included in the deformation analysis.

- The significance of $\hat{\Gamma}_m$ represents the energy dissipated/consumed by the active deformation of each mode [44, 45]. In case of human HSPC, the cell predominantly undergoes elliptic deformation ($m = 2$), followed by triangular deformation ($m = 3$). However, the deformation of higher modes ($m = 4, 5, 6\ldots$) is negligible due to the limited deformability.
- Namely, the sum of all powers (called total power), $\sum \hat{\Gamma}_m = \hat{\Gamma}_2 + \hat{\Gamma}_3$, coincides with the total energy that HSPC dissipated by active deformation, indicating the activity of HSPC. This can be used as a new quantitative index describing the influence of new mobilization agents and drugs [30, 31, 45].

3.8 Extraction of Dynamic Cellular Phenotypes (Motion)

- Plot the migration trajectory of each cell by tracking the center of mass extracted from each frame (Fig. 6a, b).
- Extract mean square displacement, MSD $= (\Delta x(t))^2$. If the migration is a stochastic random motion, MSD yields the

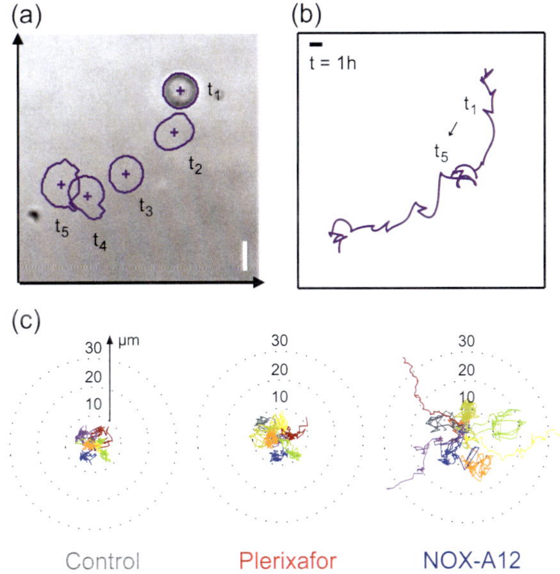

Fig. 6 (**a**) Tracking of cell center and periphery of HSPC from the phase-contrast images, as shown in Fig. 5. (**b**) Cell trajectory monitored for 1 h. Scale bar: 10 μm. (**c**) Migration trajectories of HSPC recorded over 1 h in the absence of extrinsic factors (left) and in the presence of 50 ng/mL plerixafor (middle) and NOX-A12 (right)

diffusion coefficient D, following the Einstein equation: MSD $= 4Dt$. However, a different type of diffusion shows nonzero power law coefficients, i.e., MSD $\propto t^{\beta}$ ($\beta \neq 1$). $\beta < 1$ is defined as a sub-diffusion (or obstacled diffusion), while $\beta > 1$ as super-diffusion. Especially, $\beta = 2$ is a characteristic fingerprint of a ballistic/directional motion. It is highly relevant if the mobilization agents could alter the mode of diffusion from a random walk to a ballistic motion.

- Check whether chemokines and mobilization agents cause the extension or shrinkage of migration trajectories (Fig. 6c). The persistence of the trajectory can be quantified by the persistence time τ, which is defined as [45]:

$$C(t) = \cos\zeta(0)\cos\zeta(t) + \sin\zeta(0)\sin\zeta(t) = e^{-\frac{t}{\tau}}.$$

  $\zeta(t)$ is the angle between two neighboring migration segments.

3.9 Final Remarks

Not only the quantitative information about adhesion (adhesion area and mechanical strength of adhesion) but also the dynamic phenotypes have been used as new physical biomarkers to identify the influence of SDF1α, plerixafor, NOX-A12, and some more mobilization agents. For example, we found that NOX-A12 directly interferes with adhesion mediated via the SDF1α-CXCR4 axis by neutralizing SDF1α. On the other hand, plerixafor does not cause any change in the adhesion area, implying that plerixafor does not function only as a pure antagonist to the SDF1α-CXCR4 axis. Intriguingly, both agents significantly suppressed the energy consumption by HSPC, which is distinctly different from naturally occurring SDF1α.

Recently, we have proposed an equation of motion for cell migration by relating the active deformation and motion [31]. It has been demonstrated that the migration trajectories of HSPC in the absence of SDF1α can be well reproduced by the model where the deformation and motion are linearly coupled. On the other hand, the trajectories of HSPC in the presence of SDF1α can only be recapitulated by taking the nonlinear coupling of deformation and motion into account. Though such a numerical modeling is beyond the scope of this article, physical and hence quantitative information about the dynamics of HSPC is fully complementary to the static phenotypes obtained by high-throughput microscopy, shedding light not only on the influence of mobilization agents but also on the unknown side effects of target-specific inhibitors and chemotherapy.

4 Notes

1. Reagent grade lecithins from soybean or egg yolk contain impurities such as triglycerides. Purified lecithins from the same sources are commercially available, but they contain phosphatidylcholines with different chain lengths and different degrees of chain unsaturations.
2. Ellipsometry yields the thickness and refractive index of the layer from two ellipsometric angles, Δ and Ψ [46]. As one of the two angles becomes insensitive in case Si wafers are coated with 1–2-nm-thick native oxides, the use of substrates with thermal oxides (150 nm) is recommended [47]. Neutron reflectivity yields the thickness, roughness, and scattering length density of the membrane in sub-Å accuracy [48, 49]. The film thickness appears in the form of Kiessig fringes, whose separation distance in momentum transfer Δq is given as $\Delta q = \frac{2\pi}{d}$. Since d is the thickness of the layer, the fringes from 140- to 150-nm-thick thermal oxides disturb the signal from $\approx$ 5-nm-thick lipid membranes. Thus, for neutron reflectivity, the use of silicon substrates with native oxides is recommended.
3. In case of Ca^{2+}-dependent proteins, such as cadherin, add 1 mM $CaCl_2$ to the working buffer.

4. One experiment requires approximately 3–10 × 10^4 cells to get statistically reliable data. Due to the limited access to HSPC samples, HSPC were defined as $CD34^+$ cells [28, 31, 42].

Acknowledgments

M.T. is grateful to Anthony D. Ho for insightful suggestions and continuous supports. M.T. also thanks P. Wuchter, R. Saffrich, and V. Eckstein for the long-lasting collaboration bridging clinical hematology and physics; A. Burk, C. Monzel, H. Yoshikawa, J. Thoma, and A. Yamamoto for developing the experimental and analytical platforms; and T. Ohta for the development of theoretical models. M.T. thanks the German Science Foundation (SFB873 B07), JSPS (17H00855, 16KT0070), and Nakatani Foundation for supports.

References

1. Zhang J, Niu C, Ye L, Huang H, He X, Tong W-G, Ross J, Haug J, Johnson T, Feng JQ, Harris S, Wiedemann LM, Mishina Y, Li L (2003) Identification of the haematopoietic stem cell niche and control of the niche size. Nature 425(6960):836–841
2. Mendez-Ferrer S, Michurina TV, Ferraro F, Mazloom AR, Macarthur BD, Lira SA, Scadden DT, Ma'ayan A, Enikolopov GN, Frenette PS (2010) Mesenchymal and haematopoietic stem cells form a unique bone marrow niche. Nature 466(7308):829–834
3. Puch S, Armeanu S, Kibler C, Johnson KR, Müller CA, Wheelock MJ, Klein G (2001) N-cadherin is developmentally regulated and functionally involved in early hematopoietic cell differentiation. J Cell Sci 114 (8):1567–1577
4. Hosokawa K, Arai F, Yoshihara H, Iwasaki H, Hembree M, Yin T, Nakamura Y, Gomei Y, Takubo K, Shiama H, Matsuoka S, Li L, Suda T (2010) Cadherin-based adhesion is a potential target for niche manipulation to protect hematopoietic stem cells in adult bone marrow. Cell Stem Cell 6(3):194–198
5. Wein F, Pietsch L, Saffrich R, Wuchter P, Walenda T, Bork S, Horn P, Diehlmann A, Eckstein V, Ho AD, Wagner W (2010) N-cadherin is expressed on human hematopoietic progenitor cells and mediates interaction with human mesenchymal stromal cells. Stem Cell Res 4(2):129–139
6. Aiuti A, Webb IJ, Bleul C, Springer T, Gutierrez Ramos JC (1997) The chemokine SDF-1 is a chemoattractant for human CD34(+) hematopoietic progenitor cells and provides a new mechanism to explain the mobilization of CD34(+) progenitors to peripheral blood. J Exp Med 185(1):111–120
7. Möhle R, Bautz F, Rafii S, Moore MAS, Brugger W, Kanz L (1998) The chemokine receptor CXCR-4 is expressed on CD34(+) hematopoietic progenitors and leukemic cells and mediates transendothelial migration induced by stromal cell-derived factor-1. Blood 91(12):4523–4530
8. Dar A, Goichberg P, Shinder V, Kalinkovich A, Kollet O, Netzer N, Margalit R, Zsak M, Nagler A, Hardan I, Resnick I, Rot A, Lapidot T (2005) Chemokine receptor CXCR4–dependent internalization and resecretion of functional chemokine SDF-1 by bone marrow endothelial and stromal cells. Nat Immunol 6:1038–1046
9. Zepeda-Moreno A, Saffrich R, Walenda T, Hoang VT, Wuchter P, Sánchez-Enríquez S, Corona-Rivera A, Wagner W, Ho AD (2012) Modeling SDF-1–induced mobilization in leukemia cell lines. Exp Hematol 40(8):666–674
10. To LB, Haylock DN, Simmons PJ, Juttner CA (1997) The biology and clinical uses of blood stem cells. Blood 89(7):2233–2258
11. Copelan EA (2006) Hematopoietic stem-cell transplantation. N Engl J Med 354 (17):1813–1826
12. Duong HK, Savani BN, Copelan E, Devine S, Costa LJ, Wingard JR, Shaughnessy P, Majhail N, Perales M-A, Cutler CS, Bensinger W, Litzow MR, Mohty M,

Champlin RE, Leather H, Giralt S, Carpenter PA (2014) Peripheral blood progenitor cell mobilization for autologous and allogeneic hematopoietic cell transplantation: guidelines from the American Society for Blood and Marrow Transplantation. Biol Blood Marrow Transplant 20(9):1262–1273

13. Petit I, Szyper-Kravitz M, Nagler A, Lahav M, Peled A, Habler L, Ponomaryov T, Taichman RS, Arenzana-Seisdedos F, Fujii N, Sandbank J, Zipori D, Lapidot T (2002) G-CSF induces stem cell mobilization by decreasing bone marrow SDF-1 and up-regulating CXCR4. Nat Immunol 3 (7):687–694
14. Welte K (2014) G-CSF: filgrastim, lenograstim and biosimilars. Expert Opin Biol Ther 14 (7):983–993
15. Wuchter P, Ran D, Bruckner T, Schmitt T, Witzens-Harig M, Neben K, Goldschmidt H, Ho AD (2010) Poor mobilization of hematopoietic stem cells—definitions, incidence, risk factors, and impact on outcome of autologous transplantation. Biol Blood Marrow Transplant 16(4):490–499
16. Pusic I, DiPersio JF (2010) Update on clinical experience with AMD3100, an SDF-1/CXCL12–CXCR4 inhibitor, in mobilization of hematopoietic stem and progenitor cells. Curr Opin Hematol 17(4):319–326
17. Broxmeyer HE, Orschell CM, Clapp DW, Hangoc G, Cooper S, Plett PA, Liles WC, Li X, Graham-Evans B, Campbell TB, Calandra G, Bridger G, Dale DC, Srour EF (2005) Rapid mobilization of murine and human hematopoietic stem and progenitor cells with AMD3100, a CXCR4 antagonist. J Exp Med 201(8):1307–1318
18. Dar A, Schajnovitz A, Lapid K, Kalinkovich A, Itkin T, Ludin A, Kao WM, Battista M, Tesio M, Kollet O, Cohen NN, Margalit R, Buss EC, Baleux F, Oishi S, Fujii N, Larochelle A, Dunbar CE, Broxmeyer HE, Frenette PS, Lapidot T (2011) Rapid mobilization of hematopoietic progenitors by AMD3100 and catecholamines is mediated by CXCR4-dependent SDF-1 release from bone marrow stromal cells. Leukemia 25:1286–1296
19. Wuchter P, Leinweber C, Saffrich R, Hanke M, Eckstein V, Ho AD, Grunze M, Rosenhahn A (2014) Plerixafor induces the rapid and transient release of stromal cell-derived factor-1 alpha from human mesenchymal stromal cells and influences the migration behavior of human hematopoietic progenitor cells. Cell Tissue Res 355(2):315–326
20. Hoellenriegel J, Zboralski D, Maasch C, Rosin NY, Wierda WG, Keating MJ, Kruschinski A, Burger JA (2014) The Spiegelmer NOX-A12, a novel CXCL12 inhibitor, interferes with chronic lymphocytic leukemia cell motility and causes chemosensitization. Blood 123 (7):1032–1039
21. Boutros M, Heigwer F, Laufer C (2015) Microscopy-based high-content screening. Cell 163(6):1314–1325
22. Schmitt L, Dietrich C, Tampe R (1994) Synthesis and characterization of chelator-lipids for reversible immobilization of engineered proteins at self-assembled lipid interfaces. J Am Chem Soc 116(19):8485–8491
23. Rudd AK, Valls Cuevas JM, Devaraj NK (2015) SNAP-tag-reactive lipid anchors enable targeted and spatiotemporally controlled localization of proteins to phospholipid membranes. J Am Chem Soc 137(15):4884–4887
24. Kaindl T, Rieger H, Kaschel L-M, Engel U, Schmaus A, Sleeman J, Tanaka M (2012) Spatio-temporal patterns of pancreatic cancer cells expressing CD44 isoforms on supported membranes displaying hyaluronic acid oligomers arrays. PLoS One 7(8):e42991
25. Rieger H, Yoshikawa HY, Quadt K, Nielsen MA, Sanchez CP, Salanti A, Tanaka M, Lanzer M (2015) Cytoadhesion of plasmodium falciparum–infected erythrocytes to chondroitin-4-sulfate is cooperative and shear enhanced. Blood 125(2):383–391
26. Kern W, Puotinen DA (1970) Cleaning solutions based on hydrogen peroxide for use in silicon semiconductor technology. RCA Rev 31:187–206
27. Sackmann E (1996) Supported membranes: scientific and practical applications. Science 271(5245):43–48
28. Tanaka M, Sackmann E (2005) Polymer-supported membranes as models of the cell surface. Nature 437(7059):656–663
29. Körner A, Deichmann C, Rossetti FF, Köhler A, Konovalov OV, Wedlich D, Tanaka M (2013) Cell differentiation of pluripotent tissue sheets immobilized on supported membranes displaying cadherin-11. PLoS One 8(2): e54749
30. Burk AS, Monzel C, Yoshikawa HY, Wuchter P, Saffrich R, Eckstein V, Tanaka M, Ho AD (2015) Quantifying adhesion mechanisms and dynamics of human hematopoietic stem and progenitor cells. Sci Rep 5:9370
31. Monzel C, Becker AS, Saffrich R, Wuchter P, Eckstein V, Ho AD, Tanaka M (2018) Dynamic cellular phenotyping defines specific mobilization mechanisms of human

hematopoietic stem and progenitor cells induced by SDF1α versus synthetic agents. Sci Rep 8(1):1841
32. Ludwig A, Saffrich R, Eckstein V, Bruckner T, Wagner W, Ho AD, Wuchter P (2014) Functional potentials of human hematopoietic progenitor cells are maintained by mesenchymal stromal cells and not impaired by plerixafor. Cytotherapy 16(1):111–121
33. Dexter TM, Moore MAS (1977) In vitro duplication and cure of haemopoietic defects in genetically anaemic mice. Nature 269 (5627):412–414
34. Vater A, Sahlmann J, Kröger N, Zöllner S, Lioznov M, Maasch C, Buchner K, Vossmeyer D, Schwoebel F, Purschke WG, Vonhoff S, Kruschinski A, Hübel K, Humphrey M, Klussmann S, Fliegert F (2013) Hematopoietic stem and progenitor cell mobilization in mice and humans by a first-in-class mirror-image oligonucleotide inhibitor of CXCL12. Clin Pharmacol Ther 94(1):150–157
35. Rädler J, Sackmann E (1993) Imaging optical thicknesses and separation distances of phospholipid vesicles at solid surfaces. J Phys II France 3(5):727–748
36. Limozin L, Sengupta K (2009) Quantitative reflection interference contrast microscopy (RICM) in soft matter and cell adhesion. ChemPhysChem 10:2752–2768
37. Yamamoto A, Mishima S, Maruyama N, Sumita M (1998) A new technique for direct measurement of the shear force necessary to detach a cell from a material. Biomaterials 19 (7):871–879
38. Sagvolden G, Giaever I, Pettersen EO, Feder J (1999) Cell adhesion force microscopy. Proc Natl Acad Sci U S A 96(2):471–476
39. Chen S, Springer TA (2001) Selectin receptor–ligand bonds: formation limited by shear rate and dissociation governed by the bell model. Proc Natl Acad Sci U S A 98 (3):950–955
40. Landau LD, Lifshitz EM (1987) Chapter II: Viscous fluids. In: Landau LD, Lifshitz EM (eds) Fluid mechanics, 2nd edn. Pergamon, New York, pp 44–94
41. Merkel R, Nassoy P, Leung A, Ritchie K, Evans E (1999) Energy landscapes of receptor–ligand bonds explored with dynamic force spectroscopy. Nature 397:50–53
42. Yoshikawa HY, Rossetti FF, Kaufmann S, Kaindl T, Madsen J, Engel U, Lewis AL, Armes SP, Tanaka M (2011) Quantitative evaluation of mechanosensing of cells on dynamically tunable hydrogels. J Am Chem Soc 133 (5):1367–1374
43. Yu L, Li J, Hong J, Takashima Y, Fujimoto N, Nakajima M, Yamamoto A, Dong X, Dang Y, Hou Y, Yang W, Minami I, Okita K, Tanaka M, Luo C, Tang F, Chen Y, Tang C, Kotera H, Liu L (2018) Low cell-matrix adhesion reveals two subtypes of human pluripotent stem cells. Stem Cell Rep 11(1):142–156
44. Partin AW, Schoeniger JS, Mohler JL, Coffey DS (1989) Fourier analysis of cell motility: correlation of motility with metastatic potential. Proc Natl Acad Sci U S A 86 (4):1254–1258
45. Ohta T, Monzel C, Becker AS, Ho AD, Tanaka M (2018) Simple physical model unravels influences of chemokine on shape deformation and migration of human hematopoietic stem cells. Sci Rep 8(1):10630
46. Azzam RMA, Bashara NM (1978) Ellipsometry and polarized light. Phys Today 31(11):72
47. Seitz PC, Reif MD, Konovalov OV, Jordan R, Tanaka M (2009) Modulation of substrate--membrane interactions by linear poly (2-methyl-2-oxazoline) spacers revealed by X-ray reflectivity and ellipsometry. Chem PhysChem 10(16):2876–2883
48. Fragneto-Cusani G (2001) Neutron reflectivity at the solid/liquid interface: examples of applications in biophysics. J Phys Condens Matter 13(21):4973–4989
49. Rossetti FF, Schneck E, Fragneto G, Konovalov OV, Tanaka M (2015) Generic role of polymer supports in the fine adjustment of interfacial interactions between solid substrates and model cell membranes. Langmuir 31 (15):4473–4480

Chapter 3

Colony Formation: An Assay of Hematopoietic Progenitor Cells

Romy Kronstein-Wiedemann and Torsten Tonn

Abstract

The colony-forming cell (CFC) assay is used to study the proliferation and differentiation pattern of each input hematopoietic progenitors by their ability to form colonies in a semisolid medium. The resulting colonies are consisting of more differentiated cells, and the number and the morphology of the colonies provide preliminary information about the ability of progenitors to differentiate and proliferate. To allow colonies to grow to a size which facilitates accurate counting and identification, about 14 days of culture is sufficient. In certain situations also shorter periods may be used.

Key words Ficoll, Red blood cell depletion, Hematopoietic progenitors, Methyl cellulose, Semisolid culture, Colony-forming units, Erythroid and myeloid colonies, STEMvision

1 Introduction

Mature blood cells have a limited life-span and are continuously replaced by proliferation and differentiation of a very small population of multipotent hematopoietic stem cells (HSC) in bone marrow. HSC have the ability to differentiate into all hematopoietic lineages but also retain their self-renewal capacity [1]. Hematopoietic stem cell transplantation (HSCT) utilizing bone marrow as a stem cell source has become an accepted treatment modality for a variety of metabolic, immunologic, and hematologic malign and non-malign disorders [1–3]. Most hematopoietic stem cells can be identified by expression of CD34 antigen, and these cells have a high engraftment capacity. The discovery and clinical application of granulocyte colony-stimulating growth factor (G-CSF) and granulocyte–macrophage colony-stimulating growth factor (GM-CSF) led to the observation that CD34+ bone marrow stem cells can be mobilized into the peripheral blood in large numbers [4].

The presence of primitive hematopoietic cells in adult peripheral blood (PB) has been known for more than four decades. Studies with PB showed the demonstration and quantitation of

Gerd Klein and Patrick Wuchter (eds.), *Stem Cell Mobilization: Methods and Protocols*, Methods in Molecular Biology, vol. 2017, https://doi.org/10.1007/978-1-4939-9574-5_3,

specific progenitor populations detected by colony formation assays [5–7]. This prompted the evaluation of leukapheresis harvests as an alternative source of cells for therapeutic applications where autologous or allogeneic bone marrow transplants were not feasible [8–11]. The key to the development of strategies for increasing the concentration of clonogenic progenitors in the circulation as well as to an improved understanding of factors that regulate stem cell mobilization, recruitment, and marrow colonization is the availability of a quantitative assay for cells capable of reconstituting and sustaining hematopoiesis in transplant recipients [12].

Although HSC have the capacity to proliferate and differentiate in culture, most cells detected in hematopoietic culture assays consist of hematopoietic progenitor cells which have limited self-renewal capacity and short-term hematopoietic potential. Progenitor cells detected in culture assays can either be multipotential or restricted to one or two lineages (erythrocytes, granulocytes, monocytes/macrophages, or platelets) [13–15]. While CFU-E are clusters of a total of 8–200 mature erythroid progenitors, BFU-E are more immature progenitors and produce a colony containing >200 erythroblasts. CFU-GM may contain thousands of granulocytes (CFU-G) or of macrophages (CFU-M) or cells of both lineages in single or multiple clusters. The CFU-GEMM are progenitors that can form colonies, in semisolid culture medium, comprising mixtures of granulocytes, erythroblasts, megakaryocytes, and macrophages [16]. Clonogenic assays are the most directly quantitative means of measuring human hematopoietic progenitor cells in vitro [16]. Hematopoietic colonies are essentially clones of cells produced by a single progenitor cell. The aim of colony-forming unit (CFU) assays is to define the potential of hematopoietic stem and progenitor cell populations for proliferation and lineage differentiation, and the colonies can be morphologically analyzed. With the help of short-term in vitro colony assays, a more detailed model of the compartmentalization of intermediate, lineage-restricted progenitor subsets was established (Fig. 1).

2 Materials

All procedures for cell processing and setup of CFU assays should be performed using sterile technique and universal handling precautions. Use of water-jacketed incubators with a water pan placed in the chamber is recommended. It is important to use medical-grade CO_2, as inhibition of CFU growth due to toxic substances present in the CO_2 gas source has been reported.

- Sterile serological pipettes.
- Sterile polystyrene tubes.
- Conical tubes.
- Sterile pipette tips.

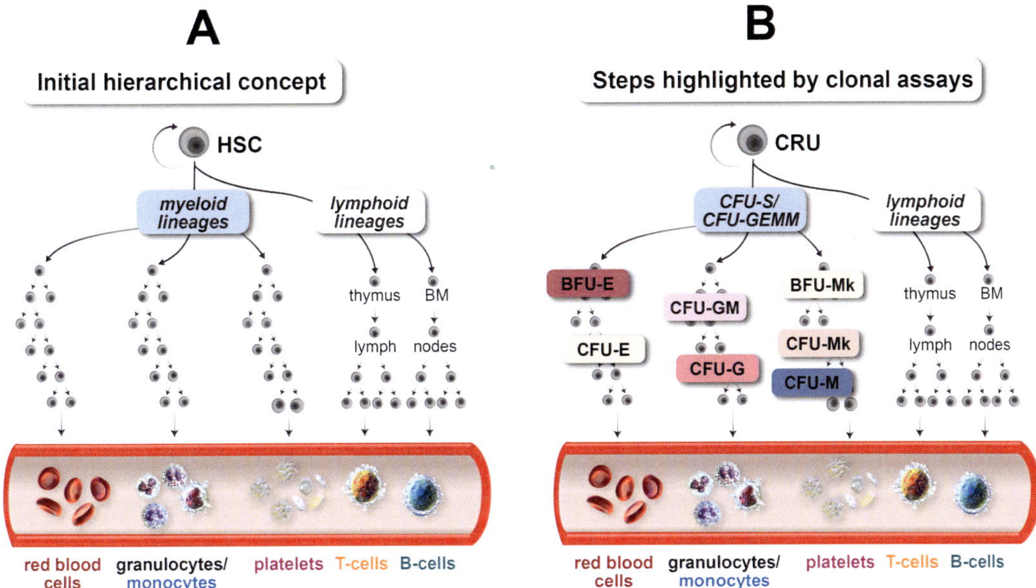

Fig. 1 Hierarchical models of HSC self-renewal and differentiation modified by Eaves 2015 [17]. (**a**) Initial view showing all lymphoid and all myeloid potentialities as the first lineage groupings to be segregated. (**b**) A more detailed view of the compartmentalization of intermediate, lineage-restricted progenitor subsets based on their behavior in short-term in vitro colony assays and properties allowing their separate isolation. *HSC* hematopoietic stem cell, *CRU* competitive repopulating unit, *CFU-S* colony-forming unit-spleen, *CFU-GEMM* colony-forming unit-granulocyte, erythroid, macrophage, megakaryocyte, *BFU-E* burst-forming unit-erythroid, *CFU-E* colony-forming unit-erythroid, *CFU-G* colony-forming unit-granulocyte, *CFU-M* colony-forming unit-macrophage, *BFU-Mk* burst-forming unit-megakaryocyte, *CFU-Mk* colony-forming unit-megakaryocyte, *BM* bone marrow

2.1 Preparation of Cell Samples

- 0.8% ammonium chloride solution.
- Ficoll Paque™ ($\rho = 1.077$ g/mL).
- Isolation buffer: Phosphate-buffered saline (PBS, pH 7.2), 2 mM EDTA, and 0.5% human serum albumin; keep buffer at 2–8 °C.
- Human CD34 MicroBead Kit (e.g., from Miltenyi): contains FcR blocking reagent and CD34 microbeads.
- LS columns.
- Cell separator (e.g., from Miltenyi).
- 0.4% trypan blue solution.

2.2 Colony-Forming Unit (CFU) Assay

- Methylcellulose: We recommend the use of complete methylcellulose media (e.g., STEMCELL Technologies or Miltenyi), which contains all cytokines for detection of CFU-E, BFU-E, CFU-GM, and CFU-GEMM.
- Iscove's Modified Dulbecco's Medium (IMDM) + 2% FBS.

- Syringes (luer lock).
- 16-gauge blunt-end needles.
- 35 mm culture dishes or SmartDish™ 6-well culture plates.
- 100 mm culture dishes or 245 mm square dishes.

3 Methods

3.1 Preparation of Cell Samples

The presence of large numbers of red blood cells (RBCs) in a colony-forming unit assay prevents hematopoietic colonies from being accurately visualized either manually or automatically. RBCs must be removed from fresh cord blood, bone marrow, and mobilized peripheral blood samples (whether whole or processed), before performing the CFU assay.

3.1.1 Red Blood Cell Depletion by Treatment with Ammonium Chloride Solution

1. Mix sample well, and remove a small volume of cells (100 μL) to perform an initial cell count using a Neubauer hemocytometer or a calibrated automated cell counter. Perform an initial nucleated cell count to establish the number and concentration of nucleated cells in the original sample.
2. Add buffered ammonium chloride solution to sample to give a minimum of 4:1 (v/v) ratio (i.e., $\geq$4 mL ammonium chloride solution: 1 mL of sample). Gently vortex the mixture immediately after adding the lysis reagent, and put on ice for 10 min with gentle vortexing or inversion once or twice during the incubation period. All of the RBCs should be lysed within 10 min.
3. Make up the volume in each tube to 12 mL with isolation buffer, and centrifuge the contents at 300 × *g* for 10 min at room temperature (15–25 °C) with the brake on.
4. Quickly, but carefully, remove and discard the supernatant so as not to remove the cell pellet, and resuspend the cell pellet, first by gentle vortexing and then after addition of 10 mL of isolation buffer by vortexing more vigorously. Cells from multiple tubes of the same sample may be pooled.
5. Centrifuge the cell suspension at 300 × *g* for 10 min at room temperature with the brake on. Discard the supernatant and wash the cells once more. Discard the supernatant from the final wash, and gently resuspend the cells in minimum 1–2 mL of isolation buffer.
6. Record the exact final volume and count live cells using a hemacytometer or light microscope.

3.1.2 Viable Cell Count Using Trypan Blue Dye Exclusion

1. Mix cell suspension thoroughly and transfer a 100 μL aliquot to a separate tube. The trypan blue dye exclusion method should be performed by diluting the cell sample with an equal volume of trypan blue (1:2 dilution). If additional dilution is required, the cell sample should be diluted in cell culture medium prior to dilution in trypan blue.
2. Mix the diluted sample well and allow the resulting solution to sit for 5 min.
3. Fill both chambers of the hemocytometer. Do not over- or underfill the chambers.
4. Count cells in four large squares (Fig. 2). The dead cells are stained blue as they have taken up the trypan blue due to a decrease in cell membrane integrity. The live cells are clear and refractile as they have not taken up the trypan blue.
5. Calculate the **viable cell count** as follows:

$$\text{Viable cell count per mL} = \text{average total viable cells per square} \times \text{dilution factor} \times 10^4$$

6. Calculate the **percent viability** as follows:

$$\%\text{Viability} = \frac{\text{Total number of viable cells}}{\text{Total number of viable} + \text{nonviable cells}} \times 100\%$$

Hematopoietic colony-forming cells are present in the mononuclear cell fraction (WBC, Fig. 3) of different hematopoietic cell

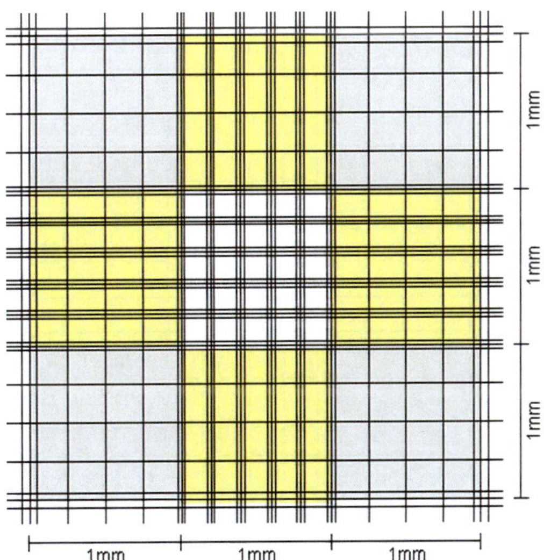

Fig. 2 Schematic draft of a "Neubauer-improved" hemocytometer counting chamber for cells. The counting grid consists of 3 × 3 large squares, each with an area of 1 mm^2

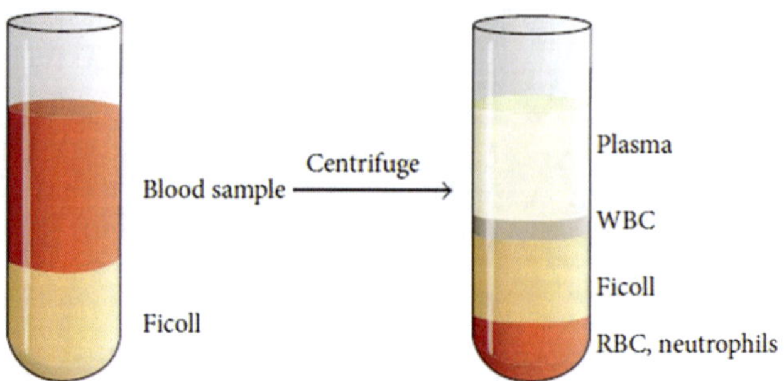

Fig. 3 Principle of density gradient centrifugation method. The blood sample is layered on top of Ficoll Paque™, a medium with higher density. Using centrifugal force, particles move through the culture medium and density gradient and are suspended at a point at which the density of the particles equals the surrounding medium [18]

3.1.3 Isolation of Light Density (Mononuclear) Cells by Ficoll Paque™

sources. Isolation of light density cells using Ficoll Paque™ enriches white blood cells and depletes mature red blood cells, nucleated erythroid progenitors, neutrophils, and dense nonviable cells [18].

1. Mix sample well, and remove a small volume of cells (100 μL) to perform an initial cell count as described in Subheading 3.1.1.
2. Dilute the cells at least 1:1 with isolation buffer, and invert gently to mix the cell suspension.
3. Add 15 mL Ficoll Paque™ to a 50 mL conical tube, and slowly layer 35 mL of cell sample or whole blood on top of the Ficoll Paque™ so that a distinct layer forms.
4. Centrifuge at 800 × g for 20 min in swinging-bucket rotor without brake.
5. Carefully transfer the mononuclear cell layer (lymphocytes, monocytes, nucleated erythroid progenitors, stem cells), localized at the interphase, to a new 50 mL conical tube.
6. Fill the conical tube with isolation buffer, mix, and centrifuge at 300 × g for 10 min at 20 °C. Carefully remove the supernatant completely.
7. For removal of platelets, resuspend the cell pellet in 50 mL of isolation buffer, and centrifuge at 200 × g for 10 min. Carefully remove the supernatant completely.
8. Resuspend the cell pellet in an appropriate amount of isolation buffer, and proceed to magnetic labeling.

3.1.4 Cell Preparation by Positive Selection of CD34+ Cells

The majority of human hematopoietic progenitor cells, including most BFU-E, CFU-GM, and CFU-GEMM, express the CD34 antigen and lack antigens present on more mature lineage-committed cells. Therefore, CFUs can be enriched from bone marrow, cord blood, and mobilized peripheral blood samples by isolation of CD34+ cells using specific monoclonal antibodies and immunoseparation technologies. CD34+ cells are magnetically labeled with CD34 MicroBeads (e.g., from Miltenyi). Then, the cell suspension is loaded onto a column which is placed in the magnetic field of a special separator. The magnetically labeled CD34+ cells are retained within the column. The unlabeled cells run through. After removing the column from the magnetic field, the magnetically retained CD34+ cells can be eluted.

1. Determine cell number (*see* **Note 1**).
2. Centrifuge cell suspension at 300 × *g* for 10 min. Aspirate supernatant completely, and resuspend cell pellet in 300 μL of buffer for up to 10^8 total cells.
3. Add 100 μL of FcR Blocking Reagent and 100 μL of CD34 MicroBeads for up to 10^8 total cells.
4. Mix well and incubate for 30 min in the refrigerator (2–8 °C).
5. Fill the conical tube with isolation buffer, and centrifuge at 300 × *g* for 10 min. Aspirate supernatant completely.
6. Resuspend up to 10^8 cells in 500 μL of buffer.
7. Place column in the magnetic field of a suitable separator.
8. Prepare LS column by rinsing with 3 mL of isolation buffer.
9. Apply cell suspension onto the column. Collect flow-through containing unlabeled cells.
10. Wash column with 3 × 3 mL of isolation buffer.
11. Remove column from the separator, and place it on a suitable collection tube.
12. Pipette 5 mL of isolation buffer onto the column. Immediately flush out the magnetically labeled cells by firmly pushing the plunger into the column.

3.2 Preparation of Methylcellulose Media

The complete methylcellulose media are formulated to allow the addition of cells to methylcellulose medium at a 1:10 (v/v) ratio, which maintains the optimal viscosity of the medium (*see* **Note 2**).

1. Thaw bottle of complete MethoCult™ medium at room temperature (15–25 °C) or overnight at 2–8 °C (*see* **Note 3**).
2. Shake vigorously for 1–2 min, and then let stand for at least 5 min to allow bubbles to rise to the top before aliquoting (*see* **Note 4**).
3. Use a 3 mL or 6 mL luer lock syringe attached to a 16-gauge blunt-end needle to dispense methylcellulose medium into 14 mL (17 × 95 mm) sterile tubes (*see* **Note 5**).

4. Dispense 3 mL per tube for 1.1 mL duplicate cultures or 4 mL per tube for 1.1 mL triplicate cultures (*see* **Notes 6** and 7).
5. Vortex tubes to mix well. Tubes of complete medium can be used immediately, stored at 2–8 °C for up to 1 month, or stored at −20 °C. After thawing aliquoted tubes of methylcellulose medium, mix well and use immediately. Do not refreeze.

3.3 CFU Assay

1. Thaw the required number of pre-aliquoted tubes of complete methylcellulose medium at room temperature (15–25 °C) or overnight at 2–8 °C (*see* **Note 3**).
2. Prepare culture dishes by placing 2 × 35 mm culture dishes with lids inside a 100 mm Petri dish with a lid. Add a third 35 mm culture dish without a lid as a water dish. This set of dishes is sufficient for one duplicate assay (*see* **Note 8**).
3. Count prepared cells.
4. Dilute the cells with IMDM with 2% FBS to 10× the final concentration(s) required for plating. Refer to Table 1 for recommended plating concentrations for cell samples from different tissues. For example, to achieve a final plating concentration of 1×10^4 cells per dish, prepare a cell suspension of 1×10^5 cells per mL (*see* **Note 9**).
5. For a duplicate assay, add 0.3 mL of diluted cells to a pre-aliquoted 3 mL methylcellulose tube. For a triplicate assay, add 0.4 mL of diluted cells to a pre-aliquoted 4 mL methylcellulose tube (*see* **Note 10**).
6. Vortex the tube vigorously for at least 4 s to mix the contents thoroughly, and let it stand for at least 5 min to allow the bubbles to rise to the top.

Table 1
Recommended plating concentrations (refer to STEMCELL Technologies)

Cell source	10× concentration to be prepared	Plating concentration (cells per 35 mm dish)*
BM, ammonium chloride-treated	5×10^5 (2×10^5–1×10^6)	5×10^4 (2×10^4–1×10^5)
BM MNCs	2×10^5 (1×10^5–5×10^5)	2×10^4 (1×10^4–5×10^4)
CB MNCs	1×10^5 (5×10^4–2×10^5)	1×10^4 (5×10^3–2×10^4)
PB MNCs	2×10^6 (1×10^6–2×10^6)	2×10^5 (1×10^5–2×10^5)
MPB MNCs	2×10^5 (1×10^5–5×10^5)	2×10^4 (1×10^4–5×10^4)
Lin-depleted ($CD34^+$ enriched BM, CB, MPB)	1×10^4 (5×10^3–2×10^4)	1000 (500–2000)
$CD34^+$ cells (BM, CB, MPB)	5×10^3 (5×10^3–2×10^4)	500 (500–2000)

* MethoCult™ containing recombinant cytokines

7. To dispense the methylcellulose mixture containing cells into culture dishes, attach a sterile 16-gauge blunt-end needle to a sterile 3 mL luer lock syringe (*see* **Notes 2** and **11**).
8. To expel the air from the syringe, place the needle below the surface of the methylcellulose medium, and draw up approximately 1 mL to remove the air from the syringe. Gently depress the plunger and expel the medium completely.
9. Draw up the methylcellulose mixture containing cells into the syringe, and dispense a volume of 1.1 mL into each 35 mm dish as follows: while holding the syringe containing the methylcellulose and cells in one hand, remove the lid of a 35 mm dish with the opposite hand. Position the syringe over the center of the dish without touching the syringe to the dish. Dispense 1.1 mL and replace the lid.
10. Distribute the medium evenly across the surface of each 35 mm dish by gently tilting and rotating the dish to allow the medium to attach to the wall of the dish on all sides.
11. Place the culture dishes into the outer dish (e.g., 100 mm Petri dish or 245 mm square dish), and add approximately 3 mL of sterile water to the uncovered 35 mm dish(es). Place lid onto outer dish.
12. Incubate at 37 °C, in 5% CO_2 with ≥95% humidity for 14 days (*see* **Note 12**).

3.4 Analysis of Hematopoietic CFCs

CFCs can be visualized either manually or automatically using STEMvision.

CFU-E: Colony-forming unit-erythroid. CFU-E are mature erythroid progenitors. They produce 1–2 clusters containing a total of 8–200 erythroblasts and require erythropoietin (EPO) for differentiation (Fig. 4a).

BFU-E: Burst-forming unit-erythroid. BFU-E are more immature progenitors than CFU-E and require EPO and cytokines with burst-promoting activity such as interleukin-3 (IL-3) and stem cell factor (SCF) for optimal colony growth. They produce a colony containing >200 erythroblasts in a single and multiple clusters (Fig. 4b).

CFU-GM: Colony-forming unit-granulocyte, macrophage. CFU-GM may contain thousands of cells in single or multiple clusters. They produce colonies containing at least 20 granulocytes (CFU-G), macrophages (CFU-M), or cells of both lineages (CFU-GM) (Fig. 4c).

CFU-GEMM: Colony-forming unit-granulocyte, erythroid, macrophage, megakaryocyte. They are multipotential progenitors that produce large colonies containing >500 cell containing erythroblast and cells of at least two other recognizable lineages (Fig. 4d).

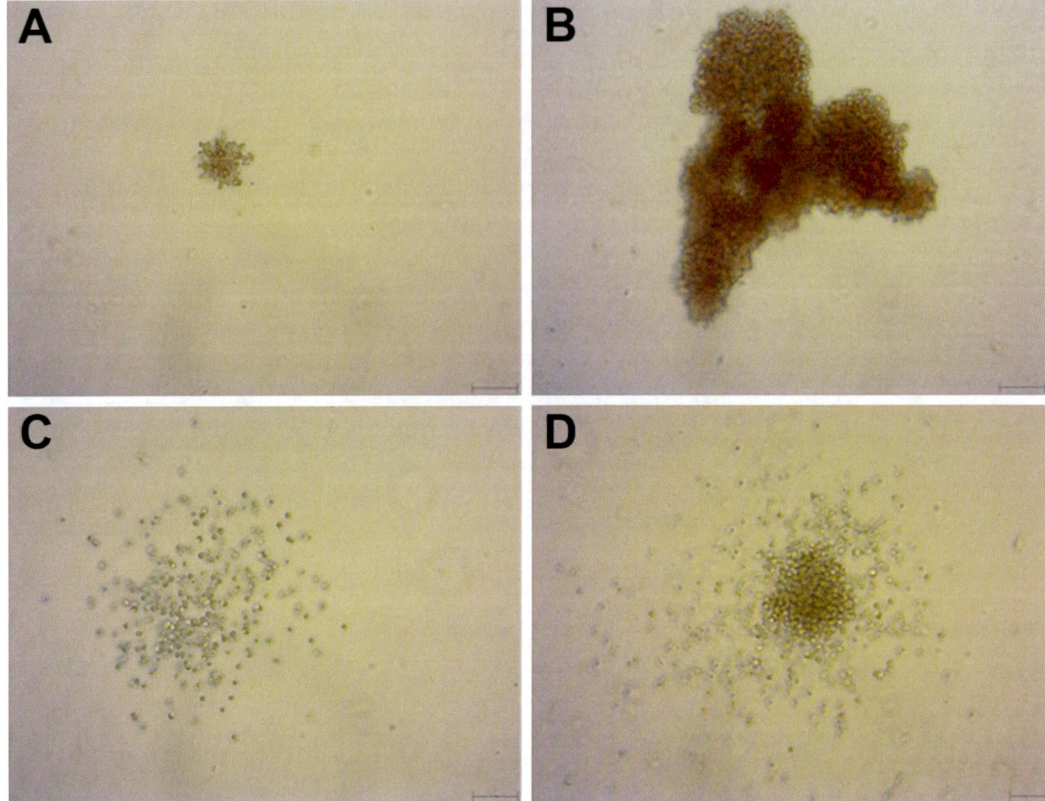

Fig. 4 Classes of human hematopoietic progenitors. (**a**) Colony-forming unit-erythroid (CFU-E). (**b**) Burst-forming unit-erythroid (BFU-E). (**c**) Colony-forming unit-granulocyte, macrophage (CFU-GM). (**d**) Small colony-forming unit-granulocyte, erythroid, macrophage, megakaryocyte (CFU-GEMM). Data information: (**a–c**) bar = 100 μm, (**d**) bar = 200 μm

4 Notes

1. Work fast, keep cells cold, and use pre-cooled solutions. This will prevent capping of antibodies on the cell surface and nonspecific cell labeling. For optimal performance, it is important to obtain a single-cell suspension before magnetic labeling. Passing cells through 30 μm nylon mesh will remove cell clumps which may clog the column.
2. Do not use pipettes to dispense methylcellulose as the volume dispensed will not be accurate. Syringes and large bore blunt-end needles should be used for accurate dispensing of viscous methylcellulose medium and to prevent needle-stick injuries.
3. Do not thaw methylcellulose medium at 37 °C.
4. For incomplete methylcellulose medium, add desired growth factors, supplements, and IMDM with 25 mM HEPES to yield a total volume of 100 mL.

5. Place the needle below the surface of the methylcellulose medium, and draw up approximately 1 mL to remove the air from the syringe. Gently depress the plunger and expel the medium completely. Repeat until no air space is visible.
6. Do not expel the medium to the “0” mark on the syringe when aliquoting. For example, measure from 3.5 mL to 0.5 mL rather than 3.0 mL to 0 mL.
7. It is preferable to dispense the entire contents of the bottle into tubes in order to avoid repeated freezing and thawing of the bottle.
8. If using SmartDish™ cultureware, add 4–8 mL of sterile water to the empty spaces between the SmartDish™ wells. Place the SmartDish™ in a 245 mm square dish, along with additional 35 mm culture dishes each containing 3–4 mL of sterile water.
9. When it is difficult to anticipate the correct plating cell concentration, use two or more cell concentrations that differ by twofold to threefold, e.g., 2×10^4 cells per dish and 1×10^4 cells per dish.
10. This 1:10 (v/v) ratio of cells/medium gives the correct viscosity to ensure optimal CFU growth and morphology.
11. For each tube plated, use a new sterile disposable 3 mL syringe fitted with a new 16-gauge blunt-end needle to prevent contamination between samples.
12. Proper culture conditions are critical for optimal CFU growth. Use of water-jacketed incubators with water pan in chamber and routine monitoring of temperature and CO_2 levels are recommended.

References

1. Kosan C, Godmann M (2016) Genetic and epigenetic mechanisms that maintain hematopoietic stem cell function. Stem Cells Int 2016:5178965
2. Barrett D, Fish JD, Grupp SA (2010) Autologous and allogeneic cellular therapies for high-risk pediatric solid tumors. Pediatr Clin N Am 57(1):47–66
3. Milanetti F, Abinun M, Voltarelli JC, Burt RK (2010) Autologous hematopoietic stem cell transplantation for childhood autoimmune disease. Pediatr Clin N Am 57(1):239–271
4. Özgüner M, Tavil B, Köksal J, Canal E, Bozkaya I, Tunç B (2011) Comparison of colony forming unit-assay results of different hematopoietic stem cell sources. Turkish J Pediatr Dis 5(4):197–201
5. Chervenick PA, Boggs DR (1971) In vitro growth of granulocytic and mononuclear cell colonies from blood of normal individuals. Blood 37(2):131–135
6. Clarke BJ, Housman D (1977) Characterization of an erythroid precursor cell of high proliferative capacity in normal human peripheral blood. Proc Natl Acad Sci U S A 74 (3):1105–1109
7. McCredie KB, Hersh EM, Freireich EJ (1971) Cells capable of colony formation in the peripheral blood of man. Science 171:293–294
8. Juttner CA, To LB, Ho JQ, Bardy PG, Dyson PG, Haylock DN, Kimber RJ (1988) Early lympho-hemopoietic recovery after autografting using peripheral blood stem cells in acute non-lymphoblastic leukemia. Transplant Proc 20(1):40–42
9. Kessinger A, Armitage JO, Landmark JD, Smith DM, Weisenburger DD (1988) Autologous peripheral hematopoietic stem cell

transplantation restores hematopoietic function following marrow ablative therapy. Blood 71(3):723–727

10. Korbling M, Dorken B, Ho AD, Pezzutto A, Hunstein W, Fliedner TM (1986) Autologous transplantation of blood-derived hemopoietic stem cells after myeloablative therapy in a patient with Burkitt's lymphoma. Blood 67 (2):529–532
11. Reiffers J, Bernard P, David B, Vezon G, Sarrat A, Marit G, Moulinier J, Broustet A (1986) Successful autologous transplantation with peripheral blood hemopoietic cells in a patient with acute leukemia. Exp Hematol 14 (4):312–315
12. Udomsakdi C, Lansdorp PM, Hogge DE, Reid DS, Eaves AC, Eaves CJ (1992) Characterization of primitive hematopoietic cells in normal human peripheral blood. Blood 80 (10):2513–2521
13. Eaves C (1995) Assays of hematopoietic progenitor cells. In: Beutler E, Lichtman MA, Coller BS, Kipps TJ (eds) Williams hematology, 5th edn. McGraw-Hill, New York, pp 22–66
14. Klug CA, Jordan CT (eds) (2002) Hematopoietic stem cell protocols, vol 63. Humana Press, Totowa, NJ
15. Miller CL, Lai B (2005) Human and mouse hematopoietic colony-forming cell assays. Methods Mol Biol 290:71–89
16. Gordon MY (1993) Human haemopoietic stem cell assays. Blood Rev 7(3):190–197
17. Eaves CJ (2015) Hematopoietic stem cells: concepts, definitions, and the new reality. Blood 125(17):2605–2613
18. Low WS, Wan Abas WA (2015) Benchtop technologies for circulating tumor cells separation based on biophysical properties. Biomed Res Int 2015:239362

Chapter 4

Mobilization and Collection of Peripheral Blood Stem Cells in Adults: Focus on Timing and Benchmarking

Katharina Kriegsmann and Patrick Wuchter

Abstract

Peripheral blood stem cells (PBSCs) are preferentially used as a hematopoietic stem cell source for autologous blood stem cell transplantation (ABSCT) upon high-dose chemotherapy (HDT) in a variety of hemato-oncologic diseases. As a prerequisite, hematopoietic stem cells have to be mobilized into the peripheral blood (PB) and collected by leukapheresis (LP). Despite continuous improvements, e.g., the introduction of plerixafor, current challenges are the further optimization regarding the leukapheresis procedure, preventing collection failures, as well as benchmarking and harmonization of mobilization approaches between institutions.

This chapter summarizes the current PBSC mobilization and collection approaches and is focusing on timely orchestration of mobilization therapy, granulocyte colony-stimulating factor (G-CSF) application, and peripheral blood (PB) $CD34^{+}$ cell assessment. Moreover, strategies for prediction and performance assessment of the PBSC collection yield are discussed.

Key words Leukapheresis, PBSC, Timing, Benchmarking

1 Introduction

High-dose chemotherapy (HDT) is an established treatment option in a variety of hemato-oncologic diseases, mainly multiple myeloma (MM) and relapsed or refractory non-Hodgkin's lymphoma (NHL) [1]. Less frequently, HDT is performed in other indications, such as sarcoma, germ cell tumors, amyloid light-chain (AL) amyloidosis and autoimmune diseases [2]. In the context of HDT, reinfused $CD34^{+}$ hematopoietic stem cells aim to restore the bone marrow (BM) function. Due to a faster hematopoietic and immune reconstitution as compared to BM stem cells, peripheral blood stem cells (PBSCs) are preferentially used as a hematopoietic stem cell source in adult MM, NHL, and Hodgkin's lymphoma patients [3–6]. As a prerequisite and due to the fact that hematopoietic stem cells are found in PB only in small numbers, they have to be mobilized from BM to peripheral blood (PB), collected by

Gerd Klein and Patrick Wuchter (eds.), *Stem Cell Mobilization: Methods and Protocols*, Methods in Molecular Biology, vol. 2017, https://doi.org/10.1007/978-1-4939-9574-5_4,

leukapheresis (LP), and stored until reinfusion [7, 8]. Despite continuous improvements, mobilization and collection failures are still frequently observed [9]. Moreover, mobilization approaches and outcomes differ considerably between collection centers [10, 11]. Therefore, the current challenges are the further optimization, benchmarking, and harmonization of mobilization approaches.

The intention of this chapter is to present current PBSC mobilization and collection strategies, focusing on timing of mobilization therapy and PB $CD34^+$ cell assessment, predicting $CD34^+$ cell collection yield, and benchmarking of LP sessions.

2 Mobilization Strategies and Regimens

Generally, two different PBSC mobilization approaches can be distinguished: steady-state mobilization with granulocyte colony-stimulating factor (G-CSF) only and chemotherapy/G-CSF mobilization (Table 1).

Table 1
Autologous PBSC mobilization strategies

A Steady-state mobilization with G-CSF only	B Chemotherapy/G-CSF mobilization	
	B'	B''
Disease-specific (chemo) therapy		Disease-specific (chemo) therapy
	Disease-specific chemotherapy as mobilization therapy	Separate mobilization chemotherapy
	↓	↓
G-CSF	G-CSF	G-CSF
↓	↓	↓
Leukapheresis	Leukapheresis	Leukapheresis

PBSC mobilization can be approached by (A) steady-state mobilization with granulocyte colony-stimulating factor (G-CSF) only or by (B) chemotherapy/G-CSF mobilization. In case of chemotherapy/G-CSF mobilization, chemomobilization can be incorporated in a disease-specific induction or salvage therapy (B′) or given as a separate mobilization chemotherapy (B″). Adopted from Mohty et al. [28]

2.1 Steady-State Mobilization

In case of steady-state mobilization, G-CSF cytokines are given subcutaneously in a dosage of 10 μg/kg per day. Currently, filgrastim (Neupogen®), lenograstim (Granocyte®), and the biosimilar Filgrastim Hexal® are approved for PBSC mobilization in Europe. Compared to chemotherapy/G-CSF mobilization, pure G-CSF mobilization is associated with fewer side effects and shows more predictable mobilization kinetics [11, 12–14]. However, variable mobilization failures and slightly lower collection yields have been reported in MM and NHL patients upon single-agent G-CSF mobilization [16–21]. Therefore, G-CSF alone might offer a first-line mobilization option in patients without risk factors for poor mobilization [13, 22]. On the other hand, for older patients or patients with considerable comorbidities, this approach is associated with less toxicity than conventional chemomobilization [23].

2.2 Chemotherapy/G-CSF Mobilization

In NHL and sarcoma patients, chemomobilization regimens are usually incorporated in induction or salvage therapy strategies and are chosen based on the disease characteristics with regard to the local clinical practice [13, 24–26]. In MM patients, cyclophosphamide-based mobilization regimens are frequently used as mobilization chemotherapy. However, the additional antitumor effect of this regimen following modern induction chemotherapy is more and more under debate [18, 27]. After chemomobilization, G-CSF is administered as daily subcutaneous injection of filgrastim (5 μg/kg body weight [bw]) or lenograstim (150 $\mu g/m^2$). Some centers tend to use filgrastim up to 10 μg/kg bw; in particular such a dose escalation is practiced if an insufficient increase of $CD34^+$ cells in PB is noted. Selected chemomobilization regimens are listed in detail in Table 2.

Compared to steady-state mobilization with G-CSF, chemomobilization is associated with higher toxicity and more side effects [12, 24]. Moreover, due to less predictable PB $CD34^+$ cell kinetics, chemomobilization requires a more comprehensive orchestration of G-CSF application, PB $CD34^+$ cell assessment, and initiation of LP sessions [11, 14, 26]. However, reduction of mobilization failure rates and higher PBSC yields has been documented as advantages of this approach [13, 21].

3 G-SCF: Originators and Biosimilars

The G-CSF originators filgrastim (Neupogen®) and lenograstim (Granocyte®) are widely used for PBSC mobilization [27]. In recent years, biosimilar agents have been approved by the European Medicines Agency for the same indication as the G-CSF originators. The biosimilar of filgrastim produced by Sandoz is currently marketed as "Filgrastim Hexal®" in Germany, as

Table 2
Chemomobilization regimens

Regimen	Substance	Dosage/application	Duration	Entity
CAD				MM [63, 66, 87–89]
	Cyclophosphamide	1000 mg/m^2/day, i.v.	d 1	AL amyloidosis [91]
	Doxorubicin	15 mg/m^2/day, i.v.	d 1–4	
	Dexamethasone	40 mg/day, p.o.	d 1–4	
Cd				AL amyloidosis [91–93]
	Cyclophosphamide	1000 mg/m^2/day, i.v.	d 1–2	
	Dexamethasone	20 mg/day, p.o.	d 1–3	
C				MS [94], SSc [94]
	Cyclophosphamide	1000 mg/m^2/day, i.v.	d 1–2	
(R)CHOP				NHL [95–97]
	Rituximab	375 mg/m^2/day, i.v.	d 0	
	Cyclophosphamide	750 mg/m^2/day, i.v.	d 1	
	Doxorubicin	50 mg/m^2/day, i.v.	d 1	
	Vincristine	1.4 mg/m^2/day, i.v. (max. 2.0 mg)	d 1	
	Prednisone	100 mg/day, p.o.	d 1–5	
(R)CHOEP				NHL [95–97]
	(R)CHOP	/	d 1–5	
	Etoposide	100 mg/m^2/day, i.v.	d 1–3	
(R)DHAP				relapsed NHL [98–101]
	Rituximab	375 mg/m^2/day, i.v.	d 0	
	Cisplatin	25 mg/m^2/day, i.v.	d 1–4	
	Cytarabine	2 × 2000 mg/m^2/day, i.v.	d 2	
	Dexamethasone	40 mg/day, p.o.	d 1–4	
(R)ICE				relapsed NHL, GCT [99, 102–105]
	Rituximab	375 mg/m^2/day, i.v.	d 0	
	Etoposide	100 mg/m^2/day, i.v.	d 1–3	
	Ifosfamide	5000 mg/m^2/day, i.v.	d 2	
	Carboplatin	AUC 5, i.v. (max. 800 mg)	d 2	
(R)AraC/TT				NHL with CNS involvement [106]
	Rituximab	375 mg/m^2/day, i.v.	d 0	
	Cytarabine	3000 mg/m^2/day, i.v.	d 1–2	
	Thiotepa	40 mg/m^2/day, i.v.	d 2	
PEI				GCT [26, 107–112]
	Cisplatin	20 mg/m^2/day, i.v.	d 1–5	
	Etoposide	75 mg/m^2/day, i.v.	d 1–5	
	Ifosfamide	1200 mg/m^2, i.v.	d 1–5	
VIDE				Sarcoma [25, 44]
	Vincristine	1.5 mg/m^2/day, i.v.	d 1	
	Ifosfamide	3000 mg/m^2/day, i.v	d 1–3	
	Doxorubicin	20 mg/m^2/day, i.v.	d 1–3	

(continued)

Table 2 (continued)

Regimen	Substance	Dosage/application	Duration	Entity
GMALL				NHL
Block C1				
	Rituximab	375 mg/m^2/day, i.v.	d 0	
	Dexamethasone	10 mg/m^2/day, p.o.	d 1–5	
	Vindesine	3 mg/m^2, i.v.	d 1	
	Methotrexate	1.5 g/m^2, i.v.	d 1	
	Etoposide	250 mg/m^2/day, i.v.	d 4–5	
	Cytarabine	2 × 2 g/m^2, i.v.	d 5	

The dosage of the chemotherapeutic agents may vary between different authors
AL Amyloidosis, light-chain amyloidosis, *CNS* Central nervous system, *d* Day, *GCT* Germ cell tumor, *i.v.* Intravenously, *MM* Multiple myeloma, *MS* Multiple sclerosis, *NHL* Non-Hodgkin's lymphoma, *p.o.* Per os, *SSc* Systemic sclerosis. Adopted from Kriegsmann et al. [27]

"Zarxio®" in the United States, and as "Zarzio®" in other countries. The approvals of G-CSF biosimilars were mainly based on the positive proof of molecular similarity to G-CSF originators. However, as the technical production process of biosimilars is not identical with that of their originators [29], some skepticism has been raised regarding the clinical application of biosimilars.

So far, autologous PBSC mobilization outcomes of G-CSF originators and biosimilars were analyzed in heterogeneous and/or small cancer patient cohorts and partially contradictory results were obtained [30–34]. Exemplarily, Ria et al. reported a higher CD34$^+$ yield using lenograstim compared with filgrastim in a heterogeneous group of lymphoma and MM patients ($n = 86$) [35], while no differences in CD34$^+$ collection results were observed by other groups [36, 37]. All three G-CSF variants—filgrastim ($n = 74$), lenograstim ($n = 45$), and the biosimilar Zarzio®/Filgrastim Hexal® ($n = 131$)—were compared retrospectively in a homogeneous group of MM patients who underwent PBSC collection after cyclophosphamide, doxorubicin and dexamethasone (CAD) chemomobilization as part of their first-line therapy. No statistically significant differences in CD34$^+$ mobilization and collection yield as well as number of required LP sessions between the filgrastim, Filgrastim Hexal® and lenograstim cohorts (Figs. 1 and 2) could be observed [65]. This result is supported by previous studies that suggested lenograstim/filgrastim, reference/biosimilar filgrastim, and lenograstim/biosimilar filgrastim are equivalent for PBSC mobilization after chemotherapy [30, 31, 34, 36, 37].

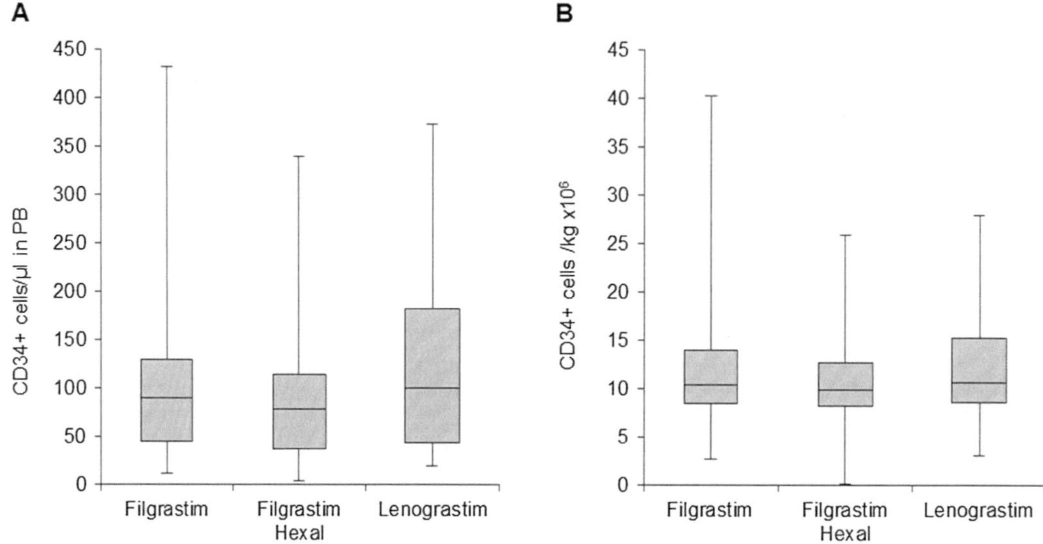

Fig. 1 Peripheral blood CD34$^+$ cell number and CD34$^+$ collection results. (**A**) The number of peripheral blood CD34$^+$ cells on the first day of leukapheresis and (**B**) the overall CD34$^+$ collection result per kg bw in multiple myeloma patients mobilized with chemotherapy and filgrastim ($n = 74$), Filgrastim Hexal® ($n = 131$), and lenograstim ($n = 45$). Adopted from Lisenko et al. [65]

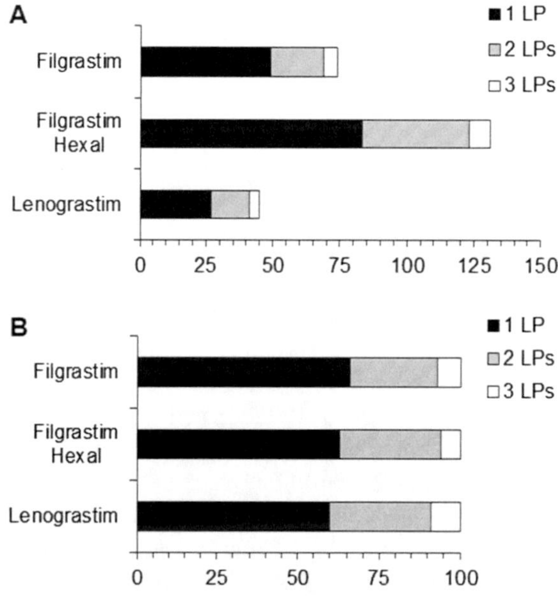

Fig. 2 Number of leukapheresis procedures. (**A**) Absolute and (**B**) relative number of LP (leukapheresis) procedures performed in multiple myeloma patients mobilized with chemotherapy and filgrastim ($n = 74$), Filgrastim Hexal® ($n = 131$), and lenograstim ($n = 45$). Adopted from Lisenko et al. [65]

4 Timing of Mobilization, PB CD34+ Assessment, and Leukapheresis Initiation

The optimal timing of mobilization chemotherapy, G-CSF application, PB CD34+ assessment, and LP initiation varies between different centers and is often subject to debate in clinical practice. However, timing is crucial for a successful and quantitatively sufficient collection of autologous PBSCs [27].

During steady-state mobilization, G-CSF (10 μg/kg bw) is given subcutaneously for 5 consecutive days. PB CD34+ assessment and, in case of adequate mobilization (PB CD34+ > 20/μl), initiation of leukapheresis should be performed at day 5 of G-CSF administration. In case of inadequate mobilization, G-CSF application can be continued for 1 or 2 more days [27].

If a chemomobilization approach is used, a protocol-specific timing of G-CSF administration and CD34+ assessment in PB should be applied. This is of particular importance, as compared to the steady-state approach, mobilization with chemotherapy and G-CSF has the limitation of a less predictable CD34+ peak, and it is thus more challenging to define the optimal timing of apheresis [27, 38–42]. Table 3 shows the specific timing of G-CSF application and the earliest time point to start assessing CD34+ cells in PB

Table 3
Timing of mobilization chemotherapy, G-CSF application, and PB CD34+ assessment

Mobilization chemotherapy		G-CSF application		PB CD34+ assessment
Name	*Duration*	*Start*	*Minimum duration (days)*	*Start*
CAD	d 1–4	d 9	4	d 13
Cd	d 1–3	d 7	4	d 11
C	d 1–2	d 5	5	d 10
(R)CHOP	d 1–5	d 6	5	d 11
(R)CHOEP	d 1–5	d 6	5	d 11
(R)DHAP	d 1–4	d 9	5	d 14
(R)ICE	d 1–3	d 6	6	d 12
(R)AraC/TT	d 1–2	d 5	5	d 10
PEI	d 1–5	d 6	6	d 12
VIDE	d 1–3	d 5	7	d 12
GMALL C1	d 1–5	d 7	7	d 14

The first day of mobilization chemotherapy application is assumed to be day 1. The day of rituximab application is assumed to be day 0, and the first day of mobilization chemotherapy application is assumed to be day 1. Adopted from Kriegsmann et al. [27]

d Day, *G-CSF* Granulocyte-colony-stimulating factor, *PB* Peripheral blood, *WBC* White blood cell. For abbreviations of chemotherapy protocols, compare to Table 2

for consecutive stem cell collection by LP, based on a recently proposed schedule [26]. Generally, in chemomobilization approaches filgrastim (5 μg/kg bw) or lenograstim (150 μg/m^2) is administered subcutaneously daily until the end of PBSC collection. If two G-CSF injections per day are required, most centers prefer to administer the larger dose in the morning. Moreover, recent data obtained from MM and lymphoma patients who were mobilized with chemotherapy followed by lenograstim showed that G-CSF injection 3 h before LP improved the efficacy of PBSC collection compared to G-CSF application half a day before LP [43].

After a chemotherapy-specific period of G-CSF application (Table 3), PB white blood cell (WBC) and thrombocyte count should be evaluated daily. At this time point a dose escalation of G-CSF up to 10 μg/kg bw/day may be considered. When the WBC count reaches $\geq 4.0 \times 10^3/\mu l$, the PB CD34$^+$ cell count should be determined by flow cytometry [44]. LP should be initiated when the PB CD34$^+$ cell count reaches $\geq 20/\mu l$ [45].

5 Risk-Adapted Rescue of Mobilization Failure

5.1 Plerixafor

Plerixafor (Mozobil®) is approved in combination with G-CSF in MM and lymphoma patients who are mobilizing PBSCs insufficiently. Stem cells adhere to their BM niche by interactions of chemokine receptor type 4 (CXCR4, expressed on CD34$^+$ cells) and stromal cell-derived factor 1 (SDF-1) [46, 47]. The mechanism of action of plerixafor is mainly based on a reversible blockade of CXCR4 [48]. Therefore, plerixafor inhibits CXCR4/SDF-1 based cell–cell interactions leading to highly efficient hematopoietic stem cell mobilization [49–53]. In case of insufficient PBSC mobilization after at least 4 days of G-CSF stimulation, plerixafor application is approved as subcutaneous injection 6–11 h before initiation of LP. The recommended dosage is 0.24 mg/kg bw per day. Plerixafor application can either be considered prior to LP initiation due to an insufficiently low PB CD34$^+$ cell number and/or other risk factors for poor mobilization (i.e., preemptive use) or after a quantitatively insufficient LP session (i.e., rescue mobilization).

5.2 Preemptive Plerixafor Application

The peak PB CD34$^+$ cell count before LP initiation is considered as one of the most relevant factors for failing the PBSC collection goal [9, 54–57], and several groups developed algorithms to trigger preemptive plerixafor administration based on PB CD34$^+$ cell counts before LP [9, 45, 58–60]. Although prospective trials for validation are lacking, a common expert consensus-based approach stratifies patients for proactive plerixafor intervention on the basis of the following peak PB CD34$^+$ cell counts: (A) in patients with peak CD34$^+$ cell count <10 cells/μl, an indication for preemptive plerixafor is given; (B) in patients with PB CD34$^+$ cell count between 10 and 20 cells/μl (gray zone), plerixafor administration should be considered taking into account other risk factors for collection failure and the PBSC collection goal; (C) no preemptive plerixafor administration is required in patients with pre-LP PB CD34$^+$ cell count of >20/μ, and LP can be initiated promptly [27, 45]. The proposed algorithm for plerixafor administration and other risk factors for poor mobilization are summarized in Table 4.

Table 4
Proposed algorithm for plerixafor administration [113–121].

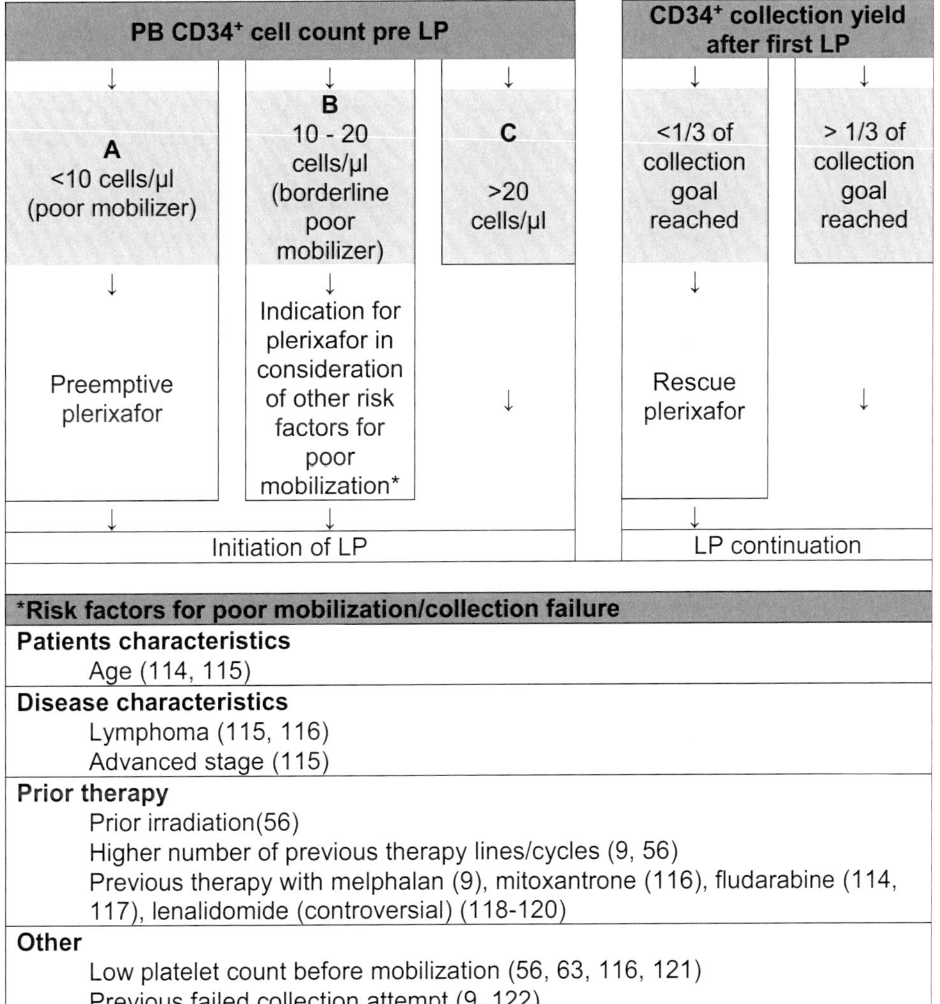

PB CD34$^+$ cell count pre LP			CD34$^+$ collection yield after first LP	
↓	↓	↓	↓	↓
A <10 cells/μl (poor mobilizer)	**B** 10 - 20 cells/μl (borderline poor mobilizer)	**C** >20 cells/μl	<1/3 of collection goal reached	> 1/3 of collection goal reached
↓	↓		↓	
Preemptive plerixafor	Indication for plerixafor in consideration of other risk factors for poor mobilization*	↓	Rescue plerixafor	↓
↓	↓		↓	
Initiation of LP			LP continuation	

*Risk factors for poor mobilization/collection failure
Patients characteristics
Age (114, 115)
Disease characteristics
Lymphoma (115, 116)
Advanced stage (115)
Prior therapy
Prior irradiation(56)
Higher number of previous therapy lines/cycles (9, 56)
Previous therapy with melphalan (9), mitoxantrone (116), fludarabine (114, 117), lenalidomide (controversial) (118-120)
Other
Low platelet count before mobilization (56, 63, 116, 121)
Previous failed collection attempt (9, 122)

Adopted from Mohty et al. [27] and Cheng et al. [45]
LP Leukapheresis, *PB* Peripheral blood

5.3 Plerixafor Application as Rescue Mobilization Strategy

If less than 1/3 of the CD34$^+$ collection goal was reached with the first LP session and therefore the collection goal will most likely not be reached within 3 LP sessions, a rescue application of plerixafor should be considered starting without delay at the same evening (Table 4). This approach aims to reach the PBSC collection goal in as few LP sessions as possible and to assure cost efficiency, patient safety, and satisfaction [27, 45, 61, 62].

6 Estimation of Daily PBSC Yields

Generally, the minimum quantity of PBSC for one transplant is defined as $\geq 2.0 \times 10^6$ CD34$^+$ cells/kg bw [27]. The individual collection goal might differ between patients in dependency of diagnosis, e.g., up to three transplants are often aimed to be collected in MM patients in first-line treatment [63–65].

The prediction of daily LP yields based on circulating PB CD34$^+$ cells that were measured shortly before the initiation of the LP procedure offers key advantages for optimal planning of PBSC collection: (A) It can be reliably estimated, whether the patient's individual collection goal will be achieved during the planned LP session or, if not, further mobilization strategies and LP session can be considered. (B) In case of sufficiently high PB CD34$^+$ cell numbers pre-LP (good mobilizer), the necessary blood volume to be processed to reach the collection goal can be calculated; therefore the LP procedure can be terminated safely without even reaching the maximum LP time (300 min) and/or maximum processed blood volume (i.e., four times the patient's total blood volume).

The authors previously presented clinical data applying the following formula [64, 66]:

$$\text{Predicted collection yield} = \frac{(\text{PB CD34 cells per L}) \times (\text{Collection efficiency of } 30\%) \times (\text{Processed blood volume in L})}{\text{Body weight in kg}}$$

This formula was previously described and validated by Rosenbaum et al. for prediction of daily CD34$^+$ collection yield per kg bw for donors with low, intermediate, and high PB CD34$^+$ cell numbers [67, 68]. The collection efficiency might vary with regard to the applied apheresis machine and protocol [67, 69–72]. In order to avoid overestimating the collection result and consecutively terminating the LP session too early, the collection efficiency coefficient should be chosen rather conservatively, as previously described [67].

If the PBSC collection goal is not reached during the first LP, LP sessions can be repeated on up to 4 consecutive days [64]. More than four LP sessions are generally not feasible.

7 Leukapheresis Benchmarking

LP sessions can be benchmarked individually by calculating a performance ratio that determines the quotient between the collected and pre-LP predicted $CD34^+$ yield [64, 67]:

$$\text{Performance ratio} = \frac{(\text{CD34}^+ \text{ cells per kg body weight collected})}{(\text{CD34}^+ \text{ cells per kg body weight predicted})} \times 100\%$$

The calculation of the performance ratio for each LP session can be used as an internal quality control. Particularly in case of a decreased performance ratio (<100%), possible technical problems should be considered. In addition, the assessment of the LP performance can constitute a benchmarking tool in case of a change in apheresis machines, tubing systems, collection regimens, etc. in larger patient cohorts. Exemplarily, the authors recently compared two apheresis systems—the refined continuous Spectra Optia mononuclear cell collection system and the original two-step Spectra Optia separation system—with regard to their performance parameters during a redeployment of apheresis systems at our institution. No significant differences in performance ratios between the two procedures in MM patients were found (median performance ratio 166% and 163%, respectively), supporting the notion that the new continuous Spectra Optia MNC system is equally efficient in collecting $CD34^+$ cells as compared to the previous two-step apheresis procedure [66].

8 Long-Term Storage of PBSC Products

PBSCs may need to be stored for months or even years until reinfusion, usually upon HDT [73]. PBSCs are commonly subjected to controlled-rate freezing and stored at liquid- or vapor-phase nitrogen temperature (≤ -140 °C) for long-term periods with cryoprotectants such as 5–10% dimethyl sulfoxide (DMSO), representing the current standard practice [74–76]. In addition, some simplified methods of cryopreservation in mechanical uncontrolled rate freezers at ≤ -80 °C had been described in the literature [77–79] but are not regularly used in clinical practice.

The question arises, whether PBSCs remain viable and maintain their hematopoietic recovery potential particularly after long-term cryostorage. Several in vitro studies on $CD34^+$ viability and clonogenic assays have shown that controlled rate freezed and cryopreserved PBSC can remain viable for safe transplantation even after up to 14 years of cryostorage [80–83]. However, the limitations of these studies include an absence of evidence for in vivo PBSC engraftment in humans and relatively small and heterogeneous patient cohorts [80–86].

In a recent study, hematopoietic reconstitution was evaluated after short- (≤12 months, n = 143), medium- (>12 and ≤60 months, n = 75), and long-term (>60 months, n = 29) PBSC cryostorage in 10% DMSO at ≤ −140 °C in a large homogenous group of MM patients who received HD melphalan as conditioning regimen. No influence of PBSC storage duration on neutrophil and platelet recovery was found, demonstrating the non-inferiority of medium- and long-term cryopreserved PBSC products on hematopoietic regeneration even in cases of second or third courses of HD chemotherapy and ABSCT [63].

References

1. Passweg JR, Baldomero H, Gratwohl A, Bregni M, Cesaro S, Dreger P et al (2012) The EBMT activity survey: 1990-2010. Bone Marrow Transplant 47(7):906–923
2. Ljungman P, Urbano-Ispizua A, Cavazzana-Calvo M, Demirer T, Dini G, Einsele H et al (2006) Allogeneic and autologous transplantation for haematological diseases, solid tumours and immune disorders: definitions and current practice in Europe. Bone Marrow Transplant 37(5):439–449
3. Passweg JR, Baldomero H, Bader P, Bonini C, Cesaro S, Dreger P et al (2016) Hematopoietic stem cell transplantation in Europe 2014: more than 40 000 transplants annually. Bone Marrow Transplant 51(6):786–792
4. Schmitz N, Linch DC, Dreger P, Goldstone AH, Boogaerts MA, Ferrant A et al (1996) Randomised trial of filgrastim-mobilised peripheral blood progenitor cell transplantation versus autologous bone-marrow transplantation in lymphoma patients. Lancet 347 (8998):353–357
5. Roberts MM, To LB, Gillis D, Mundy J, Rawling C, Ng K et al (1993) Immune reconstitution following peripheral blood stem cell transplantation, autologous bone marrow transplantation and allogeneic bone marrow transplantation. Bone Marrow Transplant 12 (5):469–475
6. To LB, Roberts MM, Haylock DN, Dyson PG, Branford AL, Thorp D et al (1992) Comparison of haematological recovery times and supportive care requirements of autologous recovery phase peripheral blood stem cell transplants, autologous bone marrow transplants and allogeneic bone marrow transplants. Bone Marrow Transplant 9 (4):277–284
7. Chen SH, Wang TF, Yang KL (2013) Hematopoietic stem cell donation. Int J Hematol 97(4):446–455
8. Pusic I, DiPersio JF (2008) The use of growth factors in hematopoietic stem cell transplantation. Curr Pharm Des 14(20):1950–1961
9. Wuchter P, Ran D, Bruckner T, Schmitt T, Witzens-Harig M, Neben K et al (2010) Poor mobilization of hematopoietic stem cells-definitions, incidence, risk factors, and impact on outcome of autologous transplantation. Biol Blood Marrow Transplant 16 (4):490–499
10. Mohty M, Ho AD (2011) In and out of the niche: perspectives in mobilization of hematopoietic stem cells. Exp Hematol 39 (7):723–729
11. Bensinger W, DiPersio JF, McCarty JM (2009) Improving stem cell mobilization strategies: future directions. Bone Marrow Transplant 43(3):181–195
12. Gertz MA, Kumar SK, Lacy MQ, Dispenzieri A, Hayman SR, Buadi FK et al (2009) Comparison of high-dose CY and growth factor with growth factor alone for mobilization of stem cells for transplantation in patients with multiple myeloma. Bone Marrow Transplant 43(8):619–625
13. Giralt S, Costa L, Schriber J, Dipersio J, Maziarz R, McCarty J et al (2014) Optimizing autologous stem cell mobilization strategies to improve patient outcomes: consensus guidelines and recommendations. Biol Blood Marrow Transplant 20(3):295–308
14. Gertz MA (2010) Current status of stem cell mobilization. Br J Haematol 150(6):647–662
15. Pusic I, Jiang SY, Landua S, Uy GL, Rettig MP, Cashen AF et al (2008) Impact of mobilization and remobilization strategies on achieving sufficient stem cell yields for autologous transplantation. Biol Blood Marrow Transplant 14(9):1045–1056
16. Narayanasami U, Kanteti R, Morelli J, Klekar A, Al-Olama A, Keating C et al

(2001) Randomized trial of filgrastim versus chemotherapy and filgrastim mobilization of hematopoietic progenitor cells for rescue in autologous transplantation. Blood 98 (7):2059–2064

17. Dazzi C, Cariello A, Rosti G, Argnani M, Sebastiani L, Ferrari E et al (2000) Is there any difference in PBPC mobilization between cyclophosphamide plus G-CSF and G-CSF alone in patients with non-Hodgkin's Lymphoma? Leuk Lymphoma 39(3–4):301–310
18. Desikan KR, Barlogie B, Jagannath S, Vesole DH, Siegel D, Fassas A et al (1998) Comparable engraftment kinetics following peripheral-blood stem-cell infusion mobilized with granulocyte colony-stimulating factor with or without cyclophosphamide in multiple myeloma. J Clin Oncol 16(4):1547–1553
19. Alegre A, Tomas JF, Martinez-Chamorro C, Gil-Fernandez JJ, Fernandez-Villalta MJ, Arranz R et al (1997) Comparison of peripheral blood progenitor cell mobilization in patients with multiple myeloma: high-dose cyclophosphamide plus GM-CSF vs G-CSF alone. Bone Marrow Transplant 20 (3):211–217
20. Bensinger W, Appelbaum F, Rowley S, Storb R, Sanders J, Lilleby K et al (1995) Factors that influence collection and engraftment of autologous peripheral-blood stem cells. J Clin Oncol 13(10):2547–2555
21. Sung AD, Grima DT, Bernard LM, Brown S, Carrum G, Holmberg L et al (2013) Outcomes and costs of autologous stem cell mobilization with chemotherapy plus G-CSF vs G-CSF alone. Bone Marrow Transplant 48 (11):1444–1449
22. Mohammadi S, Malek Mohammadi A, Nikbakht M, Norooznezhad AH, Alimoghaddam K, Ghavamzadeh A (2017) Optimizing stem cells mobilization strategies to ameliorate patient outcomes: a review of guide- lines and recommendations. Int J Hematol Oncol Stem Cell Res 11(1):78–88
23. Baertsch MA, Schlenzka J, Lisenko K, Krzykalla J, Becker N, Weisel K et al (2017) Cyclophosphamide-based stem cell mobilization in relapsed multiple myeloma patients: a subgroup analysis from the phase III trial ReLApsE. Eur J Haematol 99(1):42–50
24. Lisenko K, McClanahan F, Schoning T, Schwarzbich MA, Cremer M, Dittrich T et al (2016) Minimal renal toxicity after Rituximab DHAP with a modified cisplatin application scheme in patients with relapsed or refractory diffuse large B-cell lymphoma. BMC Cancer 16(1):267
25. Kriegsmann K, Heilig C, Cremer M, Novotny P, Kriegsmann M, Bruckner T et al (2017) Successful collection of peripheral blood stem cells upon VIDE chemomobilization in sarcoma patients. Eur J Haematol 99 (5):459–464
26. Kriegsmann K, Schmitt A, Kriegsmann M, Bruckner T, Anyanwu A, Witzens-Harig M et al (2018) Orchestration of chemomobilization and G-CSF administration for successful hematopoietic stem cell collection. Biol Blood Marrow Transplant 24(6):1281–1288
27. Mohty M, Hubel K, Kroger N, Aljurf M, Apperley J, Basak GW et al (2014) Autologous haematopoietic stem cell mobilisation in multiple myeloma and lymphoma patients: a position statement from the European Group for blood and marrow transplantation. Bone Marrow Transplant 49(7):865–872
28. Marchesi F, Mengarelli A (2016) Biosimilar filgrastim in autologous peripheral blood hematopoietic stem cell mobilization and post-transplant hematologic recovery. Curr Med Chem 23(21):2217–2229
29. Subramanyam M (2013) Clinical development of biosimilars: an evolving landscape. Bioanalysis 5(5):575–586
30. Martino M, Recchia AG, Moscato T, Fedele R, Neri S, Gentile M et al (2015) Efficacy of biosimilar granulocyte colony-stimulating factor versus originator granulocyte colony-stimulating factor in peripheral blood stem cell mobilization in de novo multiple myeloma patients. Cytotherapy 17 (10):1485–1493
31. Marchesi F, Vacca M, Gumenyuk S, Pandolfi A, Renzi D, Palombi F, Pisani F, Romano A, Spadea A, Ipsevich F, Santinelli S, De Rienzo M, Papa E, Canfora M, Laurenzi L, Foddai ML, Pierelli L, Mengarelli A (2016) Biosimilar filgrastim (Zarzio®) vs. lenograstim (Myelostim®) for peripheral blood stem cell mobilization in adult patients with lymphoma and myeloma: a single center experience.Leuk Lymphoma 57(2):489-492
32. Remenyi P, Gopcsa L, Marton I, Reti M, Mikala G, Peto M et al (2014) Peripheral blood stem cell mobilization and engraftment after autologous stem cell transplantation with biosimilar rhG-CSF. Adv Ther 31 (4):451–460
33. Lefrere F, Brignier AC, Elie C, Ribeil JA, Bernimoulin M, Aoun C et al (2011) First experience of autologous peripheral blood stem cell mobilization with biosimilar granulocyte colony-stimulating factor. Adv Ther 28 (4):304–310

34. Maul JT, Stenner-Liewen F, Seifert B, Pfrommer S, Petrausch U, Kiessling MK et al (2017) Efficacious and save use of biosimilar filgrastim for hematopoietic progenitor cell chemo-mobilization with vinorelbine in multiple myeloma patients. J Clin Apher 32 (1):21–26
35. Ria R, Gasparre T, Mangialardi G, Bruno A, Iodice G, Vacca A et al (2010) Comparison between filgrastim and lenograstim plus chemotherapy for mobilization of PBPCs. Bone Marrow Transplant 45(2):277–281
36. Kim IH, Park SK, Suh OK, Oh JM (2003) Comparison of lenograstim and filgrastim on haematological effects after autologous peripheral blood stem cell transplantation with high-dose chemotherapy. Curr Med Res Opin 19(8):753–759
37. Lefrere F, Bernard M, Audat F, Cavazzana-Calvo M, Belanger C, Hermine O et al (1999) Comparison of lenograstim vs filgrastim administration following chemotherapy for peripheral blood stem cell (PBSC) collection: a retrospective study of 126 patients. Leuk Lymphoma 35(5–6):501–505
38. Duong HK, Savani BN, Copelan E, Devine S, Costa LJ, Wingard JR et al (2014) Peripheral blood progenitor cell mobilization for autologous and allogeneic hematopoietic cell transplantation: guidelines from the American Society for Blood and Marrow Transplantation. Biol Blood Marrow Transplant 20 (9):1262–1273
39. Costa LJ, Miller AN, Alexander ET, Hogan KR, Shabbir M, Schaub C et al (2011) Growth factor and patient-adapted use of plerixafor is superior to CY and growth factor for autologous hematopoietic stem cells mobilization. Bone Marrow Transplant 46 (4):523–528
40. Tuchman SA, Bacon WA, Huang LW, Long G, Rizzieri D, Horwitz M et al (2015) Cyclophosphamide-based hematopoietic stem cell mobilization before autologous stem cell transplantation in newly diagnosed multiple myeloma. J Clin Apher 30 (3):176–182
41. Farina L, Guidetti A, Spina F, Roncari L, Longoni P, Ravagnani F et al (2014) Plerixafor "on demand": results of a strategy based on peripheral blood CD34+ cells in lymphoma patients at first or subsequent mobilization with chemotherapy+G-CSF. Bone Marrow Transplant 49(3):453–455
42. Shaughnessy P, Chao N, Shapiro J, Walters K, McCarty J, Abhyankar S et al (2013) Pharmacoeconomics of hematopoietic stem cell mobilization: an overview of current evidence and gaps in the literature. Biol Blood Marrow Transplant 19(9):1301–1309
43. Kim JE, Yoo C, Kim S, Lee DH, Kim SW, Lee JS et al (2011) Optimal timing of G-CSF administration for effective autologous stem cell collection. Bone Marrow Transplant 46 (6):806–812
44. Strauss SJ, McTiernan A, Driver D, Hall-Craggs M, Sandison A, Cassoni AM et al (2003) Single center experience of a new intensive induction therapy for ewing's family of tumors: feasibility, toxicity, and stem cell mobilization properties. J Clin Oncol 21 (15):2974–2981
45. Cheng J, Schmitt M, Wuchter P, Buss EC, Witzens-Harig M, Neben K et al (2015) Plerixafor is effective given either preemptively or as a rescue strategy in poor stem cell mobilizing patients with multiple myeloma. Transfusion 55(2):275–283
46. Lapidot T, Petit I (2002) Current understanding of stem cell mobilization: the roles of chemokines, proteolytic enzymes, adhesion molecules, cytokines, and stromal cells. Exp Hematol 30(9):973–981
47. Mohle R, Bautz F, Rafii S, Moore MA, Brugger W, Kanz L (1998) The chemokine receptor CXCR-4 is expressed on CD34+ hematopoietic progenitors and leukemic cells and mediates transendothelial migration induced by stromal cell-derived factor-1. Blood 91(12):4523–4530
48. Fricker SP, Anastassov V, Cox J, Darkes MC, Grujic O, Idzan SR et al (2006) Characterization of the molecular pharmacology of AMD3100: a specific antagonist of the G-protein coupled chemokine receptor, CXCR4. Biochem Pharmacol 72(5):588–596
49. Cashen AF, Nervi B, DiPersio J (2007) AMD3100: CXCR4 antagonist and rapid stem cell-mobilizing agent. Future Oncol 3 (1):19–27
50. Broxmeyer HE, Orschell CM, Clapp DW, Hangoc G, Cooper S, Plett PA et al (2005) Rapid mobilization of murine and human hematopoietic stem and progenitor cells with AMD3100, a CXCR4 antagonist. J Exp Med 201(8):1307–1318
51. Wuchter P, Saffrich R, Giselbrecht S, Nies C, Lorig H, Kolb S et al (2016) Microcavity arrays as an in vitro model system of the bone marrow niche for hematopoietic stem cells. Cell Tissue Res 364(3):573–584
52. Wuchter P, Leinweber C, Saffrich R, Hanke M, Eckstein V, Ho AD et al (2014) Plerixafor induces the rapid and transient release of stromal cell-derived factor-1 alpha

from human mesenchymal stromal cells and influences the migration behavior of human hematopoietic progenitor cells. Cell Tissue Res 355(2):315–326

53. Ludwig A, Saffrich R, Eckstein V, Bruckner T, Wagner W, Ho AD et al (2014) Functional potentials of human hematopoietic progenitor cells are maintained by mesenchymal stromal cells and not impaired by plerixafor. Cytotherapy 16(1):111–121
54. Sancho JM, Morgades M, Grifols JR, Junca J, Guardia R, Vives S et al (2012) Predictive factors for poor peripheral blood stem cell mobilization and peak CD34(+) cell count to guide pre-emptive or immediate rescue mobilization. Cytotherapy 14(7):823–829
55. Han X, Ma L, Zhao L, He X, Liu P, Zhou S et al (2012) Predictive factors for inadequate stem cell mobilization in Chinese patients with NHL and HL: 14-year experience of a single-center study. J Clin Apher 27(2):64–74
56. Olivieri A, Marchetti M, Lemoli R, Tarella C, Iacone A, Lanza F et al (2012) Proposed definition of 'poor mobilizer' in lymphoma and multiple myeloma: an analytic hierarchy process by ad hoc working group Gruppo ItalianoTrapianto di Midollo Osseo. Bone Marrow Transplant 47(3):342–351
57. Sinha S, Gastineau D, Micallef I, Hogan W, Ansell S, Buadi F et al (2011) Predicting PBSC harvest failure using circulating CD34 levels: developing target-based cutoff points for early intervention. Bone Marrow Transplant 46(7):943–949
58. Chen AI, Bains T, Murray S, Knight R, Shoop K, Bubalo J et al (2012) Clinical experience with a simple algorithm for plerixafor utilization in autologous stem cell mobilization. Bone Marrow Transplant 47 (12):1526–1529
59. Abhyankar S, DeJarnette S, Aljitawi O, Ganguly S, Merkel D, McGuirk J (2012) A risk-based approach to optimize autologous hematopoietic stem cell (HSC) collection with the use of plerixafor. Bone Marrow Transplant 47(4):483–487
60. Costa LJ, Alexander ET, Hogan KR, Schaub C, Fouts TV, Stuart RK (2011) Development and validation of a decision-making algorithm to guide the use of plerixafor for autologous hematopoietic stem cell mobilization. Bone Marrow Transplant 46(1):64–69
61. Hundemer M, Engelhardt M, Bruckner T, Kraeker S, Schmitt A, Sauer S et al (2014) Rescue stem cell mobilization with plerixafor economizes leukapheresis in patients with multiple myeloma. J Clin Apher 29 (6):299–304
62. Baertsch MA, Kriegsmann K, Pavel P, Bruckner T, Hundemer M, Kriegsmann M et al (2018) Platelet count before peripheral blood stem cell mobilization is associated with the need for plerixafor but not with the collection result. Transfus Med Hemother 45 (1):24–31
63. Lisenko K, Pavel P, Kriegsmann M, Bruckner T, Hillengass J, Goldschmidt H et al (2017) Storage duration of autologous stem cell preparations has no impact on hematopoietic recovery after transplantation. Biol Blood Marrow Transplant 23(4):684–690
64. Wuchter P, Hundemer M, Schmitt A, Witzens-Harig M, Pavel P, Hillengass J et al (2017) Performance assessment and benchmarking of autologous peripheral blood stem cell collection with two different apheresis devices. Transfus Med 27(1):36–42
65. Lisenko K, Baertsch MA, Meiser R, Pavel P, Bruckner T, Kriegsmann M et al (2017) Comparison of biosimilar filgrastim, originator filgrastim, and lenograstim for autologous stem cell mobilization in patients with multiple myeloma. Transfusion 57(10):2359–2365
66. Lisenko K, Pavel P, Bruckner T, Puthenparambil J, Hundemer M, Schmitt A et al (2017) Comparison between intermittent and continuous spectra optia leukapheresis systems for autologous peripheral blood stem cell collection. J Clin Apher 32 (1):27–34
67. Rosenbaum ER, O'Connell B, Cottler-Fox M (2012) Validation of a formula for predicting daily CD34(+) cell collection by leukapheresis. Cytotherapy 14(4):461–466
68. Rosenbaum ER, Wuchter P, Hundemer M, Pavel P, Witzens-Harig M, Goldschmidt H et al (2014) Validation of a predictive formula for collection of hematopoietic progenitor cells (HPC) By leukapheresis at 2 institutions using 4 different machine protocols. Blood 124(21):2458
69. Brauninger S, Bialleck H, Thorausch K, Felt T, Seifried E, Bonig H (2012) Allogeneic donor peripheral blood "stem cell" apheresis: prospective comparison of two apheresis systems. Transfusion 52(5):1137–1145
70. Flommersfeld S, Bakchoul T, Bein G, Wachtel A, Loechelt C, Sachs UJ (2013) A single center comparison between three different apheresis systems for autologous and allogeneic stem cell collections. Transfus Apher Sci 49(3):428–433
71. Reinhardt P, Brauninger S, Bialleck H, Thorausch K, Smith R, Schrezenmeier H et al (2011) Automatic interface-controlled apheresis collection of stem/progenitor cells:

results from an autologous donor validation trial of a novel stem cell apheresis device. Transfusion 51(6):1321–1330

72. Cousins AF, Sinclair JE, Alcorn MJ (2015) R HAG, Douglas KW. HPC-A dose prediction on the optia(R) cell separator based on a benchmark CE2 collection efficiency: Promoting clinical efficiency, minimizing toxicity, and allowing quality control. J Clin Apher 30 (6):321–328
73. Passweg JR, Baldomero H, Bader P, Bonini C, Cesaro S, Dreger P et al (2015) Hematopoietic SCT in Europe 2013: recent trends in the use of alternative donors showing more haploidentical donors but fewer cord blood transplants. Bone Marrow Transplant 50 (4):476–482
74. Leemhuis T, Padley D, Keever-Taylor C, Niederwieser D, Teshima T, Lanza F et al (2014) Essential requirements for setting up a stem cell processing laboratory. Bone Marrow Transplant 49(8):1098–1105
75. Berz D, McCormack EM, Winer ES, Colvin GA, Quesenberry PJ (2007) Cryopreservation of hematopoietic stem cells. Am J Hematol 82(6):463–472
76. Veeraputhiran M, Theus JW, Pesek G, Barlogie B, Cottler-Fox M (2010) Viability and engraftment of hematopoietic progenitor cells after long-term cryopreservation: effect of diagnosis and percentage dimethyl sulfoxide concentration. Cytotherapy 12 (6):764–766
77. Detry G, Calvet L, Straetmans N, Cabrespine A, Ravoet C, Bay JO et al (2014) Impact of uncontrolled freezing and long-term storage of peripheral blood stem cells at - 80 degrees C on haematopoietic recovery after autologous transplantation. Report from two centres. Bone Marrow Transplant 49(6):780–785
78. Watts MJ, Sullivan AM, Ings SJ, Barlow M, Devereux S, Goldstone AH et al (1998) Storage of PBSC at −80 °C. Bone Marrow Transplant 21(1):111–112
79. Katayama Y, Yano T, Bessho A, Deguchi S, Sunami K, Mahmut N et al (1997) The effects of a simplified method for cryopreservation and thawing procedures on peripheral blood stem cells. Bone Marrow Transplant 19 (3):283–287
80. Fernyhough LJ, Buchan VA, McArthur LT, Hock BD (2013) Relative recovery of haematopoietic stem cell products after cryogenic storage of up to 19 years. Bone Marrow Transplant 48(1):32–35
81. McCullough J, Haley R, Clay M, Hubel A, Lindgren B, Moroff G (2010) Long-term storage of peripheral blood stem cells frozen and stored with a conventional liquid nitrogen technique compared with cells frozen and stored in a mechanical freezer. Transfusion 50 (4):808–819
82. Spurr EE, Wiggins NE, Marsden KA, Lowenthal RM, Ragg SJ (2002) Cryopreserved human haematopoietic stem cells retain engraftment potential after extended (5–14 years) cryostorage. Cryobiology 44 (3):210–217
83. Valeri CR, Pivacek LE (1996) Effects of the temperature, the duration of frozen storage, and the freezing container on in vitro measurements in human peripheral blood mononuclear cells. Transfusion 36(4):303–308
84. Abbruzzese L, Agostini F, Durante C, Toffola RT, Rupolo M, Rossi FM et al (2013) Long term cryopreservation in 5% DMSO maintains unchanged CD34(+) cells viability and allows satisfactory hematological engraftment after peripheral blood stem cell transplantation. Vox Sang 105(1):77–80
85. Liseth K, Ersvaer E, Abrahamsen JF, Nesthus I, Ryningen A, Bruserud O (2009) Long-term cryopreservation of autologous stem cell grafts: a clinical and experimental study of hematopoietic and immunocompetent cells. Transfusion 49(8):1709–1719
86. Aird W, Labopin M, Gorin NC, Antin JH (1992) Long-term cryopreservation of human stem cells. Bone Marrow Transplant 9(6):487–490
87. Al-Anazi KA (2012) Autologous hematopoietic stem cell transplantation for multiple myeloma without cryopreservation. Bone Marrow Res 2012:917361
88. Fruehauf S, Klaus J, Huesing J, Veldwijk MR, Buss EC, Topaly J et al (2007) Efficient mobilization of peripheral blood stem cells following CAD chemotherapy and a single dose of pegylated G-CSF in patients with multiple myeloma. Bone Marrow Transplant 39 (12):743–750
89. Breitkreutz I, Lokhorst HM, Raab MS, Holt B, Cremer FW, Herrmann D et al (2007) Thalidomide in newly diagnosed multiple myeloma: influence of thalidomide treatment on peripheral blood stem cell collection yield. Leukemia 21(6):1294–1299
90. Lisenko K, Wuchter P, Hansberg M, Mangatter A, Benner A, Ho AD et al (2017) Comparison of different stem cell mobilization regimens in AL amyloidosis patients. Biol Blood Marrow Transplant 23 (11):1870–1878

91. Worel N, Schulenburg A, Mitterbauer M, Keil F, Rabitsch W, Kalhs P et al (2006) Autologous stem-cell transplantation in progressing amyloidosis is associated with severe transplant-related toxicity. Wien Klin Wochenschr 118(1–2):49–53
92. Perotti C, Del Fante C, Viarengo G, Perlini S, Vezzoli M, Rodi G et al (2005) Peripheral blood progenitor cell mobilization and collection in 42 patients with primary systemic amyloidosis. Transfusion 45(11):1729–1734
93. Gertz MA, Lacy MQ, Gastineau DA, Inwards DJ, Chen MG, Tefferi A et al (2000) Blood stem cell transplantation as therapy for primary systemic amyloidosis (AL). Bone Marrow Transplant 26(9):963–969
94. Blank N, Lisenko K, Pavel P, Bruckner T, Ho AD, Wuchter P (2016) Low-dose cyclophosphamide effectively mobilizes peripheral blood stem cells in patients with autoimmune disease. Eur J Haematol 97(1):78–82
95. Endo T, Sato N, Mogi Y, Koizumi K, Nishio M, Fujimoto K et al (2004) Peripheral blood stem cell mobilization following CHOP plus rituximab therapy combined with G-CSF in patients with B-cell non--Hodgkin's lymphoma. Bone Marrow Transplant 33(7):703–707
96. Shi Y, Zhou P, Han X, He X, Zhou S, Liu P et al (2015) Autologous peripheral blood stem cell mobilization following dose-adjusted cyclophosphamide, doxorubicin, vincristine, and prednisolone chemotherapy alone or in combination with rituximab in treating high-risk non-Hodgkin's lymphoma. Chin J Cancer 34(11):522–530
97. Takeyama K, Ogura M, Morishima Y, Kasai M, Kiyama Y, Ohnishi K et al (2003) A dose-finding study of glycosylated G-CSF (Lenograstim) combined with CHOP therapy for stem cell mobilization in patients with non-Hodgkin's lymphoma. Jpn J Clin Oncol 33(2):78–85
98. Lisenko K, Cremer M, Schwarzbich MA, Kriegsmann M, Ho AD, Witzens-Harig M et al (2016) Efficient stem cell collection after modified cisplatin-based mobilization chemotherapy in patients with diffuse large B cell lymphoma. Biol Blood Marrow Transplant 22(8):1397–1402
99. Gisselbrecht C, Glass B, Mounier N, Singh Gill D, Linch DC, Trneny M et al (2010) Salvage regimens with autologous transplantation for relapsed large B-cell lymphoma in the rituximab era. J Clin Oncol 28 (27):4184–4190
100. Smardova L, Engert A, Haverkamp H, Raemakers J, Baars J, Pfistner B et al (2005) Successful mobilization of peripheral blood stem cells with the DHAP regimen (dexamethasone, cytarabine, cisplatinum) plus granulocyte colony-stimulating factor in patients with relapsed Hodgkin's disease. Leuk Lymphoma 46(7):1017–1022
101. Pavone V, Gaudio F, Guarini A, Perrone T, Zonno A, Curci P et al (2002) Mobilization of peripheral blood stem cells with high-dose cyclophosphamide or the DHAP regimen plus G-CSF in non-Hodgkin's lymphoma. Bone Marrow Transplant 29(4):285–290
102. Russell N, Mesters R, Schubert J, Boogaerts M, Johnsen HE, Canizo CD et al (2008) A phase 2 pilot study of pegfilgrastim and filgrastim for mobilizing peripheral blood progenitor cells in patients with non-Hodgkin's lymphoma receiving chemotherapy. Haematologica 93(3):405–412
103. Kewalramani T, Zelenetz AD, Nimer SD, Portlock C, Straus D, Noy A et al (2004) Rituximab and ICE as second-line therapy before autologous stem cell transplantation for relapsed or primary refractory diffuse large B-cell lymphoma. Blood 103 (10):3684–3688
104. Zelenetz AD, Hamlin P, Kewalramani T, Yahalom J, Nimer S, Moskowitz CH (2003) Ifosfamide, carboplatin, etoposide (ICE)-based second-line chemotherapy for the management of relapsed and refractory aggressive non-Hodgkin's lymphoma. Ann Oncol 14 (Suppl 1):i5–i10
105. Kingreen D, Beyer J, Kleiner S, Reif S, Huhn D, Siegert W (2001) ICE--an efficient drug combination for stem cell mobilization and high-dose treatment of malignant lymphoma. Eur J Haematol Suppl 64:46–50
106. Illerhaus G, Marks R, Ihorst G, Guttenberger R, Ostertag C, Derigs G et al (2006) High-dose chemotherapy with autologous stem-cell transplantation and hyperfractionated radiotherapy as first-line treatment of primary CNS lymphoma. J Clin Oncol 24(24):3865–3870
107. Necchi A, Nicolai N, Mariani L, Raggi D, Fare E, Giannatempo P et al (2013) Modified cisplatin, etoposide, and ifosfamide (PEI) salvage therapy for male germ cell tumors: long-term efficacy and safety outcomes. Ann Oncol 24(11):2887–2892
108. Harstrick A, Schmoll HJ, Wilke H, Kohne-Wompner CH, Stahl M, Schober C et al (1991) Cisplatin, etoposide, and ifosfamide salvage therapy for refractory or relapsing germ cell carcinoma. J Clin Oncol 9 (9):1549–1555

109. Voss MH, Feldman DR, Motzer RJ (2011) High-dose chemotherapy and stem cell transplantation for advanced testicular cancer. Expert Rev Anticancer Ther 11 (7):1091–1103
110. Hildebrandt M, Rick O, Salama A, Siegert W, Huhn D, Beyer J (2000) Detection of germ-cell tumor cells in peripheral blood progenitor cell harvests: impact on clinical outcome. Clin Cancer Res 6(12):4641–4646
111. Schwella N, Beyer J, Schwaner I, Heuft HG, Rick O, Huhn D et al (1996) Impact of pre-leukapheresis cell counts on collection results and correlation of progenitor-cell dose with engraftment after high-dose chemotherapy in patients with germ cell cancer. J Clin Oncol 14(4):1114–1121
112. Siegert W, Beyer J, Strohscheer I, Baurmann H, Oettle H, Zingsem J et al (1994) High-dose treatment with carboplatin, etoposide, and ifosfamide followed by autologous stem-cell transplantation in relapsed or refractory germ cell cancer: a phase I/II study. The German Testicular Cancer Cooperative Study Group. J Clin Oncol 12(6):1223–1231
113. Waterman J, Rybicki L, Bolwell B, Copelan E, Pohlman B, Sweetenham J et al (2012) Fludarabine as a risk factor for poor stem cell harvest, treatment-related MDS and AML in follicular lymphoma patients after autologous hematopoietic cell transplantation. Bone Marrow Transplant 47(4):488–493
114. Perseghin P, Terruzzi E, Dassi M, Baldini V, Parma M, Coluccia P et al (2009) Management of poor peripheral blood stem cell mobilization: incidence, predictive factors, alternative strategies and outcome. A retrospective analysis on 2177 patients from three major Italian institutions. Transfus Apher Sci 41(1):33–37
115. Mendrone A Jr, Arrais CA, Saboya R, Chamone Dde A, Dulley FL (2008) Factors affecting hematopoietic progenitor cell mobilization: an analysis of 307 patients. Transfus Apher Sci 39(3):187–192
116. Janikova A, Koristek Z, Vinklarkova J, Pavlik T, Sticha M, Navratil M et al (2009) Efficacious but insidious: a retrospective analysis of fludarabine-induced myelotoxicity using long-term culture-initiating cells in 100 follicular lymphoma patients. Exp Hematol 37(11):1266–1273
117. Popat U, Saliba R, Thandi R, Hosing C, Qazilbash M, Anderlini P et al (2009) Impairment of filgrastim-induced stem cell mobilization after prior lenalidomide in patients with multiple myeloma. Biol Blood Marrow Transplant 15(6):718–723
118. Mazumder A, Kaufman J, Niesvizky R, Lonial S, Vesole D, Jagannath S (2008) Effect of lenalidomide therapy on mobilization of peripheral blood stem cells in previously untreated multiple myeloma patients. Leukemia 22(6):1280–1281. author reply 1-2
119. Kumar S, Dispenzieri A, Lacy MQ, Hayman SR, Buadi FK, Gastineau DA et al (2007) Impact of lenalidomide therapy on stem cell mobilization and engraftment post-peripheral blood stem cell transplantation in patients with newly diagnosed myeloma. Leukemia 21(9):2035–2042
120. Nakasone H, Kanda Y, Ueda T, Matsumoto K, Shimizu N, Minami J et al (2009) Retrospective comparison of mobilization methods for autologous stem cell transplantation in multiple myeloma. Am J Hematol 84(12):809–814
121. Lanza F, Lemoli RM, Olivieri A, Laszlo D, Martino M, Specchia G et al (2014) Factors affecting successful mobilization with plerixafor: an Italian prospective survey in 215 patients with multiple myeloma and lymphoma. Transfusion 54(2):331–339

Chapter 5

Transmigration Assays for the Determination of Molecular Interactions Between Hematopoietic Stem Cells and Niche Cells

Reinhard Henschler and Rudolf Richter

Abstract

The transmigration capacity of hematopoietic stem and progenitor cells (HSPC) is characteristically associated with their ability to home to sites of hematopoiesis in the transplanted host, to proliferate, to differentiate, and to successfully repopulate the hematopoietic system of a transplanted host. Stimulating agents shown to induce the transmigration of HSPC were often later identified to play significant roles in mobilization of HSPC or their interaction with niche cells in the hematopoietic microenvironment. Transwell migration assays through microporous membranes have been developed in various forms to determine the migration capacity of HSPC or mesenchymal stromal cells (MSCs) toward chemoattractants. We describe here a method of a multi-well reusable transmigration assay using a small volume and low numbers of HSPC, allowing the simple and reproducible determination of HSPC transmigration capacity which enable researchers to obtain rapid answers at limited costs with high reliability.

Key words Transwell, Transmigration, Chemoattraction, HSPC, Chemokines, Porous membrane

1 Introduction

In the field of hematopoietic stem cell transplantation, the transwell migration assay gained major attention when the group of Timothy Springer used this technique to purify the chemokine stromal cell-derived factor-1 alpha (SDF-1α, CXCL12) from supernatant of the murine bone marrow stromal cell line MS-5 [1]. Cells positioned in the upper wells transmigrated through micropores toward MS-5 supernatant or SDF-1α which were only present in the lower well. This indicated chemoattraction, i.e., directed migration, since no migration ensued when the factor was present also in the upper wells. Soon after, SDF-1α was found to potently attract $CD34^+$ selected HSPCs and mediate their homing to spleen after transplantation in mice [2].

In 1999, the group of Lapidot published data showing that blocking the receptor of SDF-1α on HSPCs, CXC chemokine

Gerd Klein and Patrick Wuchter (eds.), *Stem Cell Mobilization: Methods and Protocols*, Methods in Molecular Biology, vol. 2017, https://doi.org/10.1007/978-1-4939-9574-5_5,

receptor (CXCR) 4, resulted in an engraftment defect in immunodeficient mice repopulated with human CD34$^+$ HSPC, thus linking the result of HSPC transmigration assays to hematopoietic stem cell transplantation [3]. Extending this knowledge, development of inhibitors of CXCR4 led to the identification and clinical development of AMD3100, a CXCR4 blocker, to plerixafor, a drug used for poor mobilizers of HSPC for transplantation [4]. The principle of chemoattraction as a key signal to initiate the directed migration of blood-circulating cells to home to tissues, and vice versa to exit from tissues toward the bloodstream, has been found to apply to numerous hematopoietic cell types including monocytes, granulocytes, lymphocytes, and HSPC [5, 6].

We have recently described the identification of chemotactic activities for HSPC from human plasma libraries [7]. For this, we screened a total of more than 400 fractions of chromatographically separated plasma for the presence of chemotactic activities acting on HSPC using a multi-replicate reusable transmigration chamber. This study stands as an example for the use of a rapid, robust transmigration assay, which works with small numbers of cells and small volumes. Here we describe the methodology for these transmigration assays in detail.

2 Materials

2.1 Materials for Transmigration Assay Using HSPC

- PKH26 Red Fluorescent Cell Linker Kit for general cell membrane labelling (Sigma, Munich, Germany) consisting of diluent C (6 × 10 mL) and PKH26 Cell Linker in ethanol (0.5 mL).
- IMDM (Iscove's Modified Dulbecco's Medium) with L-glutamine and phenol red, without α-thioglycerol and 2-mercaptoethanol.
- 5% (w/v) bovine serum albumin (BSA) dilution in phosphate-buffered saline (PBS).
- IMDM/0.1% BSA: IMDM supplemented with 5% BSA to a final concentration of 0.1%.
- Storage of HSPC longer than 2 h: IMDM/0.1% BSA supplemented with 10 ng/mL each of recombinant human (rh) stem cell factor (SCF) and rh flt3 ligand (FL).
- AP48 chemotaxis chamber (Neuro Probe Inc., 16008 Industrial Drive, Gaithersburg, Maryland-20877, USA; www.neuroprobe.com).
- Polycarbonate membranes, PVP-free. 5 μm pore size, 25 × 80 mm (100 membrane box; Neuro Probe; *see* **Note 1**).
- Phosphate-buffered saline (PBS) solution with Ca^{2+}/Mg^{2+}.
- Fetal bovine serum (FBS), non-heat inactivated.

- Cytokines, chemokines, and other stimulants of cell migration. Usually prepared as a 10 ng/μL or 100 ng/μL stock solution in PBS/0.1% BSA and kept frozen at −70°°C in aliquots.
- Flow cytometry microbeads for cell number determination.
- Cell culture equipment (sterile workbench class II, cytocentrifuge, micropipette, 10 mL plastic tubes).

2.2 Additional Materials for Transmigration of Mesenchymal Stromal Cells (MSCs) to Detect Factors Acting on Stromal/Niche Cells

- Human plasma fibronectin, human laminin (*see* **Note 2**), human plasma vitronectin, and poly-L-lysine. Stock solutions of 1 mg/mL should be prepared in phosphate-buffered saline (PBS), aliquoted, and stored at −80 °C. Immediately before use, aliquots from stocks are thawed and added to the media as appropriate.
- Low-glucose Dulbecco's Modified Eagle Medium (DMEM).
- Fetal calf serum (FCS), heat-inactivated.
- Basic fibroblast growth factor (bFGF).
- MSC culture medium: low-glucose DMEM supplemented with 20% FBS and 25 ng/mL basic FGF.
- Trypsin-EDTA (0.5% trypsin, 6.8 mM EDTA in PBS).
- Terg-A-Zyme (Alconox Inc., White Plains, NY).
- 8 μm PVP-free chemotaxis membranes (Neuro Probe) (*see* **Note 1**).
- 100% methanol.
- 4% paraformaldehyde.
- Wiper apparatus (Neuro Probe).
- May-Grünwald-Giemsa staining solutions.
- Inverted light microscope (Olympus CK-2).
- Neubauer cell counting chamber.
- Digital imaging camera.
- Vortex shaker.
- Cell culture material (5 and 15 mL polypropylene tubes, tissue culture plastic flasks, 15 cm cell culture dishes).

3 Methods

3.1 Method for the Transmigration Assay Using HSPC

CD34$^+$ HSPCs were obtained from aliquots retained for quality control from healthy donors or patients with malignancies after treatment with rh G-CSF and isolated by leukapheresis or from human cord blood. Informed consent for use of the CD34$^+$-enriched HSPC was obtained from allogeneic stem cell donors undergoing HPC mobilization via G-CSF, and investigations were performed according to the 2000 Declaration of Helsinki and the 2008 Declaration of Istanbul. $CD34^+$ cell selection was performed by magnetic bead cell separation systems as provided by the manufacturer (Miltenyi, Bergisch Gladbach, Germany) resulting in purities between 80 and 95%. Cells were resuspended in IMDM supplemented with 0.1% bovine serum albumin. If stored for longer than 2 h on ice, the medium was supplemented with rh SCF and rh flt3 ligand.

3.1.1 Labelling of Cells with Fluorescent Dye PKH26

Work in a sterile class II safety cabinet.

1. Prepare PKH staining solution. Add 8 μL of PKH26 Cell Linker to 1 mL diluent C and keep light-protected and on ice at 4 °C, for longer storage at <30 °C.
2. Count HSPC using a live cell counting method and subsequently pellet the HSPC by centrifugation at 400 × g for 5 min in a 10 mL tube.
3. Carefully remove the supernatant of the HSPC cell suspension and resuspend the cell pellet in 50 μL of diluent C. Then add 50 μL of PKH staining solution (for up to 10^6 cells). A typical volume is 50–100 μL for labelling between 5×10^4 to 5×10^5 cells.
4. Gently agitate and incubate the mix in the dark at room temperature for 5 min.
5. Add 2 mL of FBS to the mixture, gently agitate, and incubate for 1 min at room temperature.
6. Fill up the tube to 10 mL with IMDM/0.1% BSA, mix gently by pipetting up and down, and pellet the cells by centrifugation at 400 × g for 5 min.
7. Wash cells twice adding IMDM/0.1% BSA by centrifugation. Determine live cell concentration before the last wash.
8. Resuspend the cell pellet in IMDM/0.1% BSA to achieve a cell concentration of 5×10^5 cells/mL.

3.1.2 Preparation of Upper Well Contents of the AP48 Transwell Chemotaxis Chamber

The upper wells will have to be filled with a volume 32 μL each. A HSPC suspension is prepared in IMDM/0.1% BSA.

1. Calculate HSPC numbers and volume in IMDM/0.1% BSA for the upper wells. For 5000 cells/32 μL, a concentration of 1.56×10^5/mL HSPC is required.
2. Adapt HSPC concentration by adding up IMDM/0.1% BSA to the calculated volume or wash HSPC again by centrifugation and suspend the pellet with the calculated volume of IMDM/0.1% BSA.
3. Resuspend cells by gentle agitation or pipetting up and down.
4. Keep cells on ice, designate as "upper well."

3.1.3 Preparation of Lower Well Contents of the AP48 Transwell Chemotaxis Chamber

1. Prepare the master dilutions of chemoattractant (as experimental variable; *see* **Note 3**) for lower well, e.g., dilutions of SDF-1α. For this, prepare a 96-well cell culture U-bottom plate and prepare dilutions of the chemoattractant, e.g., SDF-1α, in 4 or 8 steps. The highest concentration is 1/3.16 of the stock chemoattractant concentration.
2. Add 21 μL IMDM/0.1% BSA into the empty well into 4 or 8 wells in a row.
3. Add 10 μL of the stock solution of the chemoattractant, e.g., SDF-1α @ 100 ng/μL, into the well with highest concentration. Mix well by pipetting up and down.
4. Transfer 10 μL of this mix to the next lower dilution well. Mix well by pipetting up and down. Repeat in a row for the entire dilution series.
5. Prepare a row of six 5 mL tubes for a series of six experimental variables, for half of the field of the 48-well chamber.
6. Add 300 μL IMDM/0.1% BSA to each tube in the dilution series.
7. Add 10 μL of the master dilution series to each of the tubes 2–5. Tube 6 will be a control without chemoattractant; tube 1 will receive 10 μL directly from the chemoattractant stock.
8. Keep on ice until pipetted into lower wells.

3.1.4 Setup of the Chemotaxis Chamber

1. Check the parts of the AP48 micro-chemotaxis chamber for cleanliness under a light microscope. For cleaning, a cleaning solution is provided by the manufacturer. Remove any visible contaminants and flush with bi-distilled water. Allow to dry in the sterile workbench.
2. Place the lower well part with the orientation sign in the lower right end. This ensures the correct assembly of the chamber. The upper and lower parts of several chambers shall not be mixed up (individual fit).
3. Using a micropipette (100 μL tips), add 34 μL of the prepared media solutions for the lower wells according to a schema (example shown below, e.g., with replicates from top to bottom and dilutions/experimental variables from right to left) into each lower well. This will result in a little elevation of the fluid surface on top of the chamber surface, which is desired to avoid air bubble formation during assembly.

   Example schema for assay layout (one half of a 48-well transmigration chamber).

Experimental variables						
Replicates	1	2	3	4	5	6
	1	2	3	4	5	6
	1	2	3	4	5	6
	1	2	3	4	5	6

4. Carefully place the microporous membrane on top, e.g., from left to right, with the shining side facing downward. Avoid air bubble formation. It is recommended to train this step with a dummy assembly before starting experiments.
5. Carefully place the rubber interspacer on top.
6. Place the upper part of the chamber on top and softly tighten the nuts (*see* **Note 4**).
7. Add HSPCs (32 μL per well) to each upper well using micropipette. This step therefore also requires care and a safe handling, and experience, since the pipette tip must not perforate the membrane.
8. Incubate in a 37 °C humidified incubator for 4–6 h.

3.1.5 Preparation of Flow Cytometry 5 mL Tubes for Cell Number Determination

Before stopping the experiment, prepare a flow cytometry 5 mL tube for each well. Add a fixed fluid volume (e.g., 200 μL) of IMDM/0.1% BSA and a fixed number of flow cytometry beads (e.g., 10.000 microbeads) into each tube.

3.1.6 Termination and Workup of the Migration Assay

1. Remove chemotaxis chamber from incubator at end of incubation period. We found that 4–6 h is an optimal period for HSPC transmigration.
2. Carefully remove the medium from the upper wells using a micropipette tip. The tip shall not perforate the membranes. This step therefore also requires care and a safe handling.
3. Open the nuts carefully and remove the upper acrylic part of the chambers without removing the membrane. Keep rubber spacer and the membrane in position on the lower acrylic part. Tweezers are helpful for this.
4. Remove the rubber separator, still keeping the membrane in place.
5. Carefully remove the membrane in a “rolling” movement. This step should be performed relatively quickly within 2–3 s. Spillover of fluid (and cells) from one well to the other shall be avoided (*see* **Note 5**).
6. Transfer the contents of each of the lower wells into the corresponding cell enumeration flow cytometry tube using a micropipette. Flush each well with 20 μL IMDM/0.1 BSA to remove all cells and collect into the respective cell enumeration tube.

7. Determine the cell numbers in the lower wells by running the cell enumeration tubes in a flow cytometer. Other methods may also be used, but due to the low numbers found in the lower wells (250–1000 cells), we found that flow cytometric counting produces the best results, detecting as little as 0.5% of the initially seeded cells in the lower wells.
8. Clean the chemotaxis chamber using the protocol provided by the manufacturer, and check microscopically for cleanliness (*see* **Note 6**). The last rinses shall be performed with distilled water.

A typical result from a series of transmigration assays serially testing 144 fractions from human plasma is shown in Fig. 1.

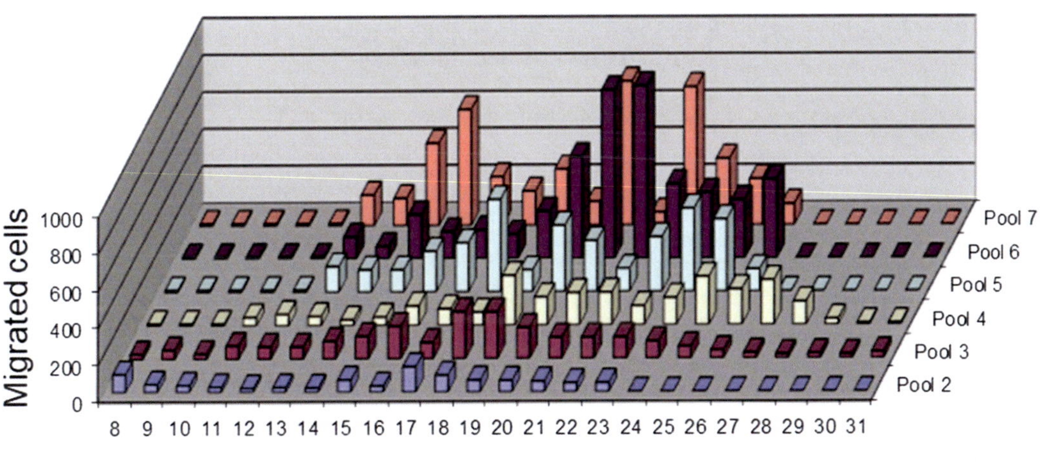

Fig. 1 HSPC transwell migration induced by fractions of human plasma, determined in the 48-well reusable migration chamber. Freeze-dried aliquots of chromatographed hemofiltrate fractions which were produced as described below were thawed and tested for transmigration in chemotaxis assays as described. Each fraction was tested in three independent experiments at 100% in the lower wells and HSPC in the upper wells as described in Subheading 3.1. Mean values are shown. For the preparation of the plasma fractions, 900 L of hemofiltrate was collected from patients with plasmacytoma and treated with alkeran and cyclophosphamide and in addition with the VACOP-B protocol and etoposide. The hemofiltrate was immediately acidified after collection and subsequently stored at -20 °C. Subsequently the hemofiltrate was ultrafiltered using a hollow fiber module UFP-50-E-A with a size exclusion membrane of 50 kDa, and was next subjected to a 3.6 L cation exchange column. Bound peptides were eluted using seven buffers with increasing pH (I, 0.1 M citric acid monohydrate, pH 3.6; II, 0.1 M acetic acid/0.1 M sodium acetate, pH 4.5; III, 0.1 M malic acid, pH 5.0; IV, 0.1 M succinic acid, pH 5.6; V, 0.1 M sodium dihydrogen phosphate, pH 6.6; VI, 0.1 M disodiumhydrogenphosphate, pH 7.4; VI,: 0.1 M ammonium carbonate, pH 9.0.) This way, seven pools were collected, and each of them was loaded on a 15 × 10 cm reverse-phase column and eluted in a gradient from 100% A (0.01 M HCl) to 60% B (80% acetonitrile, 0.01 M HCl) within 53 min [7]

3.2 Method for the Transmigration of Mesenchymal Stromal Cells (MSCs) to Detect Factors Acting on Stromal/Niche Cells

Since chemoattractants have been found to also act on stromal/niche cells, we have modified the method described above to detect migration of stromal/niche cells also [8]. This allows the detection of chemotactic activities and factors, potentially derived from hematopoietic cells or other cells in the microenvironment which act on stromal/niche cells (*see* **Note 3**).

3.2.1 Preparation of Chemotaxis Membranes

For precoating, the 8 μm PVP-free chemotaxis membranes are incubated in culture dishes with vitronectin, fibronectin, or laminin/isoforms (*see* **Note 2**) at a concentration of 10 μg/mL. Incubation of the membranes is performed for 1 h at 37 °C. Subsequently, membranes were flushed twice with PBS.

3.2.2 Preparation of MSCs

Work in a sterile class II safety cabinet.

We regularly obtained human bone marrow mesenchymal stromal cells (BM-MSCs) from patients undergoing hip joint replacement after informed consent. MSCs were prepared from the mononuclear cell fraction by density gradient centrifugation and seeded in T25 or T80 tissue culture plastic flasks at a density of $1\text{–}3 \times 10^6$ cells/cm^2 in low-glucose DMEM supplemented with 20% FCS and 25 ng/mL basic FGF. After 48 h incubation at 37 °C and 5% CO_2, non-adherent cells were removed, and the remaining cells were cultured for another 7–10 days until reaching 70–80% confluence. The cells were passaged by trypsinization. Multipotency characteristics of MSCs were confirmed by investigation of the osteoblastic, chondrogenic, and adipogenic differentiation potential [9]. Aliquots of the MSCs were frozen after the fifth passage and were used for chemotaxis assays maximally expanded for nine passages.

1. For chemotaxis assays, MSC culture medium is removed, and the cells are flushed twice with PBS within 1 h before use in the transmigration assays.
2. Subsequently, MSCs are trypsinized using trypsin-EDTA at 37 °C for 5 min.
3. When all cells were detached from the culture surface, trypsin activity is stopped by addition of 5–10 mL DMEM/10% FCS.
4. MSCs are washed once by centrifugation in low-glucose DMEM and are kept in suspension in low-glucose DMEM at 37 °C at a density of $1\text{–}5 \times 10^5$/mL in 15 mL polypropylene tubes in an upright position until seeding into the assay (max 1 h).
5. Cells are stirred on a Vortex shaker once for 1–2 s at 5 Hz before starting the assay. This is performed at all conditions because we observed that its omission can negatively influence the migration potential of MSC.

3.2.3 Preparation of the Chemotaxis Chamber

1. 37 °C pre-warmed low-glucose DMEM supplemented with 0.1% cell culture-tested BSA is used as medium for preparation of chemokine dilutions. The medium volume for four to six replicates is prepared in 5 mL polypropylene tubes for each experimental condition.
2. Lower wells of the chambers are prefilled with 34 μL pre-warmed DMEM/20% FCS.
3. After filling the lower wells, membranes are placed with the shining side upward according to the manufacturer's protocol, avoiding air bubble formation and spillover of medium toward other wells (*see* **Note 5**).
4. The chambers are closed by softly tightening the nuts. Be careful, when the pressure will get too high chamber slides would break, and too low fixation would lead to leaks in the chamber and medium outflow from the wells into the surround parts of the chamber (*see* **Note 4**).
5. The upper wells are then filled with the 37 °C pre-warmed cell suspensions in DMEM/0.1% BSA and additives as required by the experimental protocol, seeding 10,000 or 20,000/well of MSC into each well.
6. The mounting of the chambers, including the filling of all wells, should be performed within 15 min. Subsequently the chamber is placed into a humidified incubator at 5% CO_2, 37 °C.

3.2.4 Finishing the Assay and Counting the Migrated Cells

1. After a period of 4–6 h, carefully remove the medium from the upper well and avoid perforation of the membrane.
2. Carefully remove the membrane by lifting up with a forceps from a corner in a single movement within 2–3 s.
3. Immediately rinse the membrane once with PBS on both sides, but take care not to scratch the surface (*see* **Note 7**).
4. Fixation of the cells on the membrane is performed within the next 15 s by immersion in ice-cold 100% methanol for 5 min or with 4% paraformaldehyde for 15 min.
5. Wipe off the cells on the shining side with a wiper apparatus (Neuro Probe) and rinse membrane with PBS (*see* **Note 7**).
6. Stain cells on the lower part of the membranes with for 5 min with May-Grünwald solution and 20 min with Giemsa solution. Before, in between, and after staining steps, the membrane is flushed with distilled water.
7. Mount the membrane with the shining side facing upward on a glass slide. The air-dried membrane can be kept for periods up to several months in bacterial culture dishes at room temperature without deteriorating.

3.2.5 Evaluation of the Chemotaxis Assay with MSCs

Evaluation is performed under an inverted light microscope at a 40× magnification fitted with a grid scale in the ocular. Per migration experiment, cell numbers are determined in four fields. Take care: cells are on the lower membrane side, this is non-shining side. The correct focal plane has to be determined in order to ensure that only cells on the lower side of the membrane are counted (*see* **Note 7**). On the lower side of the membrane, the shape and the appearance of the pore holes are round and clearly outlined openings on the upper (shining) membrane sides, whereas pores on the lower (non-shining) side is oblique, shaped ovally with frail margins. The grid of a Neubauer cell counting chamber, underlying the membrane preparation, is used to count unit areas of the entire surface per well, and the area counted for evaluation are determined. Photographs of representative fields should be taken for documentation.

4 Notes

1. The pore size of the membranes can be chosen variably between 3, 5 and 8 μm. For granulocytes, 3 μm is usually adequate. We found 5 μm pores optimal for HSPC, both murine and human. For stromal cells like MSCs, 8 μm is required.
2. For the experiments described here, a mixture of laminins (mainly LM-111) purified from Engelbreth-Holm-Swarm murine sarcoma basement membrane was used, which does not allow conclusions regarding the individual isoform acting upon a cell. Different human laminin isoforms with different biological functions are commercially available as recombinant proteins from Biolamina, Sweden, and should be used for assessing effects of individual laminins.
3. We found several chemokines to act on HPSC, whereas MSC can routinely only be efficiently attracted by FBS. To verify whether a substance is a chemoattractant, controls shall be included where these substances are added both in the upper and lower wells or into none. Chemoattraction can then be verified as the activity which acts specifically as migration toward these substances. Dose-response and time-response curves should also be established during the optimization of the assay setup.
4. Handling of the chamber closure nuts requires care. If the pressure after tightening the nuts will get too high, this can break the chamber slides; if the pressure is too low, the fixation will lead to leaks in the chamber and medium outflow from the wells into the surrounding parts of the chamber. This requires individual training by each examiner. We found that any person

was capable of learning this; however, it can cost some broken chambers if the point is not made clear to every new user.

5. Spillover of fluid between the upper wells may result from too weak nut tightening upon mounting of the chamber or from contaminant material situated between the two acrylic chamber parts. Between lower wells, spillover can occur during the removal of the upper chamber part during the workup of the assay. We found that personal practice is able to overcome this risk once investigators are aware of the possibilities. Also, quick wiping away of droplets or small fluid areas can prohibit the overflow between wells. In doubt, the affected area (well numbers) shall be documented and omitted from the results.
6. We recommend to follow the cleaning protocols given by the manufacturer. Most researchers are nowadays used to disposable material, and securing adequate maintenance and cleanliness of the materials may require training. We found microscopical checks after cleaning and before use helpful in assessing the cleanliness of the chambers.
7. MSCs transmigrate efficiently; however, they adhere to the membranes' lower side and do not "drop" into the lower end of the well such as HSPC. To distinguish migrated cells from non-migrated cells, the non-migrated cells (on the upper side of the membrane) need therefore to be wiped away efficiently from the membrane. The procedure should be tested and verified before running routine assays. A microscopic check verifies the result. To find the upper side, the sharpness of the edge of the pores is used. The pores are stained into the membranes by a laser technology. This implies that the pores are more sharp-edged on the entry side (upper side) and have a more irregular ridge on the exit side/lower side.

Acknowledgments

We acknowledge the technical contributions by our co-workers Victoria Karpf and Bithiah Grace-Jaganathan.

References

1. Bleul CC, Fuhlbrigge RC, Casanovas JM, Aiuti A, Springer TA (1996) A highly efficacious lymphocyte chemoattractant, stromal cell-derived factor 1 (SDF-1). J Exp Med 184:1101–1109
2. Aiuti A, Webb LJ, Bleul C, Springer T, Gutierrez-Ramos JC (1997) The chemokine SDF-1 is a chemoattractant for human CD34+ hematopoietic progenitor cells and provides a new mechanism to explain the mobilization of CD34+ progenitors to peripheral blood. J Exp Med 185:111–120
3. Peled A, Petit I, Kollet O, Magid M, Ponomaryov T, Byk T, Nagler A, Ben-Hur H, Many A, Shultz L, Lider O, Alon R, Zipori D, Lapidot T (1999) Dependence of human stem cell engraftment and repopulation of NOD/S-CID mice on CXCR4. Science 283:845–848

4. De Clercq E (2003) The bicyclam AMD3100 story. Nat Rev Drug Discov 2:581–587
5. de Oliveira S, Rosowski EE, Huttenlocher A (2016) Neutrophil migration in infection and wound repair: going forward in reverse. Nat Rev Immunol 16:378–391
6. Krummel MF, Bartumeus F, Gérard A (2016) T cell migration, search strategies and mechanisms. Nat Rev Immunol 16:193–201
7. Richter R, Jochheim-Richter A, Ciuculescu F, Kollar K, Seifried E, Forssmann U, Verzijl D, Smit MJ, Blanchet X, von Hundelshausen P, Weber C, Forssmann WG, Henschler R (2014) Identification and characterization of circulating variants of CXCL12 from human plasma: effects on chemotaxis and mobilization of hematopoietic stem and progenitor cells. Stem Cells Dev 23:1959–1974
8. Rüster B, Grace B, Seitz O, Seifried E, Henschler R (2005) Induction and detection of human mesenchymal stem cell migration in the 48-well reusable transwell assay. Stem Cells Dev 14:231–235
9. Pittenger MF, Mackay AM, Beck SC, Jaiswal RK, Douglas R, Mosca JD, Moorman MA, Simonetti DW, Craig S, Marshak DR (1999) Multilineage potential of adult mesenchymal stem cells. Science 284:143–147

Chapter 6

Microfluidic Shear Force Assay to Determine Cell Adhesion Forces

Julia Hümmer, Julian Koc, Axel Rosenhahn, and Cornelia Lee-Thedieck

Abstract

Cell adhesion is implicated in many physiological settings such as the retention of hematopoietic stem cells (HSCs) in their bone marrow niches or their migration into the bloodstream. During HSC mobilization these adhesion sites are cleaved and have to be newly formed during HSC homing and engraftment. To determine the adhesive properties of HSCs on different extracellular matrix (ECM) molecules, we present a microfluidic shear force assay, where a laminar flow is used to detach a semi adherent cell population, the HSC model cell line KG-1a, from an ECM protein-coated substrate. This technique combines the high throughput of population-based assays with the ability to observe cell detachment in real time. Additionally, it is suitable for weakly adherent cells, as the setup allows cell incubation on various substrates and application of shear stress ranging several orders of magnitude in one setup without additional washing or transfer steps. As a measure for the adhesion strength of the studied cell population on the substrate, the critical shear force τ_{50} is determined which is required to remove 50% of the initially adherent cell fraction.

Key words Microfluidic assay, Shear force, Cell adhesion, Leukemic cells, Protein coating

1 Introduction

The balance between hematopoietic stem cell (HSC) retention in their bone marrow (BM) niches and HSC mobilization to the bloodstream is tightly controlled by cell adhesion complexes. In addition to their anchorage to other niche cells, such as vascular cell adhesion molecule 1 (VCAM-1)-expressing endothelial cells via integrin α4β1, HSCs are connected to a complex network of extracellular matrix (ECM) molecules, as reviewed in [1]. Via cell surface receptors such as CD44, HSCs can interact with hyaluronic acid or osteopontin which is present in the niche. Contacts to many ECM glycoproteins that are highly expressed in the HSC niche, such as fibronectin, laminins, and thrombospondins, however, are tied via integrin receptors [2]. Apart from ensuring cell anchorage points

Julia Hümmer and Julian Koc contributed equally to this work.

Gerd Klein and Patrick Wuchter (eds.), *Stem Cell Mobilization: Methods and Protocols*, Methods in Molecular Biology, vol. 2017, https://doi.org/10.1007/978-1-4939-9574-5_6, © Springer Science+Business Media, LLC, part of Springer Nature 2019

by linking the ECM with the intracellular actin cytoskeleton, these heterodimeric transmembrane receptors enable HSCs to sense biophysical cues in their environment (reviewed in [3]) and mediate bidirectional cell signaling events regulating many basic cellular behaviors such as differentiation, proliferation, and cell survival [4]. In vitro assays have shown that hematopoietic stem and progenitor cell (HSPC) cycling and differentiation are influenced by specific adhesion ligands and HSPCs interact strongly with fibronectin [5, 6]. Furthermore it was shown in vitro that the culture of murine HSPCs in the presence of fibronectin and laminin improved their engraftment capacity [7] and that fibronectin supported the growth of HSPCs [8].

Upon clinical mobilization with G-CSF, this tight anchorage of HSCs to neighboring cells and ECM molecules is lost due to proteolytic enzymes, allowing them to escape to the bloodstream [9–12]. After transplantation into a patient, HSPCs home to the BM niche from the bloodstream, a multi-step process similar to the transendothelial migration of leukocytes [13, 14]. On the surface of both, HSPCs and endothelial cells, primary adhesion molecules are involved in this process which enable the tethering and rolling of HSPCs along the endothelium [15]. When activated by chemoattractants that are secreted by endothelial cells or bound to proteoglycans in the ECM, HSPCs can firmly adhere to endothelial layers via integrins [16]. The strength of these adhesive contacts allows them to withstand the shear forces of the bloodstream and thus transendothelial migration into the BM niche.

Taken together, adhesive contacts with niche cells and ECM molecules are of utmost importance in HSC mobilization and homing. Measuring and comparing adhesive interactions between HSPCs and specific ligands or ECM molecules are therefore often faced challenges. To that end, both single-cell techniques and population-based methods have been applied to measure the adhesion strength of mostly adherent cell types to various substrates (reviewed in [17, 18]). While single-cell methods, such as cytodetachment assays using atomic force microscopy [19] or micropipette aspiration [20], allow the precise measurement of the force required to remove a single cell from its substrate, they also have some drawbacks. In addition to the costly, specialized equipment that is required for the former technique, all single-cell methods are highly time-consuming when surveying larger cell numbers. Here, population-based methods offer inherent advantages as multiple cells are assayed within one experiment, saving valuable experimental time. Population-based assays include, e.g., qualitative wash assays, where adherent cell fractions on ECM-coated surfaces can be determined after a standardized washing protocol [21, 22]. However, these methods suffer from low reproducibility and insensitivity. In addition, centrifugation assays or spinning disk assays [23–27] are used due to their high throughput; nevertheless, they do not offer real-time analysis.

Hydrodynamic shear flow assays using radial or parallel flow chambers are widely applied to study cell adhesion on defined substrates as different ranges of shear stresses can be covered in a single experiment and observation of real-time cell detachment is possible [28–31]. The versatility of this setup has been demonstrated also by the ability to study the rolling behavior of HSPCs on ECM components, such as hyaluronic acid [32, 33], which is an important step in HSPC homing. In comparison to sophisticated single-cell-based methods, these flow chambers can be easily fabricated in a parallel setup offering a high sample throughput. We show here the fabrication and assembly of a parallel plate flow chamber as well as the performance of the microfluidic detachment assay, where a laminar flow is used to detach an adherent cell population from surfaces coated with ECM proteins or peptide sequences. This method allows the application of shear stress subsequently to the incubation on various substrates in one setup, without washing or moving of the sample, which is important to study weakly adherent cells like HSPCs. Here, the applied shear stress is constantly increased in small steps, spanning several orders of magnitude. This allows the determination of the critical shear force τ_{50} which is defined as the shear force needed to remove 50% of the initially adherent cell fraction.

2 Materials

2.1 Preparation of the Plate Flow Setup (Illustrated in Fig. 1a)

1. Custom-made multi-windowed platform: 13 × 9 × 0.5 cm made of aluminum or poly(methyl methacrylate) (PMMA).
2. ECM-coated substrate (*see* Subheading 2.2).
3. Poly(dimethylsiloxane) (PDMS) channel: Mix PDMS (1:9, e.g., 1 g curing agent and 9 g polymer, Sylgard 184, Dow Corning, MI, USA) to ensure an ideal mixing of both components, and degas the mixture for 1–2 h in a desiccator (*see* **Note 1**). Pour the viscous PDMS into a polished micro-machined brass mold and cure it for 24 h at RT (Fig. 1b). The brass mold defines the channel dimensions; in this setup channel dimensions were 25 × 1.5 × 0.14 mm with a wall thickness of 1 mm (Fig. 1c). Carefully remove the channels from the brass mold (*see* **Note 2**).
4. Glass lid: Use a glass slide with the dimensions of 30 × 20 mm and 2 mm thickness. Drill two 1-mm-wide holes in a distance of 24 mm in the center of the glass lid (*see* **Note 3**).
5. PDMS seal ring: Follow the instructions described in Subheading 2.1, **item 3** using a polished micro-machined mold, to

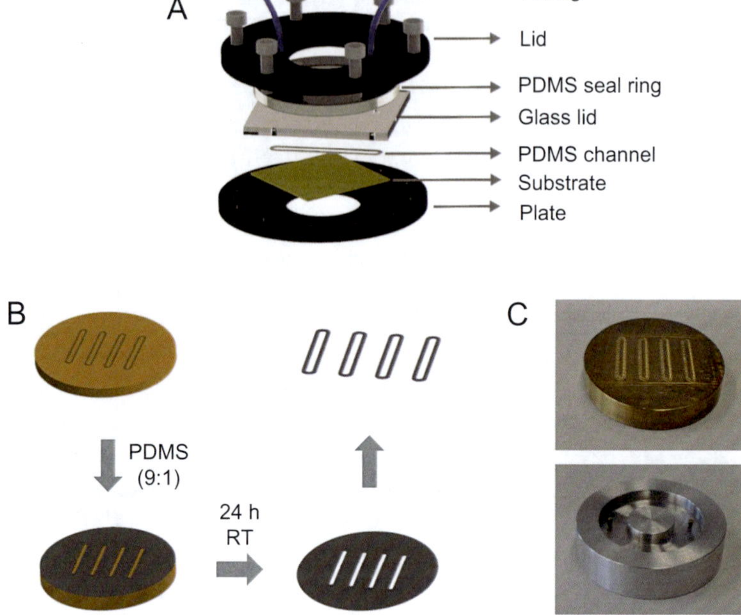

Fig. 1 (**a**) Design of the microfluidic channel with all parts needed for sealing. The upper compartment, consisting of the lid with tubing, PDMS seal ring, and glass lid, is permanently assembled. Before the experiment the PDMS channel is attached to the upper compartment and the substrate, facing the coated side upward. The assembly is fixed between the upper compartment and the plate using six M2 screws. (**b**) Schematic workflow of the PDMS channel production. Using a brass mold, a PDMS polymer solution (9:1 polymer:curing agent) is poured onto a brass mold. After solidification at RT for 24 h, the channels can be stripped off. (**c**) Photographs of the micro-machined mold for producing the PDMS channels as well as the PDMS seal ring

obtain a 4-mm-thick polymer ring with two 1.2 mm holes (Fig. 1c). After curing, remove the PDMS seal ring from the mold. Align the glass lid on the PDMS seal ring with blunt cannulas (diameter 0.55), and fix them with glue, e.g., Pattex repair extreme gel (Pattex, Henkel, Düsseldorf, Germany).

6. Lid: Use a round 3-mm-thick aluminum or PMMA disk (diameter = 4 cm), with an upper window (diameter = 1.4 cm), six outer holes for M2 screws, and two inner holes (20 mm) for the tubing.
7. Tubing: Connect a 1/16″ (1.56 mm) polyfluoroalkoxy tubing (Postnova Analytics GmbH, Landsberg am Lech, Germany) through the holes of the upper disk with the holes in the glued glass slide.
8. Manifold (VICI, Schenkon, Switzerland) with five different selectable lines (illustrated in Fig. 3b II). Connect up to four microfluidic channels with the lines of the manifold via tubing.

One line needs to be left for the manual insertion of cells. Here a tubing connected with a one-way stopcock system (Discofix®-1, B. Braun Melsungen AG, Germany) should be used.

9. Selector (Cheminert, VICI, Schenkon, Switzerland) with five different valves (Fig. 3b IV) for opening the flow path for one channel at a time. Connect the selector valves with the tubing of the flow cell via flangeless tube end fittings (VICI, Schenkon, Switzerland), consisting of polypropylene nuts and a CTFE ferrules (1/16″). Multiple flow cells can be run in parallel by connecting them to both the selector and the manifold via tubing. A useful configuration of the selector may be four channels and one bypass (tubing without microfluidic channel for equilibration of the system). The selector is connected permanently to all microfluidic channels and is located behind the flow cell plate.
10. Blunt cannulas (diameter 0.55 mm).
11. Precision wipes, e.g., Kimberly-Clark Professional Kimtech Science™ Precision Wipes™.

2.2 Preparation of Sample Surface

For surface coating with a protein or peptide, refer to standard protocols established by the distributor, as this procedure varies depending on the proteins or peptides used (*see* **Note 4**).

1. Choice of surface: standard glass microscopy slides cut to the size matching the channel dimensions (32 × 24 mm) (*see* **Note 5**).
2. Protein or peptide solution (refer to the distributer's protocols for bringing the protein or peptide of interest into solution).

2.3 Cell Culture

All cell culture work should be executed under sterile conditions in a biological safety cabinet class II. All cell culture media should be prewarmed to 37 °C before use. Here, the human acute myeloid leukemia cell line KG-1a (DSMZ, #ACC 421) was used as a model for HSPCs. Therefore, the described media are specific for these weakly adherent suspension cells.

1. Subculture of cells: cell-specific culture media – RPMI 1640 media with 20% fetal bovine serum (FBS) and 1% penicillin/streptomycin (P/S). KG-1a cells were cultured in cell-specific culture media at a density of 0.5–2.0 × 10^6 cells/ml in a humidified incubator kept at 37 °C with 5% CO_2.
2. Adhesion media: RPMI 1640 media containing 1% FBS, 1% P/S, and 25 μM $MnCl_2$.
3. Cell culture flasks for suspension cells.
4. Sterile plastic syringe (2 ml).

2.4 Microscope Setup (Fig. 4)

1. Microscope: inverted Nikon microscope TE-2000, equipped with a 4× phase-contrast objective, a cooled 5-megapixel color camera, a movable stage, and a shutter system.
2. Incubation chamber: custom made of transparent PMMA with heating cable and a temperature control unit, attached to the walls of the chamber. Gas valve and a CO_2 sensor for controlling the atmosphere (*see* **Note 6**).
3. Syringe pump: custom-built programmable syringe pump, which can be controlled via a computer using a M 403-4DG translation stage (Physik Instrumente GmbH & Co.KG, Karlsruhe, Germany). Write and use a macro (PI Mikro Move Physik Instrumente GmbH & Co.KG, Karlsruhe, Germany) whereby the pump speed is increased by 2.33% every 0.5 s. The resulting volumetric flow rate (= pump retraction area of syringe cross section) and shear stress τ will also be increased in this step range (Fig. 4b) (*see* **Note 7**). In the exemplified, presented experiment, an initial pump speed of 1×10^5%/s syringe feed was used (*see* **Note 8**).
4. Liquid reservoir with pressure control: the liquid reservoir should be equipped with a stopcock system (Discofix® C, B. Braun Melsungen AG, Germany) to regulate the flow to the manifold. The applied overpressure should be between 0.5 and 0.7 bar, depending on the system and on the maximum wall shear stress.
5. Sterile plastic syringe (volume/size depending on syringe pump used).
6. 80% ethanol p.a. in double distilled and filtrated water "type 1" (according to ISO 3696) (ddH_2O).
7. RPMI 1640 cell culture media.
8. Phosphate buffered saline (PBS).

2.5 Image Analysis and Processing Software

1. Image analysis software such as the freely available ImageJ software (https://imagej.net) with appropriate plug-ins (e.g., Fiji-bundles of plug-ins; https://fiji.sc) or NIS software (Nikon, Tokyo, Japan): For counting the attached cells and recording a digital time lapse.
2. Spreadsheet program such as Excel (Microsoft Corporation, Redmont, Washington, USA): For calculating the applied shear force at each time point.
3. Scientific data analysis program, e.g., Origin (OriginLab Corporation, Northampton, Massachusetts, USA): For plotting and statistical analysis.

3 Methods

3.1 Preparation of Sample Surface

1. As protocols for surface coatings with proteins or peptides vary depending on the protein or peptide used, refer to standard protocols established for surface coatings by the distributor. An example for a protein coating on glass is given in Subheading 4 (*see* **Note 9**).
2. As reference substrates, use uncoated glass slides.

3.2 Assembly of the Flow Channel

1. Clean the upper channel compartment (glass lid) with ethanol p.a., and remove remaining dust particles with a tissue.
2. Apply the PDMS channel to the upper compartment by wetting the channel thoroughly with ethanol, and place it on the glass using a pair of tweezers. The channel should be swimming on the glass. Fix the channel with blunt cannulas in the channel holes present in the lid in order to ensure a straight channel assembly as illustrated in Fig. 2a. Take care that the PDMS channel is applied in a flat manner. Let the chamber dry for 20 min, protected from dust. Then remove the blunt cannulas, and visually inspect the PDMS seal (Fig. 2b).
3. Place the coated glass slides, which were dried in a N_2 stream, on the lower disk with the coated surface facing upward. Attach the channel with the chamber lid by pressing the PDMS channel onto the coated surface (*see* **Note 10**). Attach the two compartments by carefully tightening the screws (*see* **Note 11**).
4. Check for the tightness of the channel by applying low pressure using ddH_2O and a syringe fixed to the tube system (*see* **Note 12**). The outlet of the tubing should be closed for this check (*see* **Note 13**). If a channel is leaky, reassemble it starting at **step 2**.

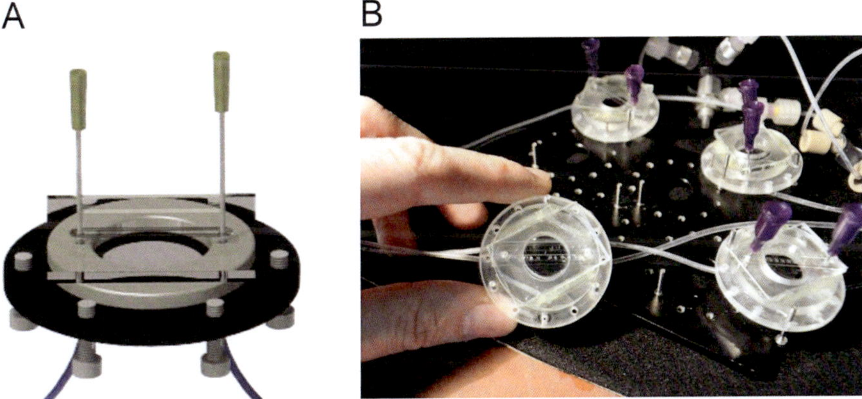

Fig. 2 (**a**) Schematic illustration and (**b**) photograph of the upper channel compartment with an attached PDMS channel. After the PDMS channel is mounted onto the glass surface by wetting with ethanol, the channel is fixed with blunt-end cannulas in the lid, and the assembly is allowed to dry completely protected from dust

3.3 Microscope Preparation

1. Adjust the temperature of the incubation chamber to 37 °C, and set the CO_2 level to 5%.
2. Place the channels inside the incubation chamber of the microscope on the sample holder, and control all channels optically (Fig. 3a). Make sure that the tubing length is sufficient, and avoid stretched tubings, as this would influence the forces in the channels later on (*see* **Note 14**). Loosely fix the tubings onto the microscope stage with tape to avoid stretching of the tubings when moving the stage.
3. Connect the manifold to the pressurized liquid reservoir as indicated in Fig. 3b.
4. Decontaminate and wash tubing: Connect an ethanol/water mixture (8:2) to the tubing system instead of the media reservoir, and inject the liquid into all channels by successively selecting the channels as well as the bypass. The ethanol solution should be kept in the system for 30 min in order to sterilize the channels and the tubing (*see* **Note 15**). Next, rinse the channels thoroughly 3× by injection of PBS and subsequently serum-free media.

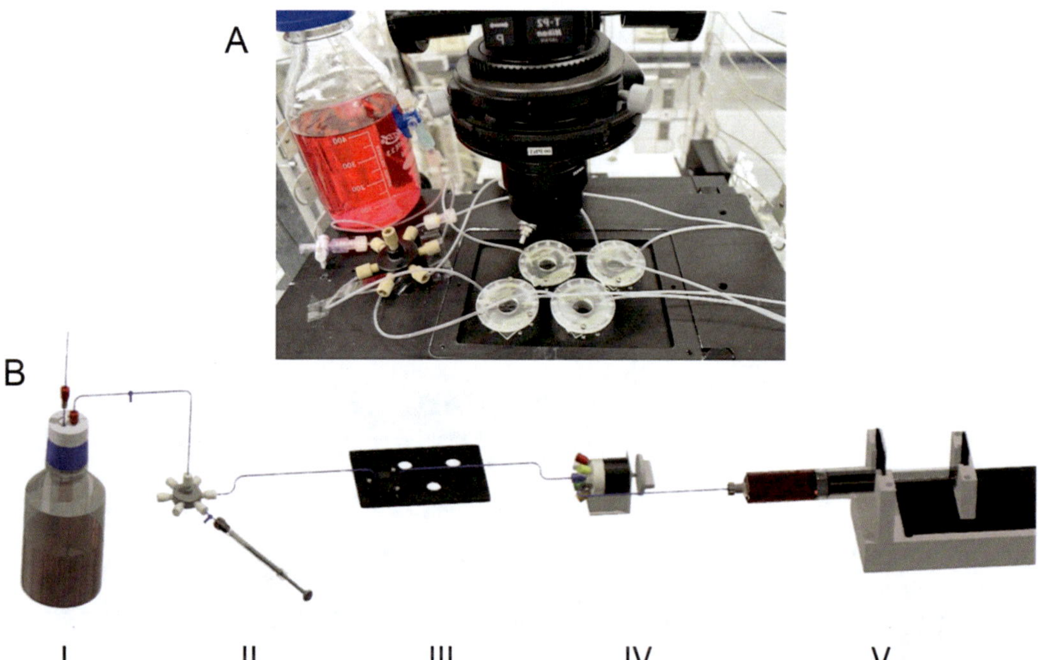

Fig. 3 (**a**) Microfluidic channels mounted onto the stage of an inverted microscope inside a custom-built incubation chamber. (**b**) Scheme depicting the assembled channel setup (simplified for one channel setup). The liquid, in the pressurized reservoir (I), is directed toward the syringe pump (V), via a manifold (II), the microfluidic channel (III), and a selection valve (IV). For cell injection via the manifold, the *T* valve to the reservoir is closed

5. Connect the syringe pump: Fill the syringe manually with liquid, then place the syringe in the syringe pump holder, and empty the syringe once by using the syringe pump. Slowly fill the syringe again by using the pump at low flow rate while selecting the bypass. The microfluidic setup after connection of all elements is shown in Fig. 3b.

3.4 Cell Injection and Cell Adhesion

1. Harvest cells from a culture flask and count the cells. Pellet the cells, and adjust the cell density to 2×10^6 cells/ml in adhesion media. Place 1 ml of cell suspension per channel in a syringe and remove all air bubbles from the syringe.
2. Close the selector and rinse the injection valve with medium to remove air bubbles. The inlet should be filled with media completely. Connect the syringe containing the cell suspension to the inlet valve (Fig. 3b II) without introducing air bubbles to the system.
3. Inject the cell suspension to the channels by subsequently choosing the channels and carefully inject approximately 1 ml of cell suspension per channel. Control the channels via the microscope to ensure an equal spread of cells in the channels (Fig. 4a). Leave the selector on bypass while incubating the cells on the surfaces, to avoid that pressure builds up. Incubate the system for 1 h to allow for cell attachment (*see* **Note 16**).

3.5 Cell Detachment Assay

1. Cell detachment: After incubation, select a channel, and simultaneously start the syringe pump software as well as the digital time-lapse recording software with one frame per second (*see* **Note 17**). As the syringe pump software is started, a defined flow of media is initiated. This flow can be converted into a shear stress that acts on the surfaces of the channel. Nonadherent cells are usually removed at low flow rates. As the shear stress is increased in small steps by increasing the volumetric flow rate, adherent cells start to be removed from the surface. Once all cells are removed, both the recording and the syringe pump software can be terminated (*see* **Note 18**). To repeat the experiment with another channel, move the microscope stage so that the channel of choice is in the field of view, and repeat the detachment step (*see* **Note 19**).
2. Determination of the channel dimensions: Remove the media reservoir, and dry the channels by pumping air through the channels. Measure the channels' width by using the microscopy software, and determine the height of the channel by using a calibrated focus stage in an optical microscope (*see* **Note 20**).

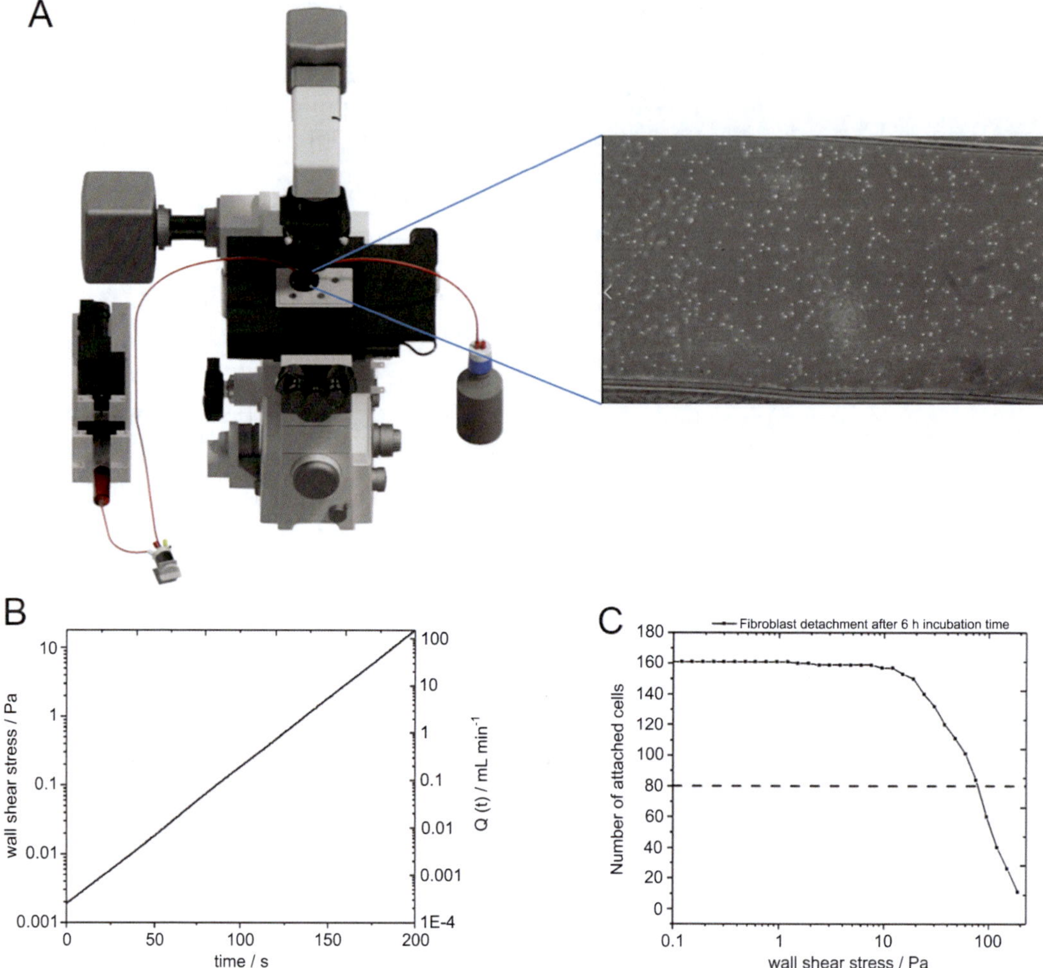

Fig. 4 (**a**) Assembled microfluidic setup after cell injection. The channel was mounted on an inverted microscope, cells were injected manually, and the channel was kept at 37 °C and 5% CO_2 during incubation. The PDMS channel with equal cell distribution before a measurement is shown on the right recorded with a 4× phase-contrast objective. (**b**) Flow profile during the microfluidic assay. The volumetric flow rate Q and thus the wall shear stress τ are increased by 2.33% every 0.5 s to span a range over five orders of magnitude. (**c**) Determination of the critical shear stress τ_{50} for a detachment assay of fibroblast cells after 6 h incubation time (graph modified from [28]). The applied wall shear stress is plotted against the number of adherent cells. The critical shear stress τ_{50} refers to the shear stress at which 50% of the initially adherent cell population is removed

3.6 Data Evaluation

1. Select the region of interest: Use an image analysis tool such as NIS to select the channel as a region of interest. If the channel is tilted relative to the field of view, rotate the rectangular region of interest, until it fits the channel area.
2. Count the attached cells: Count the cells manually or using image analysis tools.

3. Using a spreadsheet program, plot the number of cells against the applied shear stress τ. The latter results from the volumetric flow Q caused by the syringe pump, the mean height h, and the width w of the channel for each time point according to Poiseuille's model [34] with Eq. (1):

$$\tau = (6Q\mu)/(h^2 w) \tag{1}$$

where μ is the viscosity of the culture media (pure RPMI 1640 media exhibit a Newtonian behavior, and its viscosity can be approximated by the viscosity of water with ~0.67×10^{-3} kg m^{-1} s^{-1} [35]) and h and w are the measured mean height and width of the respective microfluidic channel. Using a logarithmic scale for the x-axis (Fig. 4c), the critical shear stress τ_{50}, where 50% of all cells were removed, can be derived (*see* **Note 21**).

4 Notes

1. To manufacture the PDMS channel and seal ring, the PDMS mixture needs to be thoroughly degassed according to the instructions of the manufacturer.
2. To remove the channel from the brass mold, bend the channel back using the long side of a pair of tweezers to avoid tearing or piercing the new channel.
3. To drill 1 mm holes for the tubing in the glass lid, use a diamond tip drill, and drill from both sides to avoid splintering of the glass.
4. For this assay, any kind of surface coating, which can be immobilized on glass, can be applied. However, coatings have to be rather flat to allow tight sealing of the PDMS channels.
5. Take care that the size of the substrate both matches the channel dimensions and is in accordance with the screws used. For a tight sealing of the channels, the substrate height has to be approximately 1 mm (standard microscopy glass slide) in the setup described here. Alternatively, use glass cover slips, and place on top of microscope slides to adjust the height accordingly.
6. For this experiment, any microscope incubation chamber with temperature and CO_2 control can be used.
7. Any controllable syringe pump can be implemented in the system. Always try to use the same model of the syringes by the same manufacturer to avoid changes in syringe diameter and to keep the initial stroke comparable when initiating the liquid flow.

8. The step size and/or the rate of increase of the volumetric flow rate can be adjusted for each cell type. A quick increase in shear stress (every 0.5 s by 2.33%), as applied in the described experiment, reduces the possibility of cytoskeletal rearrangement and/or strengthening of focal adhesions by the cells, which could affect cell spreading and cell adhesion forces.
9. As an example for a routine protein coating, a surface coating of glass with human fibronectin is described here. Before coating, clean glass slides via sonication in ethanol absolute for 10 min, then rinse extensively with ddH_2O, and dry in a N_2 stream. Resubstitute fibronectin from human plasma (Merck KGaA, Darmstadt, Germany) according to the manufacturer's recommendation, and dilute it to a final concentration of 20 μg/ml in PBS (Merck KGaA, Darmstadt, Germany). Spread the protein solution on the clean glass slides until the surface is fully covered, and keep in a humidified chamber at 4 °C overnight. Wash coated surfaces extensively with PBS, and store them in appropriate buffer or ddH_2O at 4 °C.
10. When assembling the channel, do not readjust the lid position once it is placed as this would damage the seal.
11. When tightening the upper compartment of the flow chamber to the lower one, tighten the screws repeatedly and subsequently with equal strength while increasing the force in very little steps to ensure a complete seal.
12. For checking the tightness of the seal, ddH_2O is manually inserted into the system at low pressure. When handling sensitive protein coatings, use an aqueous buffer such as PBS instead of water.
13. Check for noises of escaping air and bubble formation next to the channel; this could indicate a leaky channel.
14. Move the microscope stage to the position the furthest away from the syringe pump.
15. Depending on the sensitivity of the studied protein coating, sterilization with 80% ethanol for 30 min might not be feasible. Check in previous experiments if the protein coating used is affected by this sterilization method. Alternatively, other sterilization methods, such as UV irradiation or sterile assembly of the plate flow channel in a biological safety cabinet class II, have to be considered.
16. Other types of cells might require different incubation times, which should be determined in prior experiments. Turn off the light of the microscope during long-term incubations.
17. When selecting the channel, the lever at the selector should be turned rather fast from bypass configuration to the channel of choice. This prevents that pressure builds up in the system,

which could yield an initial pulse, which in turn could weaken cell adhesion or remove weakly adherent cells.

18. Before starting the cell detachment in the next flow channel, the remaining pump volume should be checked in order to avoid bubble formation when the syringe is filled completely. This would affect the shear forces present in the channel.
19. Make sure to select the bypass while moving the stage of the microscope in order to avoid that pressure builds up within the system, and move the stage as slowly as possible.
20. For each measurement of the channel height, the focus points on the lower and upper perimeter should be close to each other in *x*- and *y*-direction, so the height measurement is as precise as possible.
21. In order to compare several measurements with each other, it is more appropriate to specify the fraction of adherent cells than to compare absolute cell numbers. If the resulting curve is noisy due to low numbers of adherent cells, a sigmoidal fit of the curve can be applied to determine the estimated τ_{50} value.

References

1. Hines M, Nielsen L, Cooper-White J (2008) The hematopoietic stem cell niche: what are we trying to replicate? J Chem Technol Biotechnol 83(4):421–443
2. Ellis SJ, Tanentzapf G (2010) Integrin-mediated adhesion and stem-cell-niche interactions. Cell Tissue Res 339(1):121–130
3. Lee-Thedieck C, Spatz JP (2014) Biophysical regulation of hematopoietic stem cells. Biomater Sci 2(11):1548–1561
4. Legate KR, Wickstrom SA, Fässler R (2009) Genetic and cell biological analysis of integrin outside-in signaling. Genes Dev 23 (4):397–418
5. Franke K, Pompe T, Bornhäuser M, Werner C (2007) Engineered matrix coatings to modulate the adhesion of CD133+ human hematopoietic progenitor cells. Biomaterials 28 (5):836–843
6. Kurth I, Franke K, Pompe T, Bornhäuser M, Werner C (2011) Extracellular matrix functionalized microcavities to control hematopoietic stem and progenitor cell fate. Macromol Biosci 11(6):739–747
7. Sagar BM, Rentala S, Gopal PN, Sharma S, Mukhopadhyay A (2006) Fibronectin and laminin enhance engraftibility of cultured hematopoietic stem cells. Biochem Biophys Res Commun 350(4):1000–1005
8. Yokota T, Oritani K, Mitsui H, Aoyama K, Ishikawa J, Sugahara H, Matsumura I, Tsai S, Tomiyama Y, Kanakura Y, Matsuzawa Y (1998) Growth-supporting activities of fibronectin on hematopoietic stem/progenitor cells in vitro and in vivo: structural requirement for fibronectin activities of CS1 and cell-binding domains. Blood 91(9):3263–3272
9. Greenbaum AM, Link DC (2011) Mechanisms of G-CSF-mediated hematopoietic stem and progenitor mobilization. Leukemia 25 (2):211–217
10. Lapidot T, Petit I (2002) Current understanding of stem cell mobilization: the roles of chemokines, proteolytic enzymes, adhesion molecules, cytokines, and stromal cells. Exp Hematol 30(9):973–981
11. Petit I, Szyper-Kravitz M, Nagler A, Lahav M, Peled A, Habler L, Ponomaryov T, Taichman RS, Arenzana-Seisdedos F, Fujii N, Sandbank J, Zipori D, Lapidot T (2002) G-CSF induces stem cell mobilization by decreasing bone marrow SDF-1 and up-regulating CXCR4. Nat Immunol 3 (7):687–694
12. Klein G, Schmal O, Aicher WK (2015) Matrix metalloproteinases in stem cell mobilization. Matrix Biol 44-46C:175–183
13. Springer TA (1994) Traffic signals for lymphocyte recirculation and leukocyte emigration: the multistep paradigm. Cell 76(2):301–314
14. Sahin AO, Buitenhuis M (2012) Molecular mechanisms underlying adhesion and

migration of hematopoietic stem cells. Cell Adhes Migr 6(1):39–48
15. Zannettino A, C Berndt M, Butcher E, Butcher C, Vadas M, Simmons P (1995) Primitive human hematopoietic progenitors adhere to P-selectin (CD62P). Blood 85 (12):3466–3477
16. Mazo IB, Gutierrez-Ramos JC, Frenette PS, Hynes RO, Wagner DD, von Andrian UH (1998) Hematopoietic progenitor cell rolling in bone marrow microvessels: parallel contributions by endothelial selectins and vascular cell adhesion molecule 1. J Exp Med 188 (3):465–474
17. Khalili AA, Ahmad MR (2015) A review of cell adhesion studies for biomedical and biological applications. Int J Mol Sci 16 (8):18149–18184
18. Christ KV, Turner KT (2010) Methods to measure the strength of cell adhesion to substrates. J Adhes Sci Technol 24(13–14):2027–2058
19. Athanasiou KA, Thoma BS, Lanctot DR, Shin D, Agrawal CM, LeBaron RG (1999) Development of the cytodetachment technique to quantify mechanical adhesiveness of the single cell. Biomaterials 20 (23–24):2405–2415
20. Hochmuth RM (2000) Micropipette aspiration of living cells. J Biomech 33(1):15–22
21. Garcia AJ, Gallant ND (2003) Stick and grip: measurement systems and quantitative analyses of integrin-mediated cell adhesion strength. Cell Biochem Biophys 39(1):61–73
22. Chen Y, Lu B, Yang Q, Fearns C, Yates JR 3rd, Lee JD (2009) Combined integrin phosphoproteomic analyses and small interfering RNA--based functional screening identify key regulators for cancer cell adhesion and migration. Cancer Res 69(8):3713–3720
23. Elineni KK, Gallant ND (2011) Regulation of cell adhesion strength by peripheral focal adhesion distribution. Biophys J 101 (12):2903–2911
24. Giacomello E, Neumayer J, Colombatti A, Perris R (1999) Centrifugal assay for fluorescence-based cell adhesion adapted to the analysis of ex vivo cells and capable of determining relative binding strengths. BioTechniques 26 (4):758–762. 764–756
25. Channavajjala LS, Eidsath A, Saxinger WC (1997) A simple method for measurement of cell-substrate attachment forces: application to HIV-1 Tat. J Cell Sci 110(Pt 2):249–256
26. Horbett T, Waldburger J, Ratner B, Hoffman A (1988) Cell adhesion to a series of hydrophili–hydrophobic copolymers studies with a spinning disc apparatus. J Biomed Mat Res Part A 22(5):383–404
27. Lotz MM, Burdsal CA, Erickson HP, McClay DR (1989) Cell adhesion to fibronectin and tenascin: quantitative measurements of initial binding and subsequent strengthening response. J Cell Biol 109(4 Pt 1):1795–1805
28. Christophis C, Grunze M, Rosenhahn A (2010) Quantification of the adhesion strength of fibroblast cells on ethylene glycol terminated self-assembled monolayers by a microfluidic shear force assay. Phys Chem Chem Phys 12 (17):4498–4504
29. Goldstein AS, Dimilla PA (1997) Application of fluid mechanic and kinetic models to characterize mammalian cell detachment in a radial-flow chamber. Biotechnol Bioeng 55 (4):616–629
30. Christophis C, Sekeroglu K, Demirel G, Thome I, Grunze M, Demirel M, Rosenhahn A (2011) Fibroblast adhesion on unidirectional polymeric nanofilms. Biointerphases 6 (4):158–163
31. Lu H, Koo LY, Wang WM, Lauffenburger DA, Griffith LG, Jensen KF (2004) Microfluidic shear devices for quantitative analysis of cell adhesion. Anal Chem 76(18):5257–5264
32. Hanke M, Hoffmann I, Christophis C, Schubert M, Hoang VT, Zepeda-Moreno A, Baran N, Eckstein V, Wuchter P, Rosenhahn A, Ho AD (2014) Differences between healthy hematopoietic progenitors and leukemia cells with respect to CD44 mediated rolling versus adherence behavior on hyaluronic acid coated surfaces. Biomaterials 35(5):1411–1419
33. Christophis C, Taubert I, Meseck GR, Schubert M, Grunze M, Ho AD, Rosenhahn A (2011) Shear stress regulates adhesion and rolling of CD44+ leukemic and hematopoietic progenitor cells on hyaluronan. Biophys J 101 (3):585–593
34. Deen WM (1998) Analysis of transport phenomena. Oxford University Press, New York
35. Rinker KD, Prabhakar V, Truskey GA (2001) Effect of contact time and force on monocyte adhesion to vascular endothelium. Biophys J 80(4):1722–1732

Chapter 7

A Microcavity Array-Based 3D Model System of the Hematopoietic Stem Cell Niche

Eric Gottwald, Cordula Nies, Patrick Wuchter, Rainer Saffrich, Roman Truckenmüller, and Stefan Giselbrecht

Abstract

Despite huge advances in recent years, the interaction between hematopoietic stem and progenitor cells (HSPCs) and their niches in the bone marrow is still far from being fully understood. One reason is that hematopoiesis is a multi-step maturation process leading to HSPC heterogeneity. Subpopulations of HSPCs can be identified by clonogenic assays or in serial transplantation experiments in mice following sublethal irradiation, but it is very complex to reproduce or even maintain stem cell plasticity in vitro. Advanced model systems have been developed that allow to precisely control and analyze key components of the physiologic microenvironment for not only fundamental research purposes but, as a long-term goal, also for clinical applications. In this chapter, we describe our approach of building an artificial hematopoietic stem cell niche in the form of polymer film-based microcavities with a diameter of 300 μm and a depth of up to 300 μm and arranged in a 634-cavity array. The polymer films are provided with 3 μm pores and thus allow perfusion of the culture medium. The microcavity arrays can be inserted into a microbioreactor where a closed circulation loop can be tightly controlled with regard to medium flow and gas supply. The microcavity arrays were used for a three-dimensional (3D) co-culture of MSCs and HSPCs in a defined ratio over a time period of up to 21 days. With this setup, it could be demonstrated that the HSPCs maintained their stem cell characteristics more efficiently as compared to conventional monolayer co-culture controls.

Key words Microthermoforming, Microcavity array, MSC-HSPC co-culture, Perfusion, Microbioreactor

1 Introduction

Hematopoietic stem cells in adults reside primarily in the bone marrow that provides the microenvironment for a lifelong hematopoiesis. This is a highly controlled process of which some key cues, which maintain the orchestrated generation of blood cells, are soluble factors, dimensionality, and mechanical stability. Although there are several models that are able to explain certain aspects of the niche, the behavior of hematopoietic stem/progenitor cells (HSPCs) in their in vivo niche is not fully understood. Upon

Gerd Klein and Patrick Wuchter (eds.), *Stem Cell Mobilization: Methods and Protocols*, Methods in Molecular Biology, vol. 2017, https://doi.org/10.1007/978-1-4939-9574-5_7,

studies by Till and McCulloch [1], Worton et al. [2], and Dexter et al. [3] that led to the introduction of the niche concept by Schofield [4], the most widely accepted models are the ones of the endosteal niche [5–8], the perivascular niche [9–12], and, recently proposed, the hemosphere niche model, which combines the perivascular and endosteal niche model [13]. However, disassembling and describing an in vivo situation and developing a model that is representing the original tissue are already challenging, but building an artificial system that is capable of mimicking the physiologic situation, at least to some extent, is far more complicated. Therefore, many attempts have been made to recapitulate the in vivo bone marrow architecture and functional structure which can be subdivided with regard to, e.g., dimensionality, cellular composition, and surrounding scaffold material [14]. Heterogeneity of the approaches can be further increased by implementing medium composition, flow characteristics as well as high-throughput or upscaling capabilities [15]. Since three-dimensionality is one of the most important features to recapitulate organ function, as is impressively shown by the number of publications during the last 10–15 years, reviewed, e.g., by Shen and Nilsson [16], Choi et al. [17], or Bello et al. [18], we have tried to develop a bone marrow niche model based on osteoblast precursors, the mesenchymal stromal cells (MSCs), and human hematopoietic stem and progenitor cells (HSPCs) from cord blood [19]. The cell culture vessel is based on microsized cavities thermoformed from thin polymer films. The so-called DYNARRAYS© are arrays of such microcavities typically manufactured from thermoplastic materials such as polystyrene or polycarbonate. By complementing the central microthermoforming process with corresponding preceding or consecutive processes, the film materials can be locally surface-modified or provided with cylindrical pores in the range of several tens of nanometers to several micrometers. Since active medium flow is supposed to have an important influence in 3D cell culture models, we have set up a system in which the porous microcavity arrays are inserted into a perfused microbioreactor. This microbioreactor can be operated in different flow modes and allows for controlled gas supply of the medium reservoir making it an ideal platform for the establishment of in vivo-like flow conditions. By this means, a 3D cell culture device has been developed that is not only suited for (hematopoietic) stem cell applications but rather for nearly every culture where organotypic readouts are required.

2 Materials

2.1 Microcavity Arrays

For the 3D co-culture of MSCs and HSPCs, we used DYNARRAYS© MCA-C300-300-PC microcavity arrays. The DYNARRAYS© consist of an array of 634 microcavities with a diameter of 300 μm and up to 300 μm in depth. The spacing between adjacent

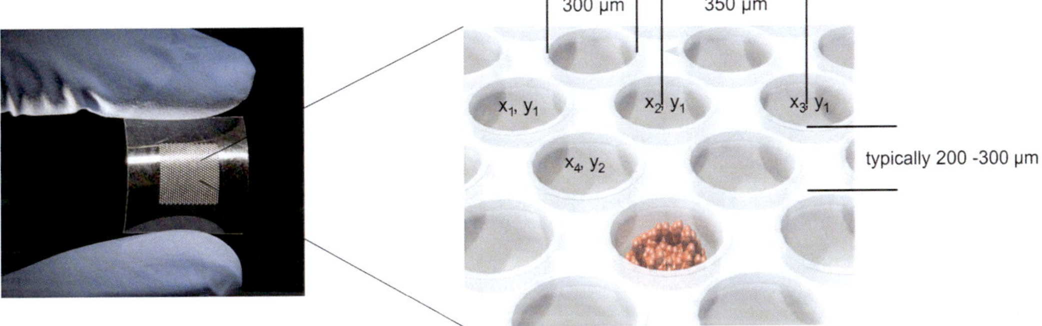

Fig. 1 Microcavity array features. Polymer film-based DYNARRAYS© microcavity array with dimensions and main size features. A 50 μm thin polycarbonate film is used to introduce the microcavities by the SMART technology (Surface Modification And Replication by Thermoforming). $X_1 Y_1$, etc. indicate positions of individual microcavities housing the three-dimensional network of cells

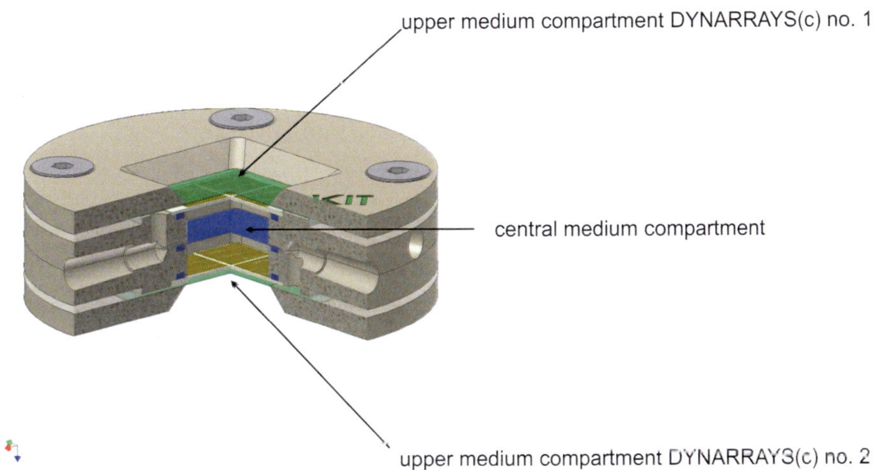

Fig. 2 Microcavity array bioreactor features. Schematic representation of the DYNARRAYS© bioreactor. The bioreactor is the size of a 5 cm petri dish and therefore fits into current microscope *x*/*y* tables. The schematic image is indicating the three medium compartments that are generated by the insertion of the DYNARRAYS© microcavity arrays, the central compartment located in between the two DYNARRAYS©, and the upper compartments located on top the microcavities, respectively

microcavities is about 50 μm. The overall size of the film chip-type culture substrate is 20 × 20 mm of which the array size takes 10 × 10 mm (Fig. 1).

2.2 Microcavity Array Bioreactor

The DYNARRAYS© have been mounted into the corresponding bioreactor that can house two DYNARRAYS© facing each other with their back sides. After the insertion of the microcavity arrays, three compartments are generated: two compartments located on the upper side of each DYNARRAYS© and one compartment located between the two DYNARRAYS© (Fig. 2). Each compartment is equipped with two inlet/and outlet ports so that a defined

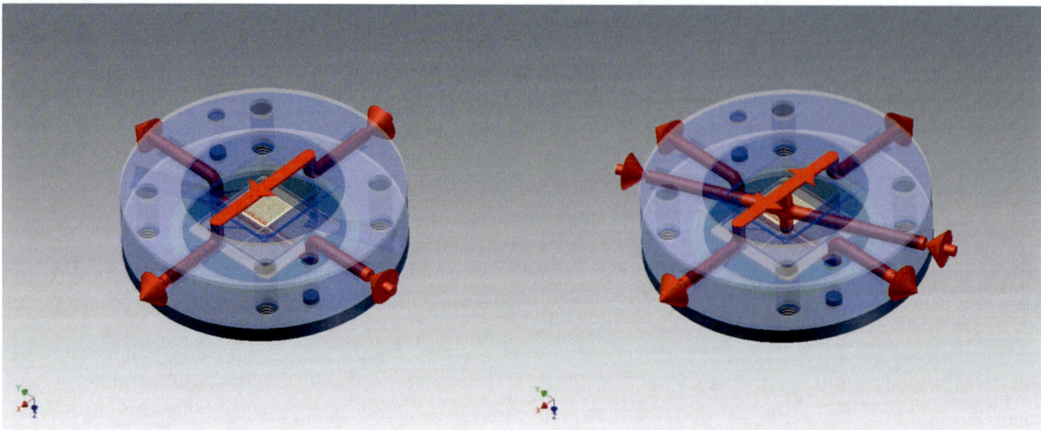

Fig. 3 Bioreactor flow modes. View into the opened bioreactor displaying the two flow modes: medium flow in superfusion (left) and perfusion mode (right) of the DYNARRAYS© bioreactor. Whereas the superfusion mode is intended to generate gradients, the perfusion mode is intended to avoid any type of gradients in the bioreactor. Reprinted with permission from [19]

medium flow can be established. The bioreactor can be operated in two flow modes: superfusion and perfusion which are displayed in Fig. 3. The main difference between the two flow modes is that during superfusion, the medium flows parallel to the microcavity surfaces, i.e., separately in the central compartment and the two upper compartments, whereas in perfusion mode, the medium is flowing perpendicularly from the central compartment through the porous DYNARRAYS© to the upper compartments. The compartments are connected to each other by tubings and connectors so that a bubble-free filling of the bioreactor is possible (Fig. 4). The filling procedure is displayed in Fig. 5. After the filling of the closed circulation loop, the loop is connected to a peristaltic pump. Since the medium reservoir is equipped with a cap that allows for the connection of a gas supply, a defined atmosphere can be established by connecting a gas mixing station and a wash bottle to the medium reservoir. To prevent airborne contaminations, sterile filters were connected to the gas inlet ports.

2.3 Isolation of HSPCs and MSCs

2.3.1 Primary Cell Material

- Human umbilical cord blood was obtained from voluntary donors after informed consent. Please note that an approval of the respective Ethics Committee is mandatory.
- Bone marrow aspirates were obtained from healthy voluntary donors after informed consent. Please note that an approval of the respective Ethics Committee is mandatory. 10–30 ml of bone marrow aspirate was collected in a syringe containing 10,000 IU heparin to prevent coagulation.

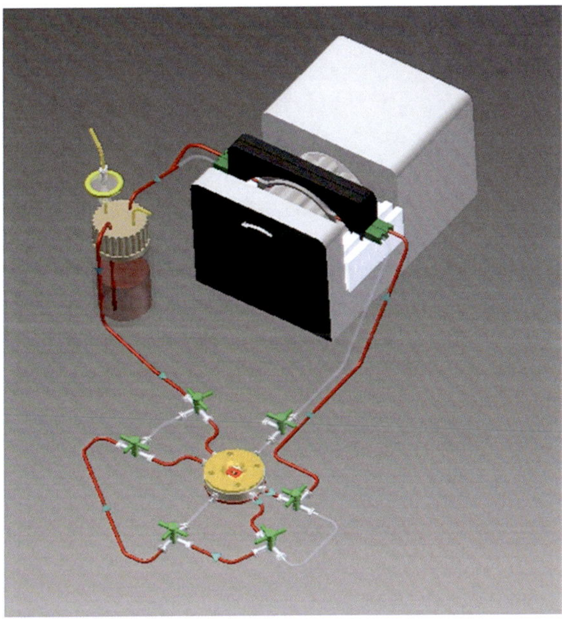

Fig. 4 Bioreactor's closed circulation loop. Minimal DYNARRAYS© bioreactor setup with medium reservoir and peristaltic pump. For long-term experiments, a wash flask (not shown) should be incorporated to prevent evaporation of liquid due to dry technical gases being introduced directly into the medium reservoir without appropriate humidification. Reprinted with permission from [19]

2.3.2 Isolation and Separation of HSPCs and MSCs

- Ficoll-Paque separation medium with a density of d = 1.077 g/cm^3.
- Isolation buffer (MACS buffer); D-PBS with 1% BSA, 2 mM EDTA, 0.09% azide.
- Manual magnetic cell separation system consisting of μMACS Separation Unit (magnet and stand) autoMACS system.
- MACS MS separation columns.
- Human CD34 MicroBead Kit for 2×10^9 to 1×10^{10} total cells.
- CD34$^+$ HSPC culture medium: Long-term bone marrow culture (LTBMC) medium, consisting of Iscove's modified Dulbecco's media (IMDM) with 12.5% fetal calf serum, 12.5% horse serum, 2 mM L-glutamine, 2 mM Pen/Strep, and 10^{-6} M hydrocortisone, w/o cytokines.
- MSC culture medium: MSCGM, Mesenchymal Stem Cell Growth Medium, ready to use, including basal medium and the necessary supplements for proliferation of human bone marrow-derived mesenchymal stem cells.
- Trypsin/EDTA (0.05%/0.02%) for MSC cell detachment.
- Cell culture flasks.
- Conical centrifuge tubes 15 ml and 50 ml.

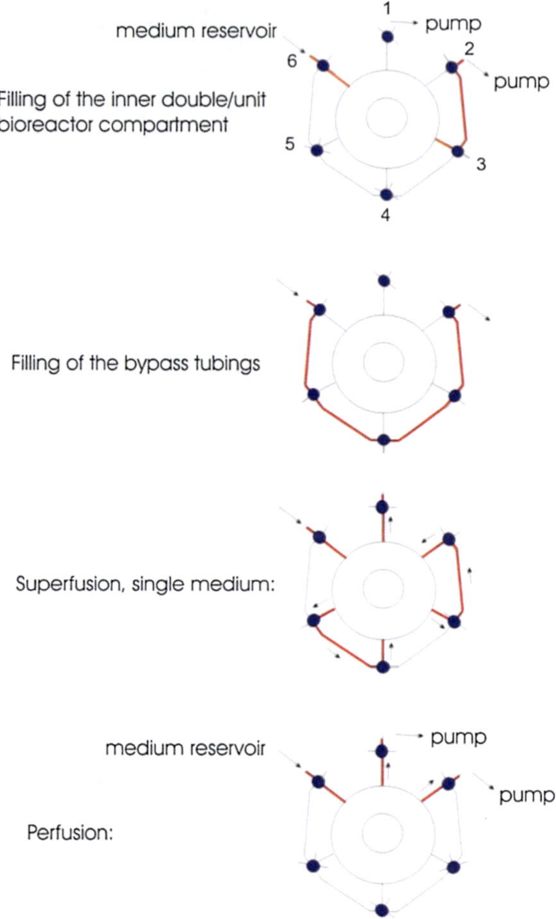

Fig. 5 Bioreactor operating modes. Scheme of the filling and operating adjustments of the three-way connectors for the run of the DYNARRAYS© bioreactor system. ●, three-way connector. Superfusion and perfusion, operating modes. Red, active medium flow

- Anti-CD34 antibody, labeled with fluorophore of choice.
- Flow cytometer for determination of CD34$^+$ cell purity.

2.3.3 Various

- Phosphate Buffered Saline (PBS).
- Neubauer cell counting chamber.
- Microscope for cell counting, observation, and analysis.
- ImageJ, image processing, and analysis software.

2.4 DYNARRAYS© Preparation

- Alcohol series with isopropanol.
- Collagen I, fibronectin, or other coating material of choice.

3 Methods

3.1 Isolation of HSPCs

1. Mononuclear cells (MNC) were isolated from umbilical cord blood by density gradient centrifugation for 20 min at $1000 \times g$ with the Ficoll-Hypaque technique.
2. $CD34^+$ cells among the MNCs were purified by positive selection with a monoclonal anti-CD34 antibody coupled to magnetic microbeads on an affinity column using a μMACS Separation Unit or alternatively an autoMACS system. Reanalysis of the isolated cells by flow cytometry should be performed to confirm a purity of >90% $CD34^+$ cells.
3. The cells were used immediately after isolation or were kept frozen at -150 °C under standard freezing conditions until use.

3.2 Isolation of MSCs

1. Bone marrow aspirates (10–30 ml) were collected in a syringe containing 10,000 IU heparin to prevent coagulation.
2. The MNC fraction was isolated by density gradient centrifugation on Ficoll-Paque and seeded in tissue culture flasks at a density of 1×10^6 cells/cm^2 in MSCGM™ medium. First colonies of plastic adherent MSCs appeared usually after 2–5 days.
3. The MSCs were expanded in MSCGM™ Mesenchymal Stem Cell Growth Medium following the manufacturer's instructions. For this step, 5000 cells/cm^2 were plated in tissue flasks without any pre-coating.
4. The culture medium was changed twice per week.
5. After reaching 80% confluence, the MSCs were trypsinized with 0.25% trypsin/1 mM EDTA for 5–7 min. Upon adding MSCGM culture medium to stopping the trypsinization reaction, the cells were counted with a Neubauer counting chamber and re-seeded at 10^4 cells/cm^2 for further expansion.
6. Sub-confluent MSC feeder layers (70–80%) of cells from passage 3 to 6 are mostly used for functional studies.
7. The capacity of the MSCs to differentiate into the osteogenic, adipogenic, and chondrogenic lineages should be confirmed as delineated in the ISCT position paper [20]. Various differentiation protocols and also commercial test kits are available for use. An exemplary description of differentiation into all three lineages can be found in [21].
8. The morphology of the MSCs in culture should be examined daily. Cell morphology, as well as intercellular connections and junctional complexes of MSCs, has been described before in detail [22].

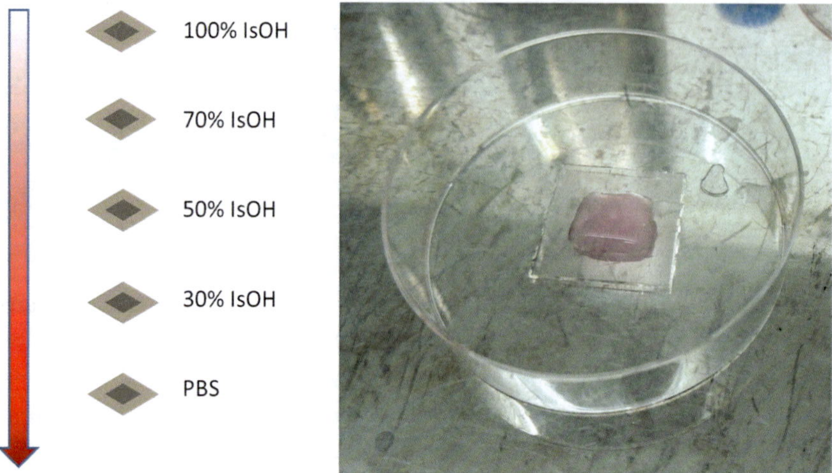

Fig. 6 Culture system preparation. Isopropanol series protocol for the hydrophilization of the DYNARRAYS© microcavity array. Upper right image: positioning of a 160 μl medium volume containing the mixture of single MSC and HSP cells leads to the formation of a droplet on top of the microcavities. Note that the dry rim of the microstructure is not wetted by the drop

9. For HSPC/MSC co-culture experiments, a 1:1 mixture of LTBMC/MSCGM medium was used. No cytokines were added to the culture medium.

3.3 DYNARRAYS© Preparation

1. For adherent cell cultures in microcavities, the array has to be hydrophilized. For this, the DYNARRAYS© are subjected to an alcohol series beginning with 100% isopropanol, moving over to 70%, 50%, 30%, and finally PBS for 10 s each (Fig. 6) (*see* **Note 1**).
2. For optional collagen coating, place a drop of 160 μl collagen I solution (200 μg/ml) on top of the microcavities, and incubate at 4 °C overnight.
3. After this, remove the collagen solution, and free the rim of the array with a pipette or vacuum device from collagen/PBS/water. Finally, a drop of 160 μl single cell suspension can be placed on top of the microcavities.

3.4 MSC/HSPC Co-Culture

1. For modelling the bone marrow niche, MSC were mixed in suspension with HSPC in 1:1 mixture of LTBMC and MSCGM medium at a ratio of 3:2 (3×10^5 MSCs and 2×10^5 HSPCs) and inoculated into the DYNARRAYS© by manually applying a 160 μl drop of single cell suspension on top of the microcavities (*see* **Note 1**).
2. As 300,000 MSCs and 200,000 HSPCs were inoculated per array, on average 480 MSCs and 320 HSPCs were inoculated per microcavity, respectively (*see* **Note 2**) (Fig. 7) which leads to a complete filling of the microcavities.

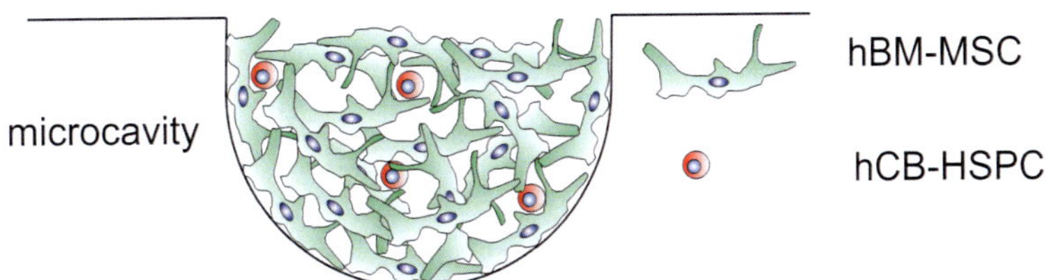

Fig. 7 3D co-culture. Schematic cross section of the MSC/HSPC co-culture inside one single microcavity. After inoculation, the cells form a complex three-dimensional network that can be observed via microscope over extended period of times (4D culture). *hBM-MSC* human bone marrow mesenchymal stromal cells, *hCB-HSPC* human cord blood hematopoietic stem and progenitor cells

3. The microcavity array was placed in an incubator for 2 h and subsequently mounted into the microbioreactor, allowing active nutrient and gas supply (*see* **Note 3**).
4. To remove the cells from the microcavities for RNA and protein isolation, the DYNARRAYS© were removed from the bioreactor, mounted in a 3 cm petri dish, and washed with PBS. Afterwards, 2 ml trypsin/EDTA (0.05%/0.02%) was added. The DYNARRAYS© were incubated this time for 10 min at 37 °C before the reaction was stopped by adding 1 ml of culture medium. The removed cells were then subjected to the downstream analysis of choice.

3.5 Bioreactor Setup and Readout

1. The DYNARRAYS© bioreactor with a connected gas supply (5% CO_2, 21% O_2, 74% N_2) was placed in an incubator and run with a flow of 400 μl/min for up to 21 days under superfusion conditions (*see* **Note 4**).
2. Every day a medium sample was drawn from the medium reservoir, and every second day, half of the medium was exchanged (*see* **Note 4**).
3. For gene expression and Western blot analysis, microscopy and colony-forming unit assays, parallel experiments should be run and analyzed according to the respective protocols.

4 Notes

1. When working with DYNARRAYS©, one might encounter empty microcavities on the chip. This can be due to entrapped air bubbles introduced and/or not removed during the alcohol series for the hydrophilization of the array. In this case, deaeration should be performed properly by increasing the duration of the alcohol series steps from 10 s to 30 s each.

2. Cells floating off the array might occur when inappropriate cell counts due to initially too high cell numbers or too long cultivation periods with proliferating cells have been used or inappropriate surface modification or extracellular matrix (ECM) deposition has occurred. To circumvent this, optimal cell numbers should be determined for each cell type separately prior to the experiment.
3. The incubation time of 2 h should not be exceeded since the cell number/medium volume ratio is very high.
4. During long-term experiments, the medium level in the reservoir will decrease. The use of dry gas mixtures will lead to an evaporation of medium through the sterile filters. To avoid evaporation of medium from the reservoir, a wash flask should be incorporated into the gas inlet tract. Small bubbles on top of the array might accumulate above the microcavities as of cellular respiration. Also, in this case, switch the three-way connectors of the bioreactor circulation to superfusion to export the bubbles.

References

1. Till JE, Mc Culloch EA (1961) A direct measurement of the radiation sensitivity of normal mouse bone marrow cells. Radiat Res 14:213–222
2. Worton RG, McCulloch EA, Till JE (1969) Physical separation of hemopoietic stem cells differing in their capacity for self-renewal. J Exp Med 130(1):91–103
3. Dexter TM, Allen TD, Lajtha LG (1977) Conditions controlling the proliferation of haemopoietic stem cells in vitro. J Cell Physiol 91 (3):335–344
4. Schofield R (1978) The relationship between the spleen colony-forming cell and the haemopoietic stem cell. Blood Cells 4(1–2):7–25
5. Lord BI, Testa NG, Hendry JH (1975) The relative spatial distributions of CFUs and CFUc in the normal mouse femur. Blood 46 (1):65–72
6. Gong JK (1978) Endosteal marrow: a rich source of hematopoietic stem cells. Science 199(4336):1443–1445
7. Calvi LM, Adams GB, Weibrecht KW, Weber JM, Olson DP, Knight MC, Martin RP, Schipani E, Divieti P, bringhurst FR, Milner LA, Kronenberg HM, Scadden DT (2003) Osteoblastic cells regulate the haematopoietic stem cell niche. Nature 425(6960):841–846
8. Zhang J, Niu C, Ye L, Huang H, He X, Tong WG, Ross J, Haug J, Johnson T, Fend JQ, Harris S, Wiedemann LM, Mishina Y, Li L (2003) Identification of the haematopoietic stem cell niche and control of the niche size. Nature 425(6960):836–841
9. Kiel MJ, Yilmaz OH, Iwashita T, Yilmaz OH, Terhorst C, Morrison SJ (2005) SLAM family receptors distinguish hematopoietic stem and progenitor cells and reveal endothelial niches for stem cells. Cell 121(7):1109–1121
10. Morrison SJ, Scadden DT (2014) The bone marrow niche for haematopoietic stem cells. Nature 505(7483):327–334
11. Yu VW, Scadden DT (2016) Hematopoietic stem cell and its bone marrow niche. Curr Top Dev Biol 118:21–44
12. Wei Q, Frenette PS (2018) Niches for hematopoietic stem cells and their progeny. Immunity 48(4):632–648
13. Ugarte F, Forsberg EC (2013) Haematopoietic stem cell niches: new insights inspire new questions. EMBO J 32(19):2535–2547
14. Mokhtari S, Baptista PM, Vyas DA, Freeman CJ, Moran E, Brovold M, Llamazares GA, Lamar Z, Porada CD, Soker S, Almeida-Porada G (2018) Evaluating interaction of cord blood hematopoietic stem/progenitor cells with functionally integrated three-dimensional microenvironments. Stem Cells Transl Med 7 (3):271–282
15. Nies C, Gottwald E (2017) Artificial hematopoietic stem cell niches–dimensionality

matters. Adv Tissue Eng Regen Med Open Access 2(5):236–247

16. Shen H, Yu H, Liang PH, Cheng H, XuFeng R, Yuan Y, Zhang P, Smith CA, Chend T (2012) An acute negative bystander effect of gamma-irradiated recipients on transplanted hematopoietic stem cells. Blood 119 (15):3629–3736
17. Choi JS, Mahadik BP, Harley BA (2015) Engineering the hematopoietic stem cell niche: Frontiers in biomaterial science. Biotechnol J 10(10):1529–1545
18. Bello AB, Park H, Lee SH (2018) Current approaches in biomaterial-based hematopoietic stem cell niches. Acta Biomater 72:1–15
19. Wuchter P, Saffrich R, Giselbrecht S, Nies C, Lorig H, Kolb S, Ho AD, Gottwald E (2016) Microcavity arrays as an in vitro model system of the bone marrow niche for hematopoietic stem cells. Cell Tissue Res 364(3):573–584
20. Dominici M, Le Blanc K, Mueller I, Slaper-Cortenbach I, Marini F, Krause D, Deans R, Keating A, Prockop DJ, Horwitz E (2006) Minimal criteria for defining multipotent mesenchymal stromal cells. The International Society for Cellular Therapy position statement. Cytotherapy 8(4):315–317
21. Wuchter P, Vetter M, Saffrich R, Diehlmann A, Bieback K, Ho AD, Horn P (2016) Evaluation of GMP-compliant culture media for in vitro expansion of human bone marrow mesenchymal stromal cells. Exp Hematol 44 (6):508–518
22. Wuchter P, Boda-Heggemann J, Straub BK, Grund C, Kuhn C, Krause U, Seckinger A, Peitsch WK, Spring H, Ho AD, Franke WW (2007) Processus and recessus adhaerentes: giant adherens cell junction systems connect and attract human mesenchymal stem cells. Cell Tissue Res 328(3):499–514

Chapter 8

Migration Assay for Leukemic Cells in a 3D Matrix Toward a Chemoattractant

Sabrina Zippel, Annamarija Raic, and Cornelia Lee-Thedieck

Abstract

In leukemia, leukemic cells hijack the hematopoietic stem cell (HSC) microenvironment in the bone marrow—the so-called stem cell niche—by flooding the niche with clonal progeny of leukemic cells. They can exploit signaling pathways which are critical for HSC development to support their own survival, homing, and maintenance. These interactions of leukemic cells with the microenvironment have an impact on therapy progress and patient outcome. Therefore, signals for homing and anchorage of leukemic cells to the bone marrow have to be investigated by using tools that allow the migration of cells toward critical signals. Here, we describe an in vitro migration assay for leukemic cells toward a chemoattractant in a 3D environment exemplified by migration of the cell line OCI-AML3 to a CXC motif chemokine ligand 12 (CXCL12) gradient. For this purpose, a chemotaxis slide is filled with a hydrogel system mimicking the extracellular matrix in vivo. The cells are encapsulated into the hydrogel network during polymerization, and a CXCL12 gradient is introduced in the enclosed chambers to trigger migration. Cell migration in the 3D network of the hydrogel is monitored by time-lapse microscopy. We describe the experimental setup and the tools for cell tracking and data analysis.

Key words Leukemic cells, Migration, 3D matrix, μ-Slides, CXCR4/CXCL12 axis, Chemokine gradient

1 Introduction

Genetic events can occur in early hematopoietic stem cells (HSCs) but also in more mature hematopoietic cells leading to a limitless self-renewal potential, uncontrolled proliferation, and reduced capacity to differentiate into mature cells. This means that HSCs and their progeny can transform into leukemic cells [1, 2]. Leukemia develops in the bone marrow where it can participate in the complex cross talk among the bone marrow cells which can result in remodeling of the HSC's supportive environment—the so-called stem cell niche. Signal cascades critical for homing, maintenance,

Sabrina Zippel and Annamarija Raic contributed equally to this work.

Gerd Klein and Patrick Wuchter (eds.), *Stem Cell Mobilization: Methods and Protocols*, Methods in Molecular Biology, vol. 2017, https://doi.org/10.1007/978-1-4939-9574-5_8,

and survival of HSCs like the CXC chemokine receptor type 4/CXC motif chemokine ligand 12 (CXCR4/CXCL12) axis are exploited by leukemic cells and can contribute to disease development and chemoresistance [3, 4]. For improvement of the therapeutic outcome of patients, investigation of the attraction of leukemic cells to CXCL12 sources in human bone marrow is urgently needed. Current protocols to study the migration of cells toward a chemoattractant include mainly well plates with integrated filter systems which allow migration of cells through a filter into a lower chamber (Boyden chamber assays). Alternatively, chemotaxis chambers which allow the observation of cell migration in a channel toward a chemoattractant source are used [5–8]. However, these methods allocate a 2D environment and are made of inflexible cell culture plastic which barely mimic the in vivo conditions. In vivo protocols enable the analysis of cell migration in a living system by using labeled cells in graft-recipient animals, but thereby the drawback of insufficient transmission to humans has to be accepted [6, 9–11]. Therefore, in vitro systems are needed which allow the migration of human cells in a 3D environment resembling the in vivo conditions. Here, we describe a protocol for the migration of leukemic cells toward a chemokine gradient embedded in a 3D polymer hydrogel substituting the extracellular matrix in the bone marrow. For this purpose, we used a commercially available migration chamber described previously by Zengel et al. in 2002 and a fast polymerizing hydrogel system (3-D Life Dextran-CD Hydrogel Kit), in which the leukemic cells can be encapsulated [12]. We illustrate the experimental setup including polymerization of the hydrogel with the leukemic cells in the migration chamber and the implementation of cytokines into the neighboring reservoirs. Furthermore, we describe the microscope settings for visualization of cell migration and cell tracking as well as the data analysis via free software available online.

2 Materials

Bring all working reagents and solutions to room temperature before usage. Store them according to the manufacturer's information. Use sterile components, e.g., cell culture equipment.

2.1 Fabrication of Cell-Laden Hydrogels and Setup of the Migration Chamber

1. μ-Slides Chemotaxis for 3D experiments (ibidi GmbH, Munich, Germany) (Fig. 1a) (*see* **Note 1**).
2. A wet chamber for the μ-Slides Chemotaxis to avoid evaporation. Use a 10 cm petri dish (Greiner Bio-One International GmbH, Frickenhausen, Germany) for this purpose, and lay it out with sterile tissue submerged in ddH_2O.
3. Depending on the aim of your experiment, use an appropriate leukemic cell line or primary leukemic cells with high

A

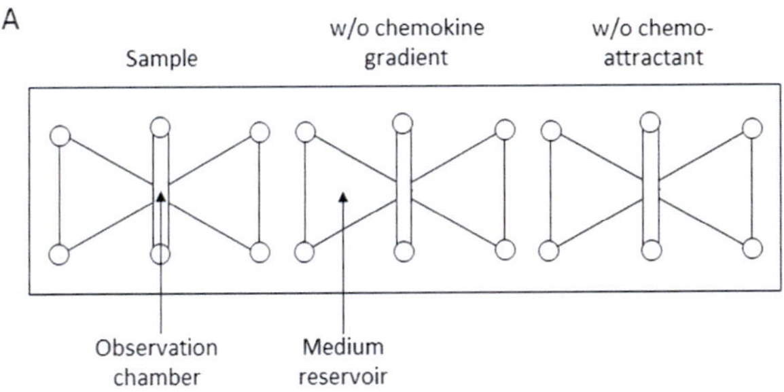

B

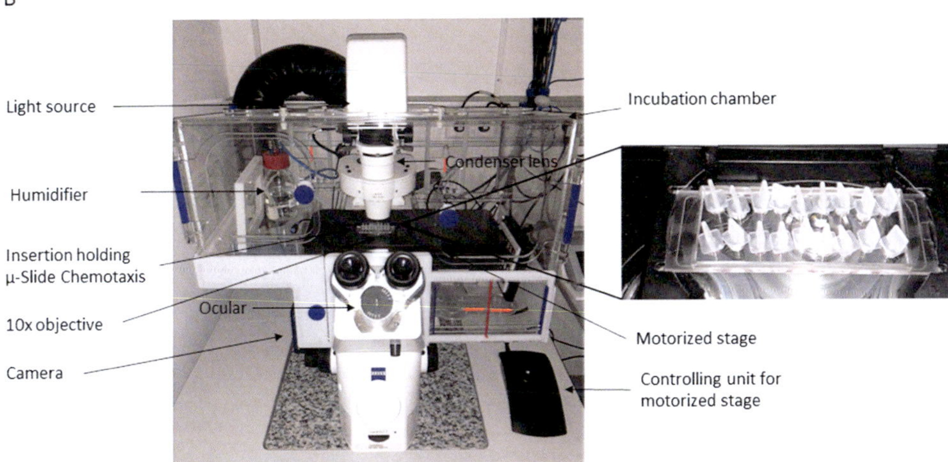

C

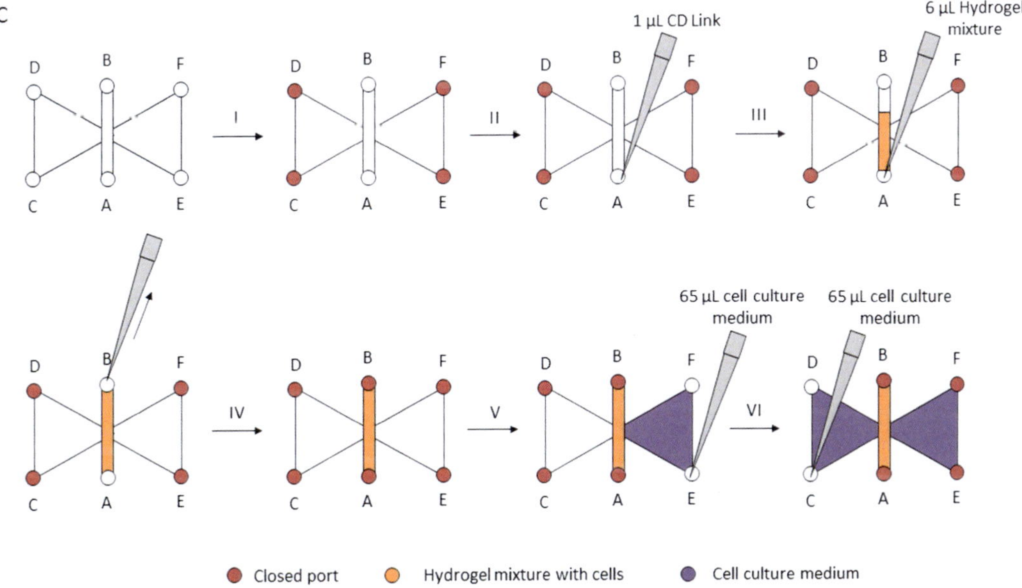

Fig. 1 (**a**) Schematic drawing of the setup of the µ-Slide Chemotaxis for 3D experiments. Three observation chambers containing the hydrogel-cell mix are adjacent to two medium reservoirs each. One medium reservoir of the sample is filled with chemoattractant leading to the formation of a chemokine gradient ("sample," left), the control without chemokine gradient ("w/o chemokine gradient," middle) contains chemokine in both medium reservoirs, and the control without chemoattractant ("w/o chemoattractant,"

chemokine receptor expression, e.g., the AML cell line OCI-AML3 with high CXCR4 expression, cultured in MEM-alpha with 10% fetal bovine serum (FBS).

4. Low-serum media: Cell culture media with 1% FBS.
5. Use a hydrogel, suitable for cell migration experiments, e.g., 3-D Life Dextran-CD Hydrogel Kit (Cellendes GmbH, Reutlingen, Germany) containing a linker (so-called CD-Link) with a peptide sequence that is cleavable by matrix metalloproteinases, a polymer (maleimide-dextran) that reacts with the linker and a suitable buffer (so-called 10× CB buffer pH 5.5). Store components at −80 °C (*see* **Note 2**).
6. Disposables: Reaction tubes 0.5 μL, pipette tips (10–200 μL), and micro pipette tips.
7. Equipment: Shaker, Bochem™ 18/10 stainless steel forceps.

2.2 Microscope Settings

1. Inverse microscope, e.g., Axio Observer Z1 (Carl Zeiss AG, Oberkochen, Germany) with incubation equipment for cell culture conditions allowing preheating to 37 °C under 5% CO_2 saturation, phase contrast mode, 10× objective, camera (AxioCam MRm, Carl Zeiss AG), motorized stage for high-precision positioning, and the appropriate software ZEN 2 (blue edition) (Carl Zeiss AG) (Fig. 1b).

2.3 Migration Experiment

1. Chemoattractant, e.g., prepare 90 μL of CXCL12 (2 μg/mL) diluted in serum-free media.

2.4 Software for Migration Analysis

1. Chemotaxis and Migration Tool 2.0 (ibidi GmbH, Munich, Germany) (*see* **Note 3**).
2. Excel (Microsoft, Dublin, Ireland).
3. ImageJ (Wayne Rasband, National Institutes of Health, Maryland, USA) (*see* **Note 4**).

Fig. 1 (continued) right) contains no chemokine. The scheme of the μ-Slide Chemotaxis is redrawn according to manufacturer's instructions (copyright ibidi GmbH, Germany). (**b**) Photograph of the experimental setup on an inverse microscope (Axio Observer Z1, Carl Zeiss AG, Oberkochen, Germany) with an incubation chamber allowing cell culture conditions (37 °C, 5% CO_2). A humidifier ensures ideal humidity to prevent evaporation by generation of warm water vapor. Further, a 10× objective, a motorized stage for high-precision positioning, and a camera are used to acquire the movies. On the right, an enlarged picture of the mounted μ-Slide Chemotaxis (all ports closed with plugs) is shown. (**c**) Schematic illustration of the process of filling the μ-Slides Chemotaxis chambers. Closed filling ports are indicated by red dots, observation chambers that are filled with hydrogel mixture and cells are indicated by orange color, and filled medium reservoirs are colored in purple. (I) The ports C–F are closed. (II) One μL CD-Link is added into filling port A. (III) Six μL of the hydrogel mixture are added to filling port A, and immediately afterward air is aspirated from filling port B by using the pipette. (IV) The ports A and B are closed. (V) The plugs from the ports E and F are removed, and 65 μL cell culture medium is filled into the medium reservoir via port E. The ports E and F are closed. (VI) The plugs from the ports C and D are removed, and 65 μL medium is added to the left observation chamber. Ports C and D are closed. The scheme of the μ-Slide Chemotaxis is adapted from ibidi GmbH, Germany

3 Methods

Place the cell culture media, the μ-Slides Chemotaxis, and plugs in the incubator (37 °C and 5% CO_2) 1 day before seeding the cells for gas equilibration to avoid air bubbles during the experiment. Work under a sterile bench and handle the plugs with forceps. Handling of the μ-Slides Chemotaxis is also described on the manufacturer's homepage online (https://ibidi.com/channel-slides/9%2D%2Dslide-chemotaxis-ibitreat.html).

3.1 Fabrication of Soft Hydrogels and Filling the Observation Chamber

1. Place the μ-Slides Chemotaxis in a wet chamber to avoid evaporation of the media.
2. Prepare approximately 50 μL cell suspension (3×10^6 cells/mL) in low-serum media.
3. Prepare the hydrogel by combining the components of the 3-D Life Dextran-CD Hydrogel Kit modified from manufacturer's instructions. For one μ-Slide Chemotaxis with three observation chambers, mix 16.6 μL ddH_2O with 2.4 μL of 10× CB buffer, pH 5.5, and 6 μL of maleimide-dextran solution in a 0.5 mL reaction tube. Mix the hydrogel mixture with a shaker.
4. Add 6 μL of the cell suspension into the hydrogel mixture, and mix gently by pipetting up and down.
5. Pipette the hydrogel mixture containing the cell suspension into the observation chambers of the μ-Slides Chemotaxis (Fig. 1c) (*see* **Note 5**). For this purpose, close the filling ports C, D, E, and F with plugs.
6. Place 1 μL of the CD-Link into filling port A to avoid formation of air bubbles (*see* **Note 6**).
7. Add 6 μL of the hydrogel solution containing the cell suspension gently on top of the CD-Link into filling port A.
8. Immediately afterward aspirate air from filling port B by using the same pipette settings of 6 μL. Press the pipette directly onto the filling port, and remove the air from the chamber. Thereby the chamber is entirely filled with the hydrogel solution.
9. Close the filling ports A and B with the plugs and check the filled chamber for air bubbles under the microscope (*see* **Note 7**).
10. Repeat **steps 5–9** for two more observation chambers (control without chemokine gradient and control without chemoattractant) on the μ-Slides Chemotaxis.
11. Fill the six reservoirs of the μ-Slides Chemotaxis with 65 μL media each (Fig. 1c). For this purpose, open the filling ports E and F, and directly place the pipette tip onto port E to avoid air

bubble formation. Close the filling ports E and F with plugs (*see* **Note 8**). Repeat the process for the filling ports C and D.

12. Incubate the μ-Slides Chemotaxis containing the hydrogel mixture with the cells at 37 °C under 5% CO_2 saturation until the microscope settings are done (*see* **Note 9**).

3.2 Setup of the Microscope

Before adding the chemoattractant and placing the μ-Slides Chemotaxis, set up the microscope. For this purpose, start the software, and prepare the method protocol in the microscope software to take a picture every 5 min of each observation chamber containing the cells for 20 h (*see* **Note 10**).

1. Place the μ-Slide Chemotaxis on the microscope stage, and immobilize it to make sure that it cannot move or bring the cells out of focus during the experiment.
2. Use phase contrast and the 10× objective.
3. Move the stage to the first observation chamber (sample) of the μ-Slides Chemotaxis, and save this position (*see* **Note 11**) (Fig. 1a).
4. Repeat **step 3** for the other two observation chambers (control without chemokine gradient and control without chemoattractant).
5. Control the *z*-positions and bring the cells into focus.
6. Carefully remove the μ-Slide Chemotaxis from the stage.

3.3 Migration Experiment

Fill the medium reservoirs of the μ-Slides Chemotaxis with chemoattractant to start the migration experiment. For this purpose, prepare the chemokine gradient with chemoattractant in one medium reservoir of the sample, the control without chemokine gradient by adding chemoattractant in both medium reservoirs, and the control without any chemoattractant in the medium reservoirs (Fig. 2a).

1. Remove the plugs of the filling ports E and F (Fig. 2b).
2. Apply 15 μL of a 0.25 μM CXCL12 solution into port E of the sample chamber.
3. Aspirate 15 μL medium from filling port F. For this purpose, press the pipette directly at the filling port. Close the filling ports with the plugs.
4. Prepare the control without chemokine gradient by repeating the **steps 1–3** for both media reservoirs.
5. Immediately, place the μ-Slides Chemotaxis carefully at the stage of the microscope.
6. Start the protocol previously prepared (*see* Subheading 3.2) to take the pictures with the microscope.

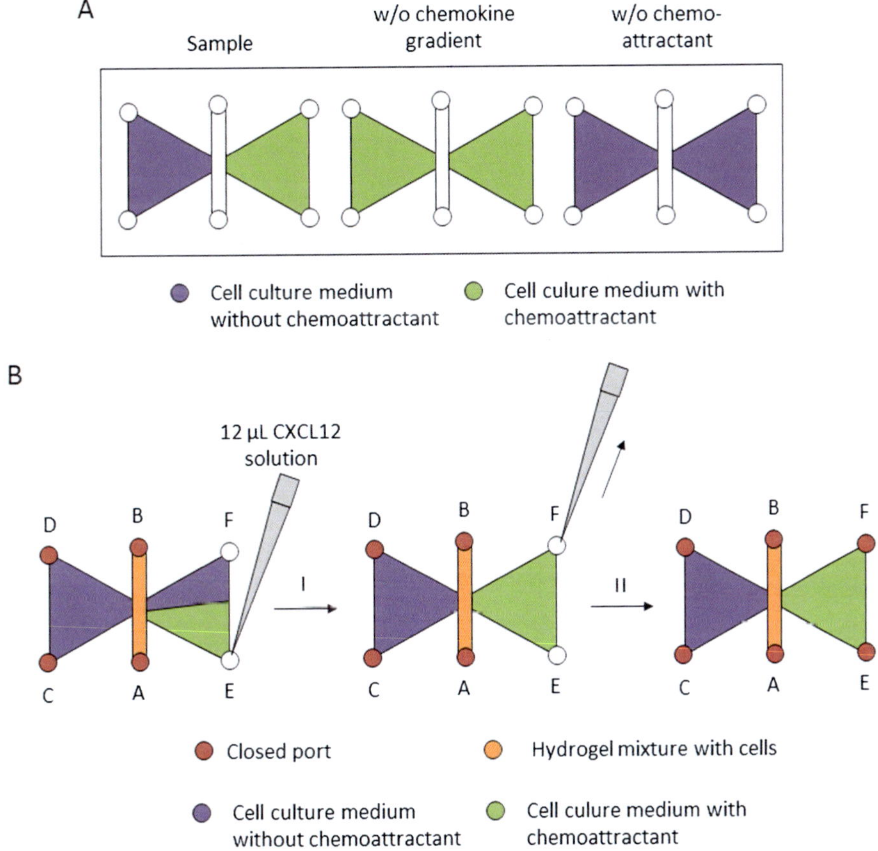

Fig. 2 (**a**) Schematic drawing of one possible experimental application using a migration assay with the µ-Slides Chemotaxis. Media reservoirs filled with cell culture medium without chemoattractant are colored in purple, and media chambers, where chemoattractant was added additionally, are indicated in green. For the sample, chemoattractant is added only to the right medium reservoir leading to the formation of a chemokine gradient. Chemoattractant is added to both medium reservoirs for the control without chemokine gradient, and no chemokine is added to the control without chemoattractant. The scheme of the µ-Slide Chemotaxis is adapted from manufacturer's instructions (ibidi GmbH, Germany). (**b**) The process of adding the chemoattractant to the medium reservoirs is illustrated in the schematic drawing. The filling ports E and F are opened, whereas A–D are closed (red dots). Both medium reservoirs are filled with cell culture medium (indicated by purple color), and the observation chamber contains the gel mixture with the cells (orange). The chemoattractant is pipetted to filling port E (green). (I) Immediately afterward, the same amount of medium is aspirated from filling port F. (II) The filling ports E and F are closed. The scheme of the µ-Slide Chemotaxis is adapted from manufacturer's instructions (ibidi GmbH, Germany)

3.4 Save the Data as Movie File

Once the experiment is finished, export the pictures taken with the microscope software to an .avi file.

1. Set the duration of the movie by selecting the frames per second, e.g., five frames per second.
2. Save the data as an .avi file.

3.5 Migration Analysis

1. Open the .avi file in ImageJ. Choose the first and the last frame of the time slot corresponding to the desired observation chamber (sample, control without chemokine gradient, and control without chemoattractant).
2. Track 30 to 40 cells per observation chamber with the plug-in "Manual tracking" (Fig. 3a).
3. Export the x and y coordinates into Excel, and convert the data from pixel into μm based on the microscope parameters.
4. Correct for drift in the movies (e.g., elicited by movement of the hydrogel due to prevalent temperatures or movement of the microscope's stage) by subtracting the drift from the recorded movement (*see* **Note 12**).
5. For plotting the data and quantitative analysis of the cell movement, use the ibidi GmbH software Chemotaxis and Migration Tool 2.0 (Fig. 3b).
6. Bring the data into a tab delimited text file by opening the data in the text editor program and saving the data as .txt file.

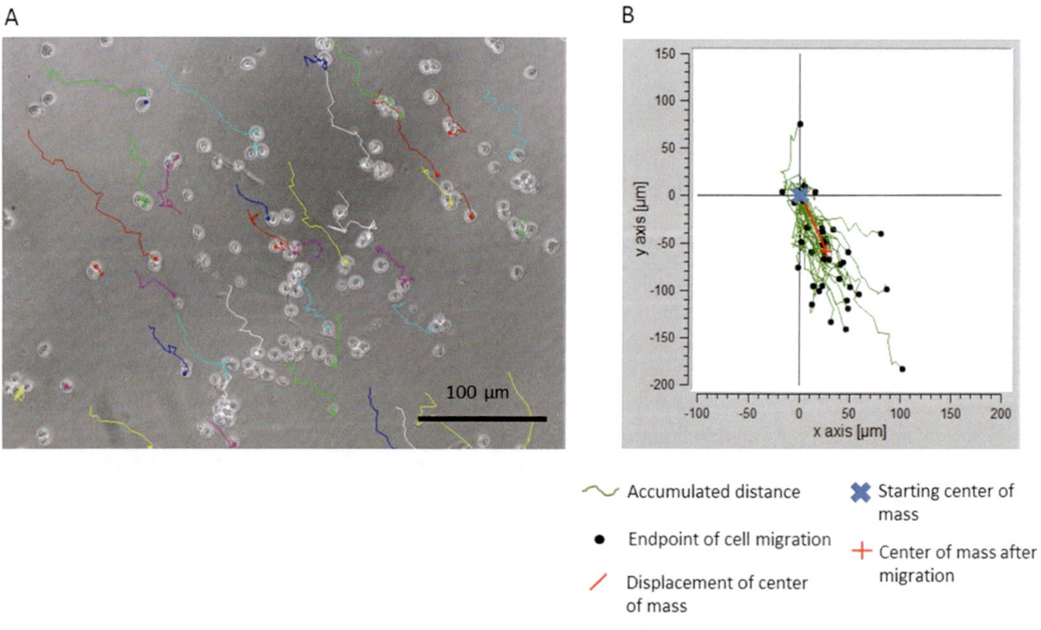

Fig. 3 (**a**) Exemplified image after manual tracking of cells with ImageJ. The tracked cells are seen at the endpoint of their migration paths (indicated as colored dots); migration paths are indicated by colored lines. (**b**) Exemplified graph resulting from the analysis with the software Chemotaxis and Migration Tool 2.0 from ibidi GmbH. In the figure, the cell migration in *x*- and *y*-direction is plotted, and the definition of the parameters for the quantitative analysis is written below the plot. The starting points of all tracked migration paths are set to the origin of the axes. Therefore, the starting center of mass is located here (indicated by a blue cross). The endpoints of the migration of cells are shown as black dots. The accumulated distance (green line) describes the route that the cells migrated. The displacement of the center of mass describes the difference between starting and end position of center of mass (red cross) and is marked by a red line

7. Import the data into the software Chemotaxis and Migration Tool 2.0.
8. Initialize the data for analysis by adding the desired number of slices of the recorded movie and setting the calibration (*see* **Note 13**).
9. Create plots and diagrams by using the corresponding menu icon. To edit the plots use the menu icon "plot settings."
10. For the quantitative analysis, the following parameters can be used by selecting the menu icons "statistics" and "measured values" of the software Chemotaxis and Migration Tool 2.0 (*see* **Note 14**).

    Forward migration index (FMI): Use *x* FMI (*see* **Note 15**) to analyze the migration of the cells. Depending on the direction of the cell migration, positive or negative values are possible. If the chemoattractant in the sample migration chamber was added to the right medium reservoir, a more positive FMI value refers to a higher migration of the cells in the direction of the chemoattractant.

    Directness can be used to characterize the straightness of the cell migration. The values vary between 0 and 1. The closer to 1, the more direct is the movement of the cells. In combination with the FMI, the directness can be a parameter to judge how direct the cells moved to the chemoattractant.

    Accumulated distance represents the total distance that the cells migrated in μm (Fig. 3b).

    Displacement of the center of mass describes the difference between the center of mass in the beginning and at the end of the migration. The center of mass represents the average coordinates of all tracked cells. The higher the value, the further the cells migrated (Fig. 3b).
11. Use the control without chemokine gradient and the control without chemoattractant to compare the quantitative values. The control without chemokine gradient contains chemoattractant in both filling reservoirs, i.e., there should be cell movement in both directions. The control without chemoattractant contains no chemoattractant, i.e., no chemoattractant related migration should occur. For the characterization of the cell migration, relate and compare the parameters described above to each other, and do not judge depending on one parameter only.

4 Notes

1. Use a μ-Slide Chemotaxis containing three observation chambers with two medium reservoirs each. One is used for the sample containing the chemokine gradient. The two others

are needed as controls: Chemoattractant is added to both reservoirs for the control without chemokine gradient, and no chemoattractant is added to the control without chemoattractant (Fig. 1a).

2. Hydrogel systems suitable for this kind of migration experiments should allow cytocompatible polymerization in the presence of cells, and they should be degradable by the embedded cells. For example, in the presented system an amino acid sequence in the linker used for hydrogel polymerization can be cleaved by matrix metalloproteinases expressed by the cells which allows migration of the cells in the 3D hydrogel.
3. Chemotaxis and Migration Tool 2.0 is freely available on the homepage of ibidi GmbH (https://ibidi.com/manual-image-analysis/171-chemotaxis-and-migration-tool.html).
4. Install the plug-in "Manual tracking" for ImageJ, freely available on the homepage of ImageJ (https://imagej.nih.gov/ij/plugins/track/track.html).
5. After placing the CD-Link in the observation chamber, perform the next steps quickly. Once the maleimide-dextran solution is added to the CD-Link, the polymerization takes only a few seconds. Always prepare the hydrogel mixture freshly. Do not store it.
6. Avoid the formation of air bubbles by directly placing the pipette on top of the filling ports.
7. Air bubbles disturb the migration assay because they interfere with the cell migration.
8. The process of opening and closing the filling ports can be facilitated by transferring a plug from one filling port to another. Thereby a filled reservoir is closed by transferring a plug from an empty reservoir that will be filled afterward.
9. There is no need to incubate the cells until they settle down due to the low height of the observation chamber which leads to the formation of only one cell layer.
10. The observation time varies depending on the cell migration capacity and the strength of the reaction to the chemoattractant. The used time specification relates to OCI-AML3 cells.
11. Make sure that you save a position which contains a wide area of the observation chamber with at least 30 to 40 cells.
12. Track eventual occurring drift in the recorded movie by tracking an artifact (e.g., bubbles or dust in the hydrogel) with ImageJ. Export the data into Excel and convert from pixel to μm. Subtract each coordinate of the artifact from the corresponding coordinates of each cell.
13. The conversion of the data from pixel into μm was already done in Excel. Therefore, perform the calibration with *X/Y*

calibration equals 1, and select μm and the time interval used. The initializing process sets the starting coordinates of each cell to zero.

14. Ibidi GmbH provides more information about possible quantitative data analysis online. We show the minimum of quantitative analysis needed to quantify your data properly.
15. The chemoattractant is added to the right medium reservoir, which leads to cell migration in positive *x*-direction. Therefore, use the *x* FMI for quantitative analysis.

Acknowledgments

Sabrina Zippel and Annamarija Raic contributed equally to this work.

References

1. Lane SW, Scadden DT, Gilliland DG (2009) The leukemic stem cell niche: current concepts and therapeutic opportunities. Blood 114 (6):1150–1157. https://doi.org/10.1182/blood-2009-01-202606
2. Bonnet D, Dick JE (1997) Human acute myeloid leukemia is organized as a hierarchy that originates from a primitive hematopoietic cell. Nat Med 3(7):730–737. https://doi.org/10.1038/nm0797-730
3. Schepers K, Campbell TB, Passegue E (2015) Normal and leukemic stem cell niches: insights and therapeutic opportunities. Cell Stem Cell 16(3):254–267. https://doi.org/10.1016/j.stem.2015.02.014
4. Colmone A, Amorim M, Pontier AL, Wang S, Jablonski E, Sipkins DA (2008) Leukemic cells create bone marrow niches that disrupt the behavior of normal hematopoietic progenitor cells. Science 322(5909):1861–1865. https://doi.org/10.1126/science.1164390
5. Mills SC, Goh PH, Kudatsih J, Ncube S, Gurung R, Maxwell W, Mueller A (2016) Cell migration towards CXCL12 in leukemic cells compared to breast cancer cells. Cell Signal 28 (4):316–324. https://doi.org/10.1016/j.cellsig.2016.01.006
6. Kitaori T, Ito H, Schwarz EM, Tsutsumi R, Yoshitomi H, Oishi S, Nakano M, Fujii N, Nagasawa T, Nakamura T (2009) Stromal cell–derived factor 1/CXCR4 signaling is critical for the recruitment of mesenchymal stem cells to the fracture site during skeletal repair in a mouse model. Arthritis Rheumatol 60 (3):813–823. https://doi.org/10.1002/art.24330
7. Aiuti A, Webb IJ, Bleul C, Springer T, Gutierrez-Ramos JC (1997) The chemokine SDF-1 is a chemoattractant for human CD34 + hematopoietic progenitor cells and provides a new mechanism to explain the mobilization of CD34+ progenitors to peripheral blood. J Exp Med 185(1):111–120. https://doi.org/10.1084/jem.185.1.111
8. Cummins TD, Wu KZL, Bozatzi P, Dingwell KS, Macartney TJ, Wood NT, Varghese J, Gourlay R, Campbell DG, Prescott A, Griffis E, Smith JC, Sapkota GP (2018) PAWS1 controls cytoskeletal dynamics and cell migration through association with the SH3 adaptor CD2AP. J Cell Sci 131(1):12. https://doi.org/10.1242/jcs.202390
9. Shanks N, Greek R, Greek J (2009) Are animal models predictive for humans? Philos Ethics Humanit Med 4:2. https://doi.org/10.1186/1747-5341-4-2
10. Knight A (2007) Animal experiments scrutinised: systematic reviews demonstrate poor human clinical and toxicological utility. ALTEX 24(4):320–325. https://doi.org/10.14573/altex.2007.4.316
11. Mestas J, Hughes CCW (2004) Of mice and not men: differences between mouse and human immunology. J Immunol 172 (5):2731–2738. https://doi.org/10.4049/jimmunol.172.5.2731
12. Zengel P, Nguyen-Hoang A, Schildhammer C, Zantl R, Kahl V, Horn E (2011) μ-Slide chemotaxis: a new chamber for long-term chemotaxis studies. BMC Cell Biol 12:21–21. https://doi.org/10.1186/1471-2121-12-21

Chapter 9

Intravital Imaging of Blood Flow and HSPC Homing in Bone Marrow Microvessels

Jonas Stewen and Maria Gabriele Bixel

Abstract

Two-photon intravital microscopy (2P-IVM) is an advanced imaging technique that allows the visualization of dynamic cellular behavior deeply inside tissues and organs of living animals. Due to the deep tissue penetration, imaging of highly light-scattering tissue as the bone becomes feasible at subcellular resolution.

To better understand the influence of blood flow on hematopoietic stem and progenitor cell (HSPC) homing to the bone marrow (BM) microvasculature of the calvarial bone, we analyzed blood flow dynamics and the influence of flow on the early homing behavior of HSPCs during their passage through BM microvessels. Here, we describe a 2P-IVM approach for direct measurements of red blood cell (RBC) velocities in the BM microvasculature using repetitive centerline scans at the level of individual arterial vessels and sinusoidal capillaries to obtain a detailed flow profile map. Furthermore, we explain the isolation and enrichment of HSPCs from long bones and the transplantation of these cells to study the early homing behavior of HSPCs in BM sinusoids at cellular resolution. This is achieved by high-resolution spatiotemporal imaging through a chronic cranial window using transgenic reporter mice.

Key words Intravital imaging, Bone marrow microvessels, Blood flow velocities, Hematopoietic stem cells, Stem cell homing, Two-photon microscopy

1 Introduction

Intravital microscopy (IVM) includes all optical and high-resolution imaging techniques that analyze biological processes in vivo [1]. Combined with two-photon excitation (2P) microscopy, it provides a powerful tool to perform functional studies in various in vivo models including the murine bone marrow (BM) [2–4]. Compared to confocal laser-scanning microscopy, 2P microscopy shows significantly reduced photobleaching, reduced phototoxicity, and substantially increased tissue penetration even in highly scattering tissue as the bone [5–7]. 2P microscopy allows to visualize fluorescent chromophores and additionally excites second-harmonic generation (SHG) of anisotropic biological structures like collagen fibers in trabecular and compact

Gerd Klein and Patrick Wuchter (eds.), *Stem Cell Mobilization: Methods and Protocols*, Methods in Molecular Biology, vol. 2017, https://doi.org/10.1007/978-1-4939-9574-5_9,

bone [8, 9]. Using 2P-IVM, we analyzed the microarchitecture of the BM vasculature and performed quantitative studies of blood flow dynamics in arterial and sinusoidal microvessels in the murine skull. Furthermore, we imaged the early homing behavior of hematopoietic stem and progenitor cells (HSPCs) to BM sinusoids [3].

Postnatal hematopoiesis is confined primarily to the BM and requires a specialized microenvironment, the hematopoietic niche, that controls the maintenance and self-renewal of hematopoietic stem cells (HSCs) by providing survival signals from various surrounding cell types [10–13]. The BM microvasculature is composed of small-caliber arterial vessels and highly interconnected sinusoidal capillaries, which are located within the BM cavities of long and flat bones [14, 15]. Thin-walled sinusoidal capillaries with fenestrations and a discontinuous basement membrane are highly specialized for cell trafficking and passage of soluble substances [16].

Blood flow and the resulting wall shear stress have a critical impact on the early homing behavior of HSPCs in BM microvessels. Only when the wall shear stress is below a critical value HSPCs successfully roll and adhere to the luminal surface of BM sinusoids to subsequently transmigrate to the BM cavities [3, 17, 18]. To better understand the impact of blood flow and to predict putative sites of HSPC adhesion and transmigration, we performed direct real-time measurements of red blood cell (RBC) flow using repetitive centerline scans [3]. 2P imaging and real-time recordings using the cranial window technique are feasible through a thin layer of cortical bone of the murine calvarium, the frontoparietal bone of the skull, despite the fact that bone is a highly light-scattering tissue [2]. In vivo imaging of the calvarium offers a unique advantage over long bones, which require surgical procedures including removal of muscle tissue and thinning of the cortical bone resulting in acute inflammatory responses of the tissue [19].

This chapter provides a detailed technical overview of 2P imaging of blood flow dynamics and HSPC homing in BM microvessels in living mice. The key to successful 2P-IVM of the BM microvasculature and its cell dynamics is (1) the surgical procedure of the chronic cranial window, (2) the isolation and enrichment of HSPCs, and (3) intravital imaging of BM microvessels, quantitative blood flow measurements, and the multistep process of HSPC homing.

2 Materials

2.1 Animal and Surgical Tools

1. Mice of both sexes (preferentially female) between 8 and 12 weeks.
2. Anesthetics: 100 mg/mL ketamine and 100 mg/mL xylazine (*see* **Note 1**).

3. Eye ointment: 5% dexpanthenol.
4. 1 mg/mL dexamethasone.
5. Stereotactic device.
6. Electric heating pad.
7. Sterile PBS.
8. 70% ethanol.
9. Surgical instruments: operation scissors, straight forceps, disposable straight scalpels (*see* **Note 2**).
10. Sterile swaps.
11. Coverslips (6 mm).
12. Brush tip holder and brush tips.
13. Cyano Veneer Powder, Cyano Veneer fast.
14. Custom-made titanium ring.
15. 50 mg/mL tramadol.

2.2 Fluorescent Probes

1. 25 mg/mL Dextran, Texas Red, 70,000 MW.
2. Anti-Endomucin antibodies (clone V.5C7) coupled to Alexa594 [3].

2.3 Isolation and Enrichment of HSPC

1. Mice of both sexes between 6 and 7 weeks.
2. Surgical instruments: operation scissors, straight forceps.
3. Mortar and pestle.
4. Ca^{2+}/Mg^{2+}-free PBS with 2% heat-inactivated bovine serum.
5. Sterile 70-μm filter.
6. Lineage cell depletion kit.
7. PKH67 green fluorescent linker kit.
8. DMEM supplemented with 25% FCS.
9. Sterile PBS.

2.4 Animal Imaging Devices

1. Custom-made head immobilization device with electric heating pad (Fig. 2b, c).

2.5 Microscope

1. Multiphoton microscope setup (TriM Scope II) with single-beam scanning and non-descanned detectors from LaVision BioTec, as described in [3] (Fig. 2a).
2. Olympus BX51 WI microscope stand equipped with high sensitive NDD detectors (PMTs).
3. Coherent Scientific Chameleon Ultra II Ti:Sapphire laser (tuning range 680–1080 nm, 120 fs pulse width, 80 MHz repetition rate).

4. Coherent Chameleon Compact OPO (automated wavelength extension from 1000 nm to 1600 nm, 200 fs pulse width, 80 MHz repetition rate).
5. IR objective lens (20× Olympus XLUMPlanFl 20×/1.0 W, WD 2.0 mm, or 16× Nikon CFI LWD Plan Fluorite, 0.8 NA, 3.0 mm WD).
6. LaVision BioTec ImSpector Software for image acquisition.
7. Pair of x-y galvanometric mirrors for scanning the sample at a scanning speed of up to 1200 lines/s.
8. Dichromatic mirrors and band-pass filters: blue (420/40 nm), green (525/50 nm), red (620/60 nm).
9. Photomultiplier tubes (PMT; Hamamatsu H67080-01 (blue channel), H67080-20 (green and red channels)).

3 Methods

3.1 Animal and Anesthesia

1. Mice are housed in a controlled environment in the institute animal facility. Few days prior to the surgery, mice are housed in a Scantainer cabinet in the lab to acclimate. Water and food are provided ad libitum.
2. Weigh mice prior to the anesthesia.
3. Inject a mixture of 100 mg/kg ketamine and 16 mg/kg xylazine in PBS (*see* **Note 1**) into the peritoneal cavity. Once the animal is unconscious, check for absence of pedal reflex (firm toe pinch) before proceeding.
4. Apply eye ointment to prevent eye dryness.
5. Place mouse on an electrical heating pad to maintain normal body temperature.

3.2 Cranial Window Surgery

1. Place the anesthetized animal in a stereotactic device, and use an electric heating pad to control the body temperature.
2. Inject 50 μL 0.2 mg/kg dexamethasone s.c. to prevent inflammatory responses.
3. Wipe down the scalp with 70% ethanol and remove excess of ethanol.
4. Lift up a small area of the skin with dull forceps at the midline, and make a small incision along the sagittal suture (Fig. 1a, *see* **Note 2**).
5. Use scissors to elongate the incision to a length of ~1.5 cm.
6. Lift up the skin at the posterior end of the incision with dull forceps, and use scissors to make a 0.5-cm-long cut to the right and to the left (Fig. 1a).

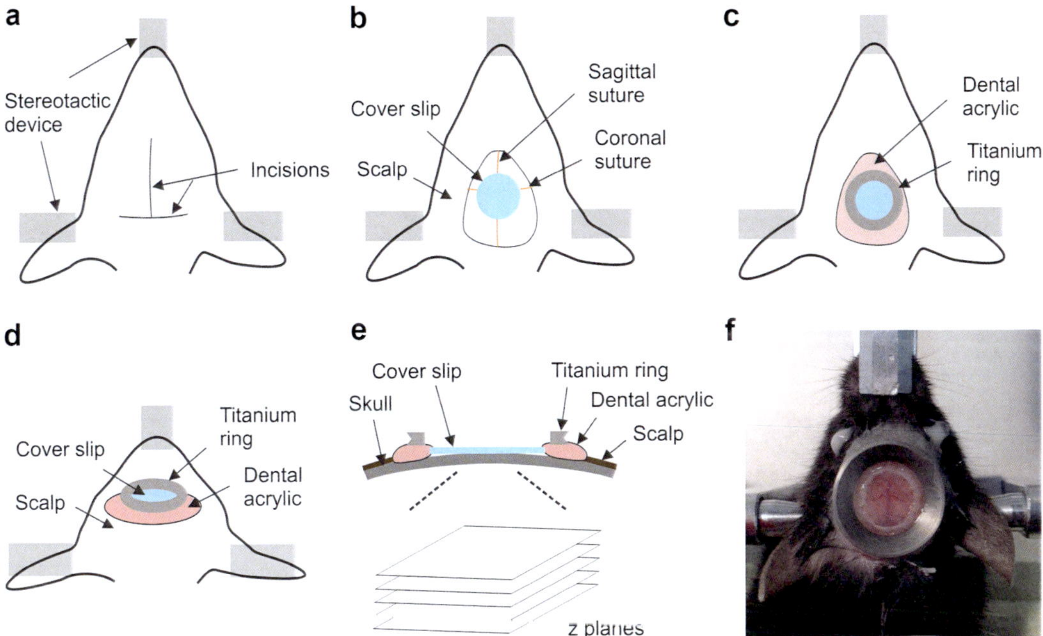

Fig. 1 Chronic cranial window surgical procedure. (**a**) Schematic showing incision lines of the skin to expose underlying calvarial bone surface. (**b**) Coverslip position in relation to the sagittal and coronal sutures. (**c**) The optical window is sealed onto the skull, and the exposed bone surface is covered with dental acrylic. A titanium ring is embedded into the dental acrylic for head fixation during imaging. (**d**) Side view of the cranial window with glass coverslip and titanium ring embedded in dental acrylic on the calvarial bone surface. (**e**) Schematic cross section through the cranial window that allows imaging of *z*-stacks of the underlying calvarial bone. (**f**) Image of a chronic cranial window after surgery is completed

7. Cut of the skin flaps on both sides with the scissors.
8. Use thin forceps to remove the thin translucent layer on the calvarial bone exposing the bone surface (*see* **Note 3**).
9. Use a scalpel to roughen the bone surface at peripheral regions outside the area of the optical window by inserting small parallel scratches.
10. Apply a thin layer of transparent dental acrylic glue with a tiny dental brush on the bone area outside of the optical window region.
11. Prepare a 6 mm glass coverslip with a sterile hanging drop of PBS, and place it carefully with a forceps on the area of interest (optical window region) (Fig. 1b, *see* **Note 4**).
12. Mix transparent component and powder of the dental acrylic until it shows a viscous consistency, and use immediately.
13. Seal the optical window onto the skull with dental acrylic, covering the edges of the glass coverslip and the area of the exposed bone surface (Fig. 1c, *see* **Note 5**).

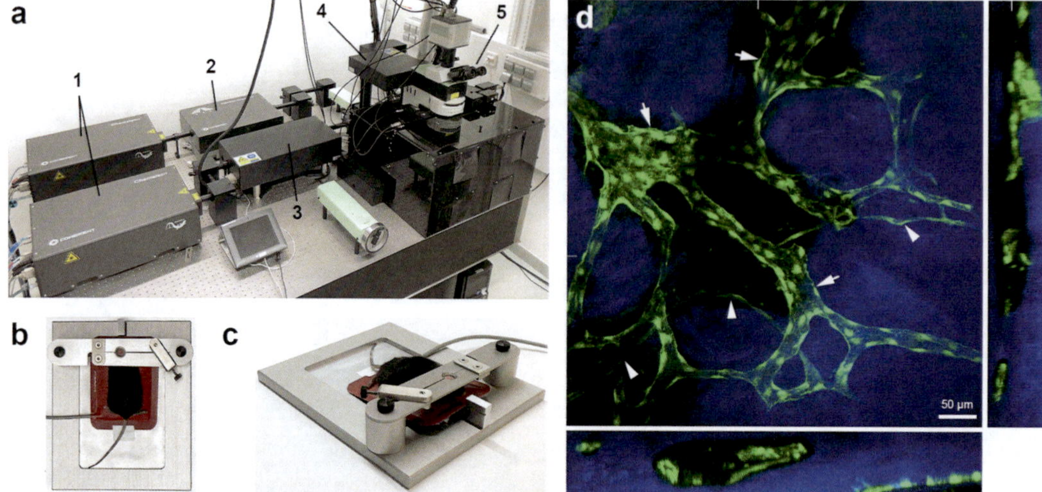

Fig. 2 Intravital imaging of BM vasculature using a chronic cranial window. (**a**) Intravital multiphoton imaging setup: (1) mode-locked Ti:Sapphire lasers, (2) optical parametric oscillator, (3) electrooptic modulator (beam shaper), (4) scan head, and (5) microscope body with dark panel box covering objective and intravital table. (**b**, **c**) Top (**b**) and side view (**c**) of the custom-made immobilization device. The anesthetized mouse is placed on a heating pad with the titanium ring fixed in the immobilization device to eliminate movement artifacts during imaging. (**d**) Representative image showing GFP^{+} (green) arterial vessels (arrow heads) and sinusoidal capillaries (arrows) and SHG^{+} (blue) bone tissue of a *Flk1-GFP* transgenic mouse (z-stack extended view (20 μm), positions of the orthogonal planes are indicated)

14. Immediately mount a custom-made titanium ring around the glass coverslip by gluing it onto the dental acrylic (Fig. 1c–f).
15. Carefully seal all gaps between the titanium ring, the bone surface, and the optical window with dental acrylic (Fig. 2f, *see* **Note 6**).
16. Allow dental acrylic to harden, and then release animal from stereotactic device, typically after 5–10 min.
17. Inject 50 μL 15 mg/kg tramadol s.c. directly after the surgery.
18. Allow animal to recover in a warm environment, and monitor for at least 4 h.
19. Provide tramadol (1 mg/mL) in the drinking water on two following days.

3.3 Isolation and Enrichment of HSPC

1. Sacrifice two mice.
2. Collect femurs and remove adherent tissue, i.e., muscle tissue.
3. Isolate BM cells by crushing femurs with a mortar and pestle in 2 mL Ca^{2+}/Mg^{2+}-free PBS with 2% heat-inactivated bovine serum.
4. Use circular movements of the pestle to crush, and then pulverize the bones to obtain a white to transparent material.
5. Pass the BM cell suspension with a syringe several times through a 25 G needle.

6. Filter the BM cell suspension with a 70-μm filter on top of a 50 mL tube.
7. Centrifuge cells (300 × *g*, 10 min) and wash once in DMEM.
8. Lineage deplete the BM cell suspension with a lineage cell depletion kit.
9. Count cells using a hemocytometer (*see* **Note 7**).
10. For fluorescent labeling, suspend 5×10^6 cells in 1 mL of diluent solution C of PKH67 green fluorescent linker kit.
11. Add 1 mL PKH67 (2×10^{-3} M) and mix carefully.
12. Incubate for 7 min at RT.
13. Inactivate excess of dye by adding 1 mL DMEM supplemented with 25% FCS.
14. Wash cells twice (300 × *g*, 10 min) and suspend in sterile ice-cold PBS (Fig. 4a).

3.4 Intravital Imaging

3.4.1 Imaging the BM Vasculature

1. Place the anesthetized animal on a heating pad to control body temperature.
2. Use a custom-made immobilization device and the mounted titanium ring for head fixation (Fig. 2b, c).
3. Use an IR objective lens (20×, Olympus, or 16×, Nikon), and place a water droplet on the optical window inside the titanium ring.
4. Place the animal under the microscope with the optical window perpendicular to the optical axis.
5. Use epifluorescence illumination to focus on the BM vasculature, and identify a region of interest.
6. Change to 2P microscopy for high magnification imaging. Determine excitation laser wavelengths and laser power (*see* **Note 8**).
7. Determine scan area, scan speed, and number of pixels acquired versus scan area.
8. Start image acquisition once all parameters are adjusted (*see* **Notes 9** and **10**).
9. Use Ti:Sapphire laser as infrared multiphoton light source at 850 nm for combined fluorescence and SHG imaging.
10. Use a red-sensitive PMT and a 620/60 band-pass filter to visualize TexasRed dextran (Fig. 3b, c) or anti-Endomucin antibodies coupled to Alexa594 (Fig. 4b, c).
11. Use a green-sensitive PMT and a 525/50 band-pass filter to visualize green chromophores, such as GFP (Fig. 2d).
12. Use a blue-sensitive PMT and a 420/40 band-pass filter for SHG imaging to visualize collagen fibers of trabecular and compact bone (Figs. 2d and 3c).

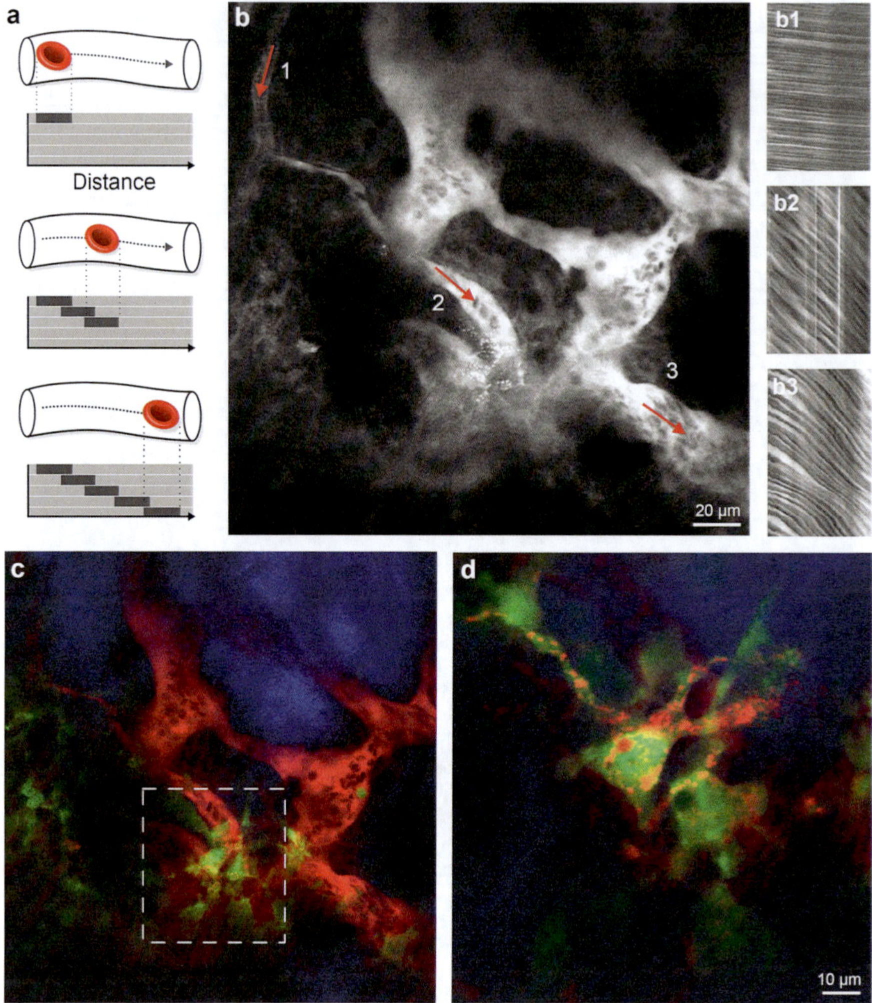

Fig. 3 Measurement of blood flow velocities in BM microvessels of the murine calvarium. (**a**) Schematic illustrating the principle of centerline scans to measure blood flow velocities. Centerline scans at a scan rate faster than the moving RBCs track their position within the vessel segment. Linear scan data are plotted as time-space image, and moving RBCs appear as diagonal streaks. RBC velocity data are calculated from the slope of the streaks. (**b**) Bone marrow microvasculature visualized by 2P microscopy after intravenous injection of TexasRed dextran that labels the blood plasma but is excluded from RBCs. Line scans of indicated vessel segments 1–3 are shown (**b1–b3**). Small-caliber arterial vessels show low-angle slopes (**b1**), while wider sinusoidal vessels show high-angle slopes (**b2**, **b3**) indicating high and low RBC velocities, respectively. (**c**) Bone marrow vasculature of a *CX3CR1-GFP* transgenic mouse after intravenous injection of TexasRed dextran indicating the BM microvessels (red) and perivascular macrophages (green). (**d**) GFP^+ perivascular macrophages in close proximity to sinusoidal vessels rapidly take up dextran dye into intracellular vesicles, decreasing the imaging quality over time (magnification of boxed area from **c**)

13. Record fluorescence and SHG signals simultaneously (Fig. 2d).
14. Allow the animal to recover in a warm environment after imaging, and monitor for at least 2 h.

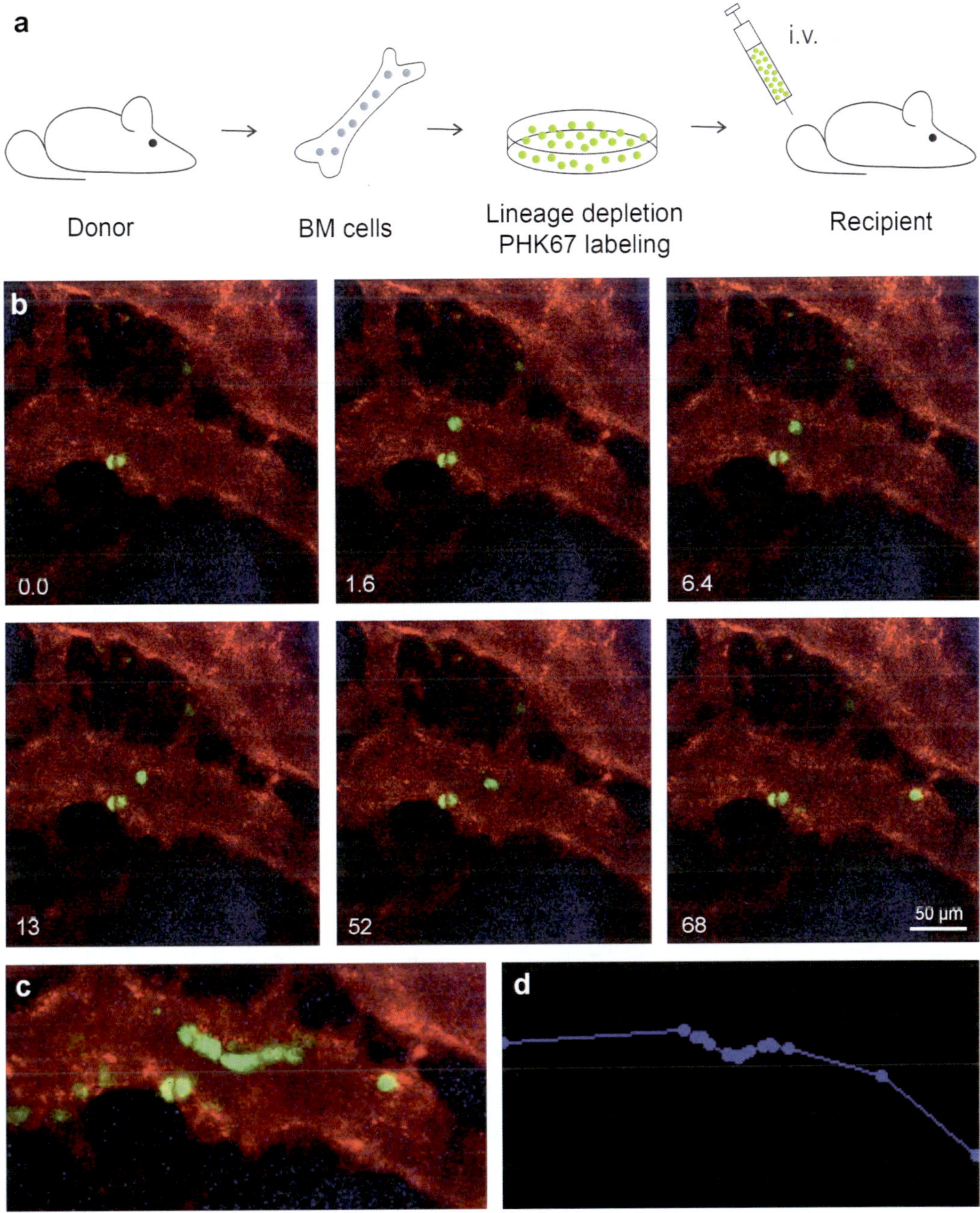

Fig. 4 HSPC homing to BM sinusoidal capillaries. (**a**) Schematic illustrating isolation of BM cells from murine long bone, lineage-depletion to enrich HSPCs and fluorescence labeling with PKH67. Labeled HSPCs are transplanted by intravenous tail vein injection. (**b**) HSPC homing in BM sinusoid was analyzed using intravital 2P imaging. Representative HSPC (green) shows rolling and transient adhesion on Endomucin$^+$ (red) BM sinusoidal capillaries. Two luminal adherent HSPCs are seen close to the vessel wall (time is in seconds). (**c**) HSPC movement pattern of in total 24 time points over a 1.2 min interval including the time points shown in **b**. (**d**) HSPC track from **c** exported from the time-lapse movie with TrackMate (ImageJ)

3.4.2 Blood Flow Measurement

1. Anesthetize the animal, and place it on a heating pad to control body temperature.
2. Inject 50 μL TexasRed dextran i.v. into the tail vein.
3. Proceed with **steps 1–13** (Subheading 3.4.1) to record fluorescence and SHG signals.
4. Capture an overview *z*-stack of the area of interest (Fig. 3b, c, *see* **Notes 11** and **12**).
5. Change the imaging mode to line scans.
6. Place a scan line (typically 50 μm) along the centerline of a depicted vessel.
7. Use repetitive line scans (500 lines) to capture moving RBCs (Fig. 3a, b).
8. Use a line scan rate of 0.5–2 kHz to obtain a time-space image with diagonal streaks (*see* **Note 13**).
9. Repeat centerline scans with remaining vessels (*see* **Notes 14** and **15**).
10. Allow the animal to recover in a warm environment after imaging, and monitor for at least 2 h.
11. Export RCB velocity data from the acquired time-space images using a python script implemented in the acquisition software.
12. Determine flow velocities of a selected vessel segment from slopes of >10 representative streaks acquired at two different scan rates, and calculate mean values.

3.4.3 Homing of HSPC in BM Microvessels

1. Intravenously inject 50 μL anti-Endomucin antibody coupled to Alexa594 (0.5–1 mg/kg) into the tail vein of a recipient animal to stain the endothelium, and wait 30 min (*see* **Note 12**).
2. Anesthetize the animal, and place it on a heating pad to control body temperature.
3. Intravenously inject 2.5×10^6 PKH67-labeled HSPCs in 200 μL PBS into the tail vein (Fig. 4a).
4. First HSPCs appear 5–10 min after transplantation. Use epifluorescence imaging to identify regions with PKH67-labeled HSPCs in the BM microvasculature (*see* **Note 16**).
5. Immediately change to two-photon imaging, and proceed with **steps 2–13** (Subheading 3.4.1) to record fluorescence and SHG signals.
6. Capture an image of the best focal plane containing the HSPC.
7. Acquire further overview images containing more details about the cellular surrounding, i.e., architecture of BM microvessels and endosteal surface by recording a *z*-stack with images of 1–5 μm step size.

8. Record real-time movies using a single plane to capture early homing behavior of HSPCs in BM sinusoidal vessels (Fig. 4b).
9. Allow the animal to recover in a warm environment after imaging, and monitor for at least 2 h.
10. Manually track individual HSPC in recorded real-time movies with TrackMate (ImageJ), and export HSPC velocities (Fig. 4c, d).

4 Notes

1. Prepare anesthetics freshly by diluting ketamine and xylazine in sterile PBS, and mix subsequently. Use 10 μL/g for intraperitoneal injection.
2. Surgical instruments should be cleaned and sterilized with 70% EtOH before use.
3. If bleedings occur from skin and/or skull vessels during the surgical procedure, soak up the blood with a sterile swap before applying dental acrylic.
4. Air bubbles under the glass coverslip result in a poor imaging depth. Ensure that PBS completely covers the gap between glass coverslip and bone surface before mounting the glass coverslip with dental acrylic.
5. When sealing the optical window onto the skull, avoid dental acrylic to drop on the glass coverslip or to flow in between glass coverslip and bone surface, thereby strongly reducing the optical transparency of the window.
6. The titanium ring, in addition to the head fixation, retains the water droplet between objective and optical window during the imaging session. An absent water droplet results in loss of the imaging signal.
7. Use flow cytometric quantification to determine enrichment of HSPCs from total BM cells using sca1, c-kit, and lineage cocktail antibodies.
8. Phototoxicity results from too much laser power. Ensure that laser power and detector gain are set properly and that all optics are properly aligned.
9. Sudden and large motion artifacts during imaging are observed when the animal is starting to wake up from anesthesia. Terminate imaging session or re-anesthetize the animal with a third of the original dose before recommencing with imaging.
10. Increasing opacity occurs due to inflammatory reactions resulting in faint fluorescence signals few days after the cranial window surgery.

11. Intravenously injected dextran dye labels the blood plasma. RCB exclude the dye and appear as dark objects against a bright fluorescent background.
12. If blood vessels are only weakly detectable, fluorescent dye or antibody was not injected properly.
13. Arterial vessels with high RBC velocities generate lines that are almost horizontal. Using a higher scanning frequency can overcome this problem; however, there is a limitation to the maximal speed that can be measured. Vertical lines often indicate stationary cells that adhere to the vessel lumen. Nonlinear line scan steaks could indicate to a vessel segment that is not completely within the imaging plane. Readjust the position of the line scan to ensure that a sufficient line scan length is in the centerline of the vessel. Irregular nonlinear streaks could point to turbulent or pulsatile flow pattern.
14. After i.v. injection, TexasRed dextran continuously leaks out of the vasculature into the surrounding tissue. Leakage is observed already at early time points due to highly permeable and fenestrated BM sinusoids. Leakage out of arterial vessels is less pronounced. Perivascular macrophages rapidly internalize TexasRed dextran resulting in highly fluorescent cells lining the BM microvasculature (Fig. 3d). For high-quality overview images, it is advisable to record images at a very early time point. Dye leakage decreases the signal intensity; however, influence on the quality of recorded line scans is less pronounced.
15. To ensure that blood flow measurements are recorded at a stable physiological condition during the imaging session, select few vessels and measure the vessel segments twice at different time points, i.e., at the beginning and the end of an experiment. Furthermore, record individual vessel segments in random order and not along a selected vessel in the direction of flow.
16. If no or only very few HSPCs are observed, HSPCs were not injected properly.

Acknowledgments

We thank F. Winkler for sharing his expertise on the imaging setup for head immobilization. This work was supported by the Max Planck Society, the University of Münster, the DFG cluster of excellence "Cell in Motion," and the European Research Council (AdG 339409 AngioBone).

References

1. Weigert R, Sramkova M, Parente L, Amornphimoltham P, Masedunskas A (2010) Intravital microscopy: a novel tool to study cell biology in living animals. Histochem Cell Biol 133(5):481–491. https://doi.org/10.1007/s00418-010-0692-z
2. Lo Celso C, Fleming HE, Wu JW, Zhao CX, Miake-Lye S, Fujisaki J, Cote D, Rowe DW, Lin CP, Scadden DT (2009) Live-animal tracking of individual haematopoietic stem/progenitor cells in their niche. Nature 457 (7225):92–96. https://doi.org/10.1038/nature07434
3. Bixel MG, Kusumbe AP, Ramasamy SK, Sivaraj KK, Butz S, Vestweber D, Adams RH (2017) Flow dynamics and HSPC homing in bone marrow microvessels. Cell Rep 18 (7):1804–1816. https://doi.org/10.1016/j.celrep.2017.01.042
4. Zhao Y, Bower AJ, Graf BW, Boppart MD, Boppart SA (2013) Imaging and tracking of bone marrow-derived immune and stem cells. Methods Mol Biol 1052:57–76. https://doi.org/10.1007/7651_2013_28
5. Denk W, Strickler JH, Webb WW (1990) Two-photon laser scanning fluorescence microscopy. Science 248(4951):73–76
6. Helmchen F, Denk W (2005) Deep tissue two-photon microscopy. Nat Methods 2 (12):932–940. https://doi.org/10.1038/nmeth818
7. Andresen V, Alexander S, Heupel WM, Hirschberg M, Hoffman RM, Friedl P (2009) Infrared multiphoton microscopy: subcellular-resolved deep tissue imaging. Curr Opin Biotechnol 20(1):54–62. https://doi.org/10.1016/j.copbio.2009.02.008
8. LaComb R, Nadiarnykh O, Carey S, Campagnola PJ (2008) Quantitative second harmonic generation imaging and modeling of the optical clearing mechanism in striated muscle and tendon. J Biomed Opt 13(2):021109. https://doi.org/10.1117/1.2907207
9. Genthial R, Beaurepaire E, Schanne-Klein MC, Peyrin F, Farlay D, Olivier C, Bala Y, Boivin G, Vial JC, Debarre D, Gourrier A (2017) Label-free imaging of bone multiscale porosity and interfaces using third-harmonic generation microscopy. Sci Rep 7(1):3419. https://doi.org/10.1038/s41598-017-03548-5
10. Zhang J, Niu C, Ye L, Huang H, He X, Tong WG, Ross J, Haug J, Johnson T, Feng JQ, Harris S, Wiedemann LM, Mishina Y, Li L (2003) Identification of the haematopoietic stem cell niche and control of the niche size. Nature 425(6960):836–841. https://doi.org/10.1038/nature02041
11. Morrison SJ, Scadden DT (2014) The bone marrow niche for haematopoietic stem cells. Nature 505(7483):327–334. https://doi.org/10.1038/nature12984
12. Ding L, Saunders TL, Enikolopov G, Morrison SJ (2012) Endothelial and perivascular cells maintain haematopoietic stem cells. Nature 481(7382):457–462. https://doi.org/10.1038/nature10783
13. Gao X, Xu C, Asada N, Frenette PS (2018) The hematopoietic stem cell niche: from embryo to adult. Development 145(2). https://doi.org/10.1242/dev.139691
14. Lassailly F, Foster K, Lopez-Onieva L, Currie E, Bonnet D (2013) Multimodal imaging reveals structural and functional heterogeneity in different bone marrow compartments: functional implications on hematopoietic stem cells. Blood 122(10):1730–1740. https://doi.org/10.1182/blood-2012-11-467498
15. Kusumbe AP, Ramasamy SK, Adams RH (2014) Coupling of angiogenesis and osteogenesis by a specific vessel subtype in bone. Nature 507(7492):323–328. https://doi.org/10.1038/nature13145
16. Abboud CN (1995) Human bone marrow microvascular endothelial cells: elusive cells with unique structural and functional properties. Exp Hematol 23(1):1–3
17. Pries AR, Secomb TW, Gaehtgens P, Gross JF (1990) Blood flow in microvascular networks. Experiments and simulation. Circ Res 67 (4):826–834
18. Pries AR, Secomb TW, Gaehtgens P (1995) Structure and hemodynamics of microvascular networks: heterogeneity and correlations. Am J Phys 269(5 Pt 2):H1713–H1722
19. Kohler A, Schmithorst V, Filippi MD, Ryan MA, Daria D, Gunzer M, Geiger H (2009) Altered cellular dynamics and endosteal location of aged early hematopoietic progenitor cells revealed by time-lapse intravital imaging in long bones. Blood 114(2):290–298. https://doi.org/10.1182/blood-2008-12-195644

Chapter 10

Assessing Cellular Hypoxic Status In Situ Within the Bone Marrow Microenvironment

Ute Suessbier and César Nombela-Arrieta

Abstract

Hematopoietic stem cells are maintained and regulated in spatially confined microenvironments within the bone marrow, in which oxygen availability is hypothesized to be very limited. The hypoxic nature of HSC niches is proposed to play a fundamental role in the preservation of fundamental stem cell properties through the induction of a distinct glycolytic metabolic profile in HSCs. Thus, the capacity to determine oxygen levels or cellular oxygenation status in specific tissue locations is essential to deepen our understanding of HSC biology. We here describe a methodology to indirectly quantify the hypoxic status of individual cells in situ within histological sections of bone marrow tissues. We employ the well-characterized nitroimidazole probe, pimonidazole, which acts as an oxygen mimetic and irreversibly incorporates into cellular proteins only under hypoxic conditions. The use of fluorescently labeled antibodies that recognize pimonidazole epitopes then enables the indirect assessment of the intracellular hypoxic status and its relationship to cell positioning within the complex tissue topography of the bone marrow.

Key words Hypoxic status, HSC niche, Imaging, Microenvironment, Metabolic state

1 Introduction

The remarkable lifelong capacity of bone marrow (BM) tissues to generate a massive cellular output of all blood lineages on a daily basis relies on the ability to host and maintain a rare but constant pool of hematopoietic stem cells (HSCs) [1]. As founders of the hematopoietic system, multipotent HSCs give rise to mature blood cells while undergoing continuous self-renewal [2]. Unlike their embryonic counterparts, adult HSCs mostly remain in a quiescent state, only infrequently entering cell cycle [3]. The preservation of the fundamental properties of HSCs is critically dependent on the signals provided by neighboring cells residing in specific sites of the BM microenvironment [4]. Recent work has suggested that albeit being highly a vascularized organ, low oxygen availability/hypoxia is a hallmark of BM tissues and in particular of HSC niches [5]. Decreased levels of oxygen stabilize hypoxia-inducible factors

Gerd Klein and Patrick Wuchter (eds.), *Stem Cell Mobilization: Methods and Protocols*, Methods in Molecular Biology, vol. 2017, https://doi.org/10.1007/978-1-4939-9574-5_10,

(HIFs) in HSCs, which in turn drive a transcriptional program that favors a distinct metabolic profile characterized by the preferential usage of anaerobic glycolytic pathways over mitochondrial oxidative phosphorylation [6–9]. This characteristic metabolism is linked to the minimal production of intracellular reactive oxygen species (ROS) derived from oxidative mitochondrial activity, which have been shown to induce DNA damage and reduce HSC fitness [10, 11]. Thus, the existence of metabolic antioxidant mechanisms, which are partly imposed by the hypoxic nature of HSC niches, is of utmost importance to the maintenance of functional HSCs [12, 13].

Recent progress in the identification of intracellular hypoxia as prerequisite for the maintenance of HSCs has strongly fostered the interest of multiple groups in the development of advanced methodologies to measure local oxygen pressures in situ in tissues, intracellular oxygenation, and/or ROS levels in cells while preserving anatomical information, to understand how spatial localization influences cellular status. For instance, the ability to precisely assess the lifetime of oxygen-sensitive phosphorescence probes using intravital two-photon microscopy of the calvarial BM cavity in mice recently provided the first spatially resolved measurements of oxygenation in the vicinity of different perivascular niches [5]. Such analyses are definitively critical to gain insight of the specific stimuli that preserve HSCs in physiological conditions and lead to their functional decline through perturbations of the native microenvironment in which they reside in the context of hematological pathologies.

While not generating a dynamic picture of tissue oxygenation, other strategies have been nevertheless extremely useful in mapping the hypoxic status of individual cells and its relationship to the relative positioning with respect to blood vessels. These methodologies are mostly based on the use of chemical probes that incorporate into cells in an oxygen-sensitive manner, which can be detected and measured, and thus provide indirect and surrogate assessments of the intracellular oxygenation status [14]. Among the most widely employed is the nitroimidazole hypoxic marker, pimonidazole (Pimo). Similar to O_2, Pimo molecules act as acceptors of electrons of the electron transport chain, albeit with lower affinity. Thus, under conditions of limited O_2 availability, Pimo molecules efficiently compete for terminal electrons. The reduction reaction leads to the formation of alkylating agents that form adducts with thiol groups in proteins in an irreversible fashion. The presence and accumulation of these moieties can be detected in individual cells with labeled monoclonal antibodies and quantified with fluorescence-based techniques (imaging or flow cytometry) to generate measurements that inversely correlate with intracellular oxygen pressure [14, 15]. We describe here a protocol to visualize and potentially quantify levels of Pimo incorporation in individual cells within tissue sections of non-decalcified murine BM, in which other cellular or structural components can be additionally detected. The methodology presented is easily amenable to the study of intracellular oxygenation in other tissues, as well as in cellular suspensions.

2 Materials

All reagents mentioned can be prepared at room temperature (RT) and should be stored at 4 °C unless stated otherwise. Follow all specific waste disposal regulations when disposing reagents.

2.1 Hypoxyprobe-1™ Kit

1. Pimonidazole-hydrochloride (Pimo, Hypoxyprobe-1™). Molecular weight of 290.1 and water solubility of 116 mg/ml; provided as a powder. In this form Pimo may be stored for prolonged periods of time at room temperature or at 4 °C, always protected from light.
2. DyLight549-conjugated IgG1 mouse monoclonal antibody (FITC-MAb1) anti-Pimo.

2.2 Fixation Buffer: Paraformaldehyde-Lysine-Periodate (PLP)

1. Monobasic buffer: 0.1 M NaH_2PO_4 in double distilled H_2O.
2. Dibasic buffer: 0.1 M Na_2HPO_4 in double distilled H_2O.
3. PB buffer: prepare by adding three parts dibasic buffer to one part monobasic.
4. PLP: for preparation of 125 ml, weigh 1.785 g L-lysine, and dissolve in 42.5 ml of ultrapure water. Add 31.25 ml of 4% paraformaldehyde, and bring up to 100 ml with PB buffer. Add 0.265 g of sodium periodate. Adjust pH of the solution with dibasic if $pH < 7.0$ or monobasic if $pH > 7.2$. Bring final volume to 125 ml using PB buffer. Keep solution protected from light and store at 4 °C. PLP should be freshly prepared before use.

2.3 Cryopreservation and Sectioning

1. 30% Sucrose/PBS: weigh 30 g sucrose and transfer to a glass container. Fill up with sterile PBS to 100 ml, and stir until sucrose is dissolved. Store at 4 °C.
2. Benchtop liquid nitrogen dewar.
3. Forceps.
4. Disposable plastic cryomolds (24 mm × 24 mm × 5 mm, or similar).
5. Cryopreserving medium optimal cutting temperature (OCT).
6. Cryostat.

7. CryoJane tape transfer system: includes long wave (360 nm) UV light source for pulsed UV illumination, adhesive tape windows, adhesive glass slides, and pressure roller.
8. Steel blade, profile D* (*see* **Note 1**).

2.4 Immunohistology

1. Blocking buffer: 10% donkey serum in PBS. For long-/medium-range storage, keep at 4 °C.
2. Streptavidin/biotin blocking kit.
3. Washing buffer: 0.1% Tween20 in PBS. Store at 4 °C.
4. Humidifying chamber/staining tray.
5. Slide Wheaton Coplin staining jars or staining dishes/racks.
6. Dehydrating ethanol (EtOH) gradient: solutions 70%, 85%, 95%, and 100% EtOH in PBS.
7. Xylene ($\geq$75% xylene isomers).
8. Shaker.
9. Hydrophobic barrier PAP pen.
10. 22 mm × 50 mm (24 mm × 32 mm) glass cover slips.
11. Vectashield™ antifade mounting media.
12. 4′,6-Diamidino-2-phenylindole (DAPI).

2.5 Microscopy and Image Analysis

1. Fluorescence-based imaging system (confocal, fluorescence microscope, or laser scanning cytometer) equipped for detection of DyLight549.

3 Methods

3.1 Bone Marrow Harvesting and Tissue Processing

1. Prepare injectable Pimo solution. Recommended and typically used dosages range from 60 to 120 mg/kg, which for a 25 g mouse would correspond to 1.5–3 mg. We thus advise that Pimo solutions are freshly prepared at 15 mg/ml in PBS, short before injection.
2. 90 min prior to euthanasia, inject mice intraperitoneally with 60–120 mg/kg Pimo.
3. Prepare PLP fixation buffer (*see* **Notes 2** and **3**).
4. Euthanize mice by CO_2 asphyxiation, and proceed immediately to achieve rapid in situ fixation of tissues (*see* **Note 4**).
5. Slowly and gradually perfuse the euthanized animal with PLP via direct cardiac puncture or injection into vena cava (*see* **Note 5**).
6. Collect femoral bones from the kneecap to the hip, and clean thoroughly (*see* **Note 6**).

7. Directly incubate entire femoral bones for 8–12 h in PLP at 4 °C for complete fixation of BM tissues.
8. Thoroughly wash bones by prolonged and repeated incubations in PBS. We recommend at least three washes of 1 h under gentle agitation at 4 °C.
9. After repeated washing, incubate fixed femoral bones in the 30% sucrose rehydrating solution, and incubate for 48–96 h at 4 °C.
10. Take bones out of the rehydrating solution using forceps, and gently dry excess liquid by placing bones in precision wipes or absorbent paper. Place bones in cryomolds with the flattest side of the femur facing down and longitudinally along the longest axis of the mold (*see* **Note 7**). Add OCT slowly to avoid the formation of air bubbles until the whole bone is covered in freezing medium. If bubbles are formed, typically between the bones and the surface of the cryomold, carefully remove them by lifting and gently moving the bones until bubbles are displaced toward the edge of the cryomold.
11. Once the bones are optimally placed, use long metal forceps to grab the cryomold, and place it such that only the lower surface is in contact with the liquid nitrogen (*see* **Note 8**).
12. Once frozen, place the samples in dry ice for temporal preservation, and/or keep at −80 °C for long-term storage.

3.2 Generation of Histological Sections of BM Tissues

1. Set temperature of the cryostat and sample holder to −25 °C. Place knife, adhesive tapes, and slides inside the cryostat at least 30 min before sectioning to allow cooling of the material to appropriate temperatures.
2. Add one drop of OCT on the metal adaptor that will then be fixed to the sample holder, and immediately place sample block on top. The drop will rapidly freeze and glue the sample to the adaptor. Once OCT is solidified, place metal adaptor and sample in the corresponding cryostat sample holder.
3. Start facing block in 30 μm steps until reaching the bone. Make sure the block is cut as homogeneously as possible throughout the entire surface. If the bone has been optimally placed flat in the mold, this will ensure that the sample is initially faced evenly throughout its longitudinal axis (Fig. 1a). Once the bone is reached, reduce the step size (to 10 μm), and gradually cut into the bone discarding all tissue sections until the BM cavity is longitudinally and evenly exposed, including both the metaphyses and the diaphysis.
4. Once the sample is properly faced, BM is exposed, and a clean section is ready to be cut, place the adhesive tape directly on the block making sure that it is straight and no wrinkles are formed.

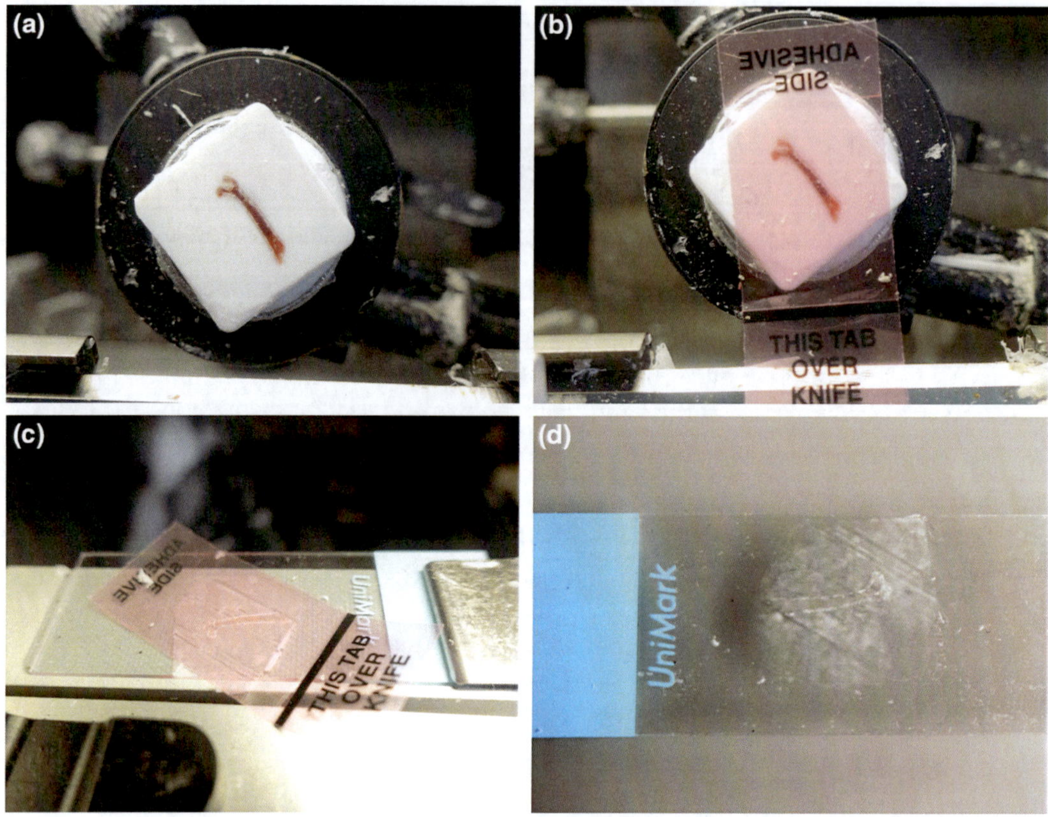

Fig. 1 Generation of histological sections of BM using CryoJane tape transfer system. Representative images of the different steps needed to obtain entire longitudinal BM sections of non-decalcified murine femurs. (**a**) Femurs are placed on the sample holders of the cryostat and iteratively sectioned until an even surface of the entire BM cavity is exposed (red marrow shown); (**b**) a piece of adhesive tape is placed on top of the tissue block, and gentle pressure is applied with a pressure roller to ensure firm adhesion; (**c**) once the tissue is cut with a 8 μm thickness, the section remains adhered to the tape and needs to be transferred to a slide which is pre-coated with an adhesive that is activated by UV light (**d**)

With the use of a roller, apply gentle pressure to ensure that the adhesive tape is properly attached and eliminate trapped bubbles, and facilitate the transfer of the tissue section (Fig. 1b).

5. Hold the loose and longer end of the adhesive tape while cutting a 8 μm (single-cell thick) longitudinal tissue section, which will remain attached to the adhesive tape. Place the tape with the tissue facing down directly on frosted adhesive-coated glass slides. Once again apply pressure with the roller to ensure that the tissue section homogeneously adheres to the surface of the slide and to facilitate the transfer of the tissue section (Fig. 1c).

6. Place the slide on the flash tray, and give 3–5× UV light pulses with intervals of 20 s (approx.). UV will activate the adhesive coated on the slides and will facilitate that the tissue section is entirely transferred from the tape to the slide. Remove the adhesive tape with a continuous and very gentle movement to complete the transfer (Fig. 1d).
7. Repeat as many times as necessary depending on the amount of sections needed for immunostaining. Keep tissue sections in dry ice or −80 °C until the staining procedure is ready to be initiated.

3.3 Immunostaining

1. Place the slides in a container with PBS, and incubate for 2 min followed by incubation in 0.1% Tween20 in PBS for another 2 min to rehydrate the samples.
2. Carefully dry the slides around the slide-mounted sample with a thin precision wipe without touching the tissue section, and draw a circle around the sample with a PAP pen. The hydrophobic barrier formed will keep the liquid within a droplet covering all tissue during the entire staining procedure (*see* **Note 9**).
3. Pipet 100–150 μl of blocking buffer to cover the sample completely, and incubate for 1 h at RT in a humidifying chamber (*see* **Note 10**).
4. If biotinylated antibodies are used, the following blocking step has to be included in the protocol. Cover the sample with streptavidin solution, and incubate for 15 min; rinse the slides with PBS, and then incubate with biotin solution for 15 min.
5. Drip off the solution (do not wash), and gently add 150–200 μl antibody buffer containing the primary antibody mix. The selected antibody cocktail employed in these steps will depend of the purpose of the experiment and will typically include marker combinations for the specific cell types for which the hypoxic status is being investigated, as well as markers for tissue landmarks such as vascular structures. An appropriate negative control slide should always be prepared (*see* **Note 11** and Fig. 2a). Place slides in a humidifying chamber on a shaker under gentle rotation. Incubate for 1 h at RT or overnight at 4 °C.
6. Place slides in staining jar, and wash samples in 0.1% Tween20 in PBS three times for 15 min.
7. Add 150–200 μl staining buffer containing secondary antibody cocktail solution. Include here DyLight549-conjugated IgG1 mouse monoclonal antibody (FITC-MAb1) anti-Pimo (1 μg/ml) (*see* **Note 12**). Incubate for 1 h at RT in humidifying chamber under gentle agitation.

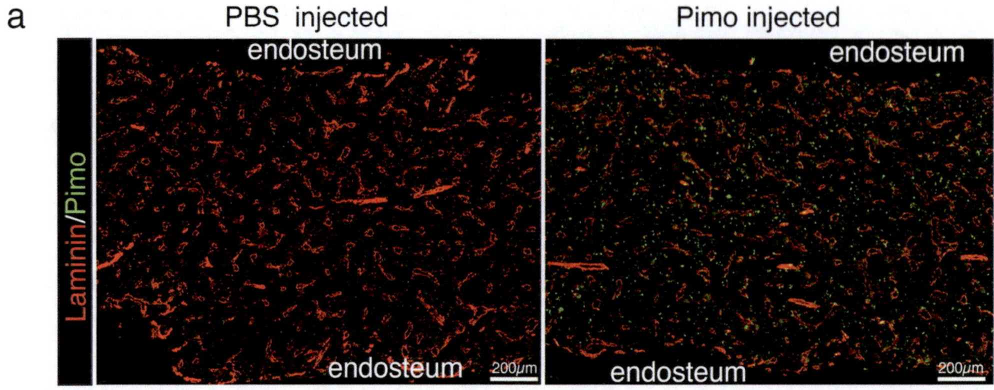

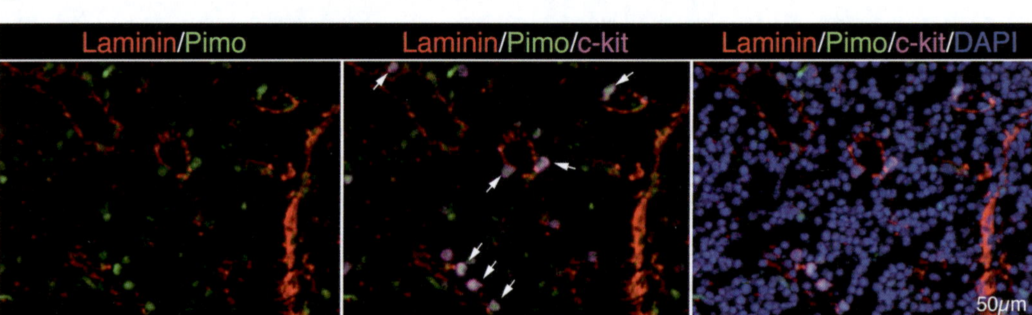

Fig. 2 Representative images of BM tissues stained for the hypoxic label Pimo. (**a**) Low-magnification image of the diaphyseal region of the femoral cavity from control (left) and Pimo-injected mice (right). Sections were stained with anti-Pimo as described here and anti-Laminin to label the outline of BM blood vessels. (**b**) High-magnification pictures of immunostained BM regions obtained with a laser scanning cytometer. Pimo-labeled cells scattered throughout BM parenchyma, the majority of which are c-kit$^+$ perivascular hematopoietic progenitors. Images were originally published in [8]

8. Repeat washing steps in 6.
9. If an additional step using tertiary antibodies is needed, repeat the procedure (**steps** 7 and **8**).
10. Incubate in 0.5–1.0 μM DAPI for 15 min at RT.
11. Place slides in staining jar, and wash samples in 0.1% Tween20 in PBS three times for 15 min.
12. For dehydration use an EtOH gradient: Prepare containers with 70%, 85%, 95%, and 100% EtOH solutions. Slides should be incubated 3 min in each alcohol solution at increasing concentrations. These solutions can be reused, but attention should be placed so that containers are tightly closed to avoid rapid evaporation of alcoholic buffers.
13. Rinse BM sections in Xylene twice for 5 min. Let dry, and remove the remaining fragments of the hydrophobic barrier (dissolved by the Xylene) with a precision wipe.

14. Place a drop of mounting medium on top of the tissue section, and adjust glass cover slip preventing the formation of air bubbles. Slides can be stored at 4 °C until imaged/scanned.

3.4 Imaging and Quantification of Immunostained Pimo Adducts

Imaging will depend on the available microscope setup. In principle any fluorescence-based imaging system for slides will work, preferentially with a motorized stage for the automatic acquisition of multiple fields of view. Visualization of the Pimo-specific signal can be performed and analyzed at different levels depending on the type of readouts sought. Depending on the resolution of the microscope objectives, the multiplexing capacity to analyze multiple fluorescent parameters, as well as on the sophistication of the image analysis pipeline, the analysis will range from the organ-scale distribution patterns of $Pimo^{+}$ cells to the quantification of the Pimo-specific signal within individual cells belonging to defined subsets in local microscopic niches (*see* **Notes 13** and **14**).

4 Notes

1. Frequent maintenance of the steel blade is critical. Blades should be repeatedly sharpened, as the presence of small nicks, dents or scratches dramatically affects the quality of the sections. Wholes, wrinkles or uneven regions in the tissue sections will typically result in artifacts, increased autofluorescence and unspecific deposition of antibodies.
2. As for any fixation procedure, the choice and preparation of reagents will have a critical impact on the quality of the histological sections and immunostaining. We normally employ and recommend pre-diluted PFA at 16% from electron microscopy sciences to prepare PLP. Alternatively, freshly prepared PFA solutions should be used. PLP is a widely employed fixative due to its good preservation of ultrastructure and optimal conservation of carbohydrate moieties of glycoproteins. Proper adjustment of the pH of the fixative between 7.0 and 7.2 is critical, especially when aiming to visualize genetically encoded fluorescent proteins. Although we have not systematically tested the suitability of other fixatives, in our experience fixation with freshly prepared 2% PFA in PBS provides similar quality and results to that of PLP.
3. Lysine is light sensitive and therefore should be stored protected from light. We found that PLP buffer can be stored for only 2–4 days. Fresh preparation of PLP for every experiment is anyways recommended.
4. Although we had the initial concern that CO_2 asphyxiation could result in rapid tissue anoxia and affect the levels of Pimo, we did not observe that the method of euthanasia

influenced Pimo incorporation and profiles in BM cells, as measured by flow cytometry or quantitative imaging.

5. Perfusion of fixative through the lower limbs results in rapid tensioning and stiffening of the muscles. Observation of this phenomenon allows to infer that the fixative has appropriately reached the femoral BM circulation.
6. It is important not to damage the bones during harvesting, as well as to remove surrounding muscle and connective tissues as much as possible. Use scalpel and/or precision wipes to carefully remove connective tissue as this is a typical source of autofluorescence and artifacts during sample preparation and imaging.
7. The epiphyses and metaphyses of the femoral bones are voluminous and stick out compared to the diaphysis making it challenging to find a flat surface along which to place the femur. However, finding the optimal position to position the bone is important for subsequent sectioning steps as it will facilitate exposing the entire BM cavity uniformly in order to get complete tissue sections.
8. Avoid complete dipping of the cryomold as this will result in the formation of air bubbles and uneven freezing. Partial dipping will rapidly result in freezing and solidification of the OCT medium.
9. The use of a humidifying chamber or stain tray is recommended. The PAP pen allows for incubation of samples in minimal volumes that maximize the use of expensive reagents. However, small incubating volumes can rapidly evaporate if samples are incubated in dry environments.
10. All the steps involving the transfer of the tissue section from the adhesive tape to the slide should be performed inside the cryostat chamber to preserve low temperatures.
11. It is absolutely essential to prepare a negative control sample to set baseline background signal and detect the potential appearance of autofluorescence. The most appropriate negative control recommended is a BM section from a control mouse that was not injected Pimo. The bone and tissue section should be prepared in the same experiment and immunostained with the same antibody cocktail as the experimental samples (including anti-Pimo antibody).
12. Although anti-Pimo antibodies are available conjugated to a variety of fluorochromes, we here advise the use of the DyLight549-conjugated antibody because we observed the lowest levels of autofluorescence/background signals in BM tissues within this fluorescence channel.

13. It is important to note that as in any fluorescence-based technique, signal intensity will depend not only on multiple factors related to the preparation of the sample, fixative employed, affinity of the antibody, and intensity of the fluorochrome but also to the sensitivity of the microscopy setup and detection system employed. Using the protocol here outlined, we achieved significant intensities of the Pimo-specific signal in the BM that could be easily detected and quantified above background noise. When possible we nevertheless recommend the acquisition of images using high-sensitivity last-generation detectors found in advanced confocal microscopy setups.

14. Recent studies uncovered that oxygen tensions in BM tissues ranged from 1% to 4%, which are certainly very low when compared to those found in experimental settings in the laboratory but considered as relative normoxic in tissues. Under these conditions we found that levels of Pimo incorporation varied within cell types but were not influenced by location or relative positioning with respect to vascular structures (Fig. 2). This type of Pimo staining pattern differs strongly from that found in solid tumors in which large volumes of avascular tissue are found. In these samples of pathological tissues, Pimo signal inversely correlates with distance to blood vessels, and thus large clusters of cells strongly positive for Pimo can be found in areas devoid of vascularization, in which perfusion is low and oxygenation can presumably reach levels close to those defined as anoxia.

Acknowledgments

This work was supported by the Swiss National Research Foundation (grant number 31003A_159597/1) and an FP7 Marie Curie Career Integration Grant (PCIG13-GA-2013- 618633) from the European Union.

References

1. Nombela-Arrieta C, Manz MG (2017) Quantification and three-dimensional microanatomical organization of the bone marrow. Blood Adv 1(6):407–416
2. Takizawa H, Boettcher S, Manz MG (2012) Demand-adapted regulation of early hematopoiesis in infection and inflammation. Blood 119:2991–3002
3. Nakamura-Ishizu A, Takizawa H, Suda T (2014) The analysis, roles and regulation of quiescence in hematopoietic stem cells. Development 141:4656–4666
4. Crane GM, Jeffery E, Morrison SJ (2017) Adult haematopoietic stem cell niches. Nat Rev Immunol 17(9):573–590
5. Spencer JA, Ferraro F, Roussakis E, Klein A, Wu J, Runnels JM, Zaher W, Mortensen LJ, Alt C, Turcotte R, Yusuf R, Côté D, Vinogradov SA, Scadden DT, Lin CP (2014) Direct measurement of local oxygen concentration in the bone marrow of live animals. Nature 508:269–273
6. Takubo K, Goda N, Yamada W, Iriuchishima H, Ikeda E, Kubota Y, Shima H,

Johnson RS, Hirao A, Suematsu M, Suda T (2010) Regulation of the HIF-1alpha level is essential for hematopoietic stem cells. Cell Stem Cell 7:391–402

7. Parmar K, Mauch P, Vergilio J-A, Sackstein R, Down JD (2007) Distribution of hematopoietic stem cells in the bone marrow according to regional hypoxia. Proc Natl Acad Sci U S A 104:5431–5436
8. Nombela Arrieta C, Pivarnik G, Winkel B, Canty KJ, Harley B, Mahoney JE, Park S-Y, Lu J, Protopopov A, Silberstein LE (2013) Quantitative imaging of haematopoietic stem and progenitor cell localization and hypoxic status in the bone marrow microenvironment. Nat Cell Biol 15:533–543
9. Lassailly F, Foster K, Lopez-Onieva L, Currie E, Bonnet D (2013) Multimodal imaging reveals structural and functional heterogeneity in different bone marrow compartments: functional implications on hematopoietic stem cells. Blood 122:1730–1740
10. Ito K, Hirao A, Arai F, Takubo K, Matsuoka S, Miyamoto K, Ohmura M, Naka K, Hosokawa K, Ikeda Y, Suda T (2006) Reactive oxygen species act through p38 MAPK to limit the lifespan of hematopoietic stem cells. Nat Med 12:446–451
11. Takizawa H, Fritsch K, Kovtonyuk LV, Saito Y, Yakkala C, Jacobs K, Ahuja AK, Lopes M, Hausmann A, Hardt WD, Gomariz Á, Nombela-Arrieta C, Manz MG (2017) Pathogen-induced TLR4-TRIF innate immune signaling in hematopoietic stem cells promotes proliferation but reduces competitive fitness. Cell Stem Cell 21:225–240.e5
12. Zhang CC, Sadek HA (2014) Hypoxia and metabolic properties of hematopoietic stem cells. Antioxid Redox Signal 20:1891–1901
13. Nombela Arrieta C, Silberstein LE (2014) The science behind the hypoxic niche of hematopoietic stem and progenitors. Hematology Am Soc Hematol Educ Program 2014:542–547
14. Krohn KA, Link JM, Mason RP (2008) Molecular imaging of hypoxia. J Nucl Med 49:129S–148S
15. Koch CJ (2008) Importance of antibody concentration in the assessment of cellular hypoxia by flow cytometry: EF5 1and pimonidazole. Radiat Res 169:677–688

Chapter 11

Analysis of Biomechanical Properties of Hematopoietic Stem and Progenitor Cells Using Real-Time Fluorescence and Deformability Cytometry

Angela Jacobi, Philipp Rosendahl, Martin Kräter, Marta Urbanska, Maik Herbig, and Jochen Guck

Abstract

Stem cell mechanics, determined predominantly by the cell's cytoskeleton, plays an important role in different biological processes such as stem cell differentiation or migration. Several methods to measure mechanical properties of cells are currently available, but most of them are limited in the ability to screen large heterogeneous populations in a robust and efficient manner—a feature required for successful translational applications. With real-time fluorescence and deformability cytometry (RT-FDC), mechanical properties of cells in suspension can be screened continuously at rates of up to 1,000 cells/s—similar to conventional flow cytometers—which makes it a suitable method not only for basic research but also for a clinical setting. In parallel to mechanical characterization, RT-FDC allows to measure specific molecular markers using standard fluorescence labeling. In this chapter, we provide a detailed protocol for the characterization of hematopoietic stem and progenitor cells (HSPCs) in heterogeneous mobilized peripheral blood using RT-FDC and present a specific morpho-rheological fingerprint of HSPCs that allows to distinguish them from all other blood cell types.

Key words Mechanical phenotyping, Hematopoietic stem and progenitor cells, Cell mechanics, Microfluidics, Flow cytometry

1 Introduction

Hematopoietic stem and progenitor cells (HSPCs) identified in the bone marrow (BM) have been shown to regenerate all blood cells and are thus used to cure leukemias or lymphomas by BM transplantation [1–7]. The success of such transplantation depends on the reliable identification and isolation of HSPCs from BM aspirates. Flow cytometry is the state-of-the-art method for HSPC identification with surrogate markers, such as the surface antigen CD34. However, there is no causal relation between the presence of these markers and HSPC functions relevant in the context of successful transplantation, such as circulation through the vasculature,

Gerd Klein and Patrick Wuchter (eds.), *Stem Cell Mobilization: Methods and Protocols*, Methods in Molecular Biology, vol. 2017, https://doi.org/10.1007/978-1-4939-9574-5_11,

extravasation, homing to the bone marrow niche, and differentiation [8, 9]. Specific modulation of cell mechanics by cytoskeletal drugs has previously been shown to influence blood cell migration [10]. Moreover, cell mechanics has been demonstrated to directly correlate with HSPC differentiation [11, 12] and to be predictive of differentiation potential of stem cells [13], the latter being not reliably mirrored by surface protein expression itself [14]. This, in addition to the fact that cell mechanics is an inherent property—which obviates the need for time- and cost-consuming labeling with fluorescent and magnetic antibodies, and their potentially adverse effects after transplantation—renders mechanical phenotyping a very promising proposition for improved functional characterization and sorting of HSPCs. Here we discuss how global, mechanical properties, in combination with conventional surface antigens, can be used as specific and functional HSPC markers.

Several experimental techniques to study mechanical properties of cells, such as atomic force microscopy (AFM) [15], micropipette aspiration [16], or the optical stretcher (OS) [17, 18], are currently available. These well-established and validated methods allow for the detection of changes in cellular shape under application of forces of different magnitudes for varying timescales. The force-induced deformation can be recorded and further used to calculate viscoelastic parameters such as the Young's modulus or steady-state viscosity of single cells [19]. Measuring mechanical properties with the OS is sensitive, label-free, and contact-free, at the single-cell level, and has a throughput (100 cells/h) higher than other aforementioned methods. Nonetheless, this measurement rate does not meet clinical application requirements. Therefore, we focus here on the implementation of a much faster method suitable for clinical needs called real-time deformability cytometry (RT-DC) [20]. This technique is based on flowing cells through a narrow constriction in a microfluidic channel, designed to deform cells by shear and normal stress under laminar flow. The hydrodynamic deformation of the cells is detected using a bright-field microscope equipped with a high-speed camera. Images are analyzed in real time at a rate reaching 1,000 cells/s. In addition, an analytical model [19] and numerical simulations [21] were developed to extract an apparent Young's modulus value for every measured cell and thus enable determination of cell elasticity using RT-DC.

In this chapter, we describe the characterization of HSPCs with real-time fluorescence and deformability cytometry (RT-FDC) [22]—an extension of RT-DC, combining 1D-resolved, fluorescence-based parameters with morpho-rheological properties. This extended functionality allows for the analysis of the mechanical phenotype and specific molecular markers of HSPCs at the same time. We present the morpho-rheological fingerprint of HSPCs from healthy human G-CSF-mobilized peripheral blood (PB) obtained with RT-FDC and explain how it can be used to distinguish HSPCs from any other blood cell types.

2 Materials

2.1 Material for RT-FDC Setup Preparation

Consumables necessary for setting up an RT-FDC experiment are listed in Table 1.

The microfluidic chip represents the main part of RT-FDC as illustrated in Fig. 1b. The chip is made from polydimethylsiloxane (PDMS, SYLGARD, 188 Dow, Corning Inc., NY, USA) as described by Herbig et al. [23]. These chips are commercially available from Zellmechanik Dresden with constriction widths ranging from 10 μm to 40 μm (Flic10, Flic15, Flic20, Flic30, Flic40).

2.2 Material for HSPC Sample Preparation and Measurement

1. PBS (phosphate-buffered saline).
2. Ethanol absolute.
3. DEPC water.
4. Ficoll—density 1.077 g/ml.
5. $CD34^{+}$ progenitor isolation kit.
6. APC-antihuman CD34 antibody.
7. Methyl cellulose—measurement buffer (MB): phosphate-buffered saline without magnesium/calcium (PBS−)

Table 1
Consumables necessary for setting up an RT-FDC experiment

Article	Product name, company	Order no.
FEP Tubing	FEP Tubing 1/16″ OD, 0.030″ ID Postnova Analytics, Germany	1520XL
Syringe connector part 1	PEEK Union for 1/16″ OD Tubing, Postnova Analytics, Germany	P-702
Syringe connector part 2	F Luer to 1/4-28 FB, F, Postnova Analytics, Germany	P-658
Sheath/sample fluid syringe	BD Luer-Lok™ 1 ml syringe, BD Biosciences, NJ, USA	613-4971
Syringe for tubing cleaning	BD Disposable Luer-Lok™ tip 5 ml syringe, Henke Sass Wolf, Germany	613-2043
Syringe needle	Blunt Fill Needle 18G, BD Biosciences, NJ, USA	BDAM305180
Syringe filter unit	Millex-GV, 0.22 μm, PVDF, Merck Millipore, Germany	SLGV004SL
Microfluidic chip	Flic10/Flic15/Flic20/Flix30/Flic40 /Zellmechanik Dresden, Germany	–
RT-FDC	Zellmechanik Dresden, Germany [22]	Accelerator + Fluorescence Module

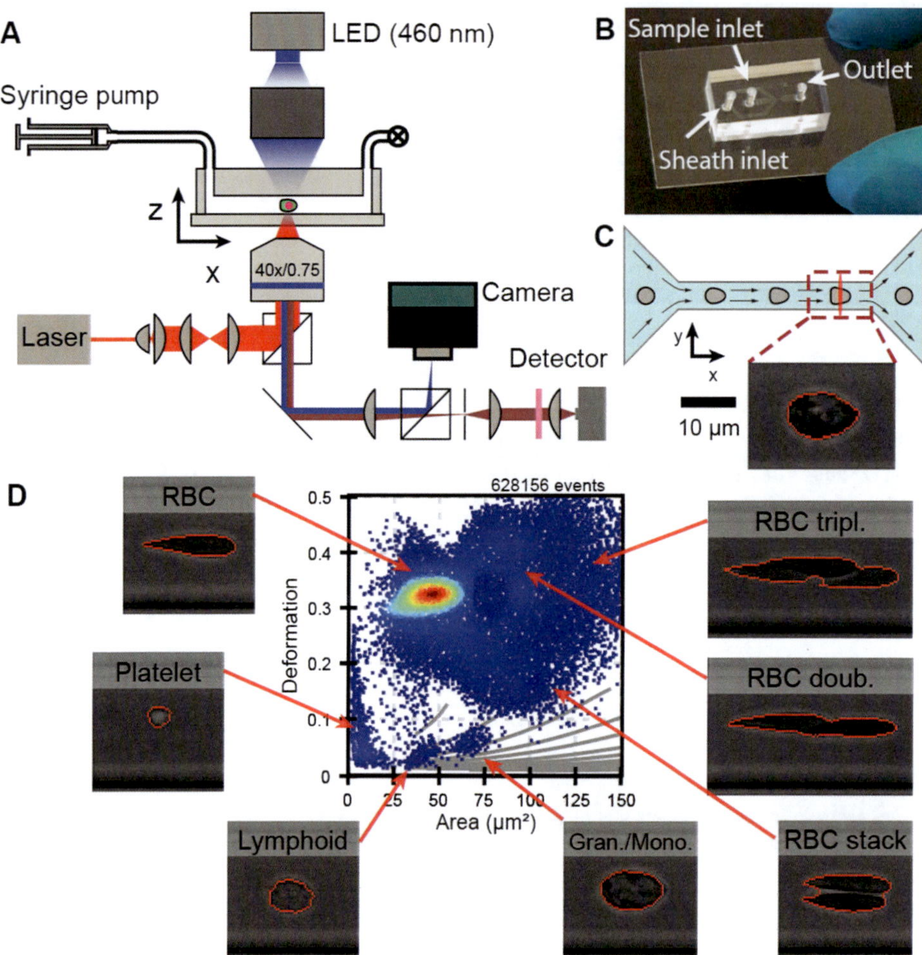

Fig. 1 Real-time fluorescence and deformability cytometry for peripheral blood measurements. (**a**) Experimental setup. Cells are flushed through a microfluidic channel, geometrically designed to deform cells by shear flow. The deformation of the cells is tracked by a high-speed camera in real time at about 1,000 cells/s, and the images can be recorded by a bright-field microscope. Lasers excite and photodiodes measure the fluorescence signal. (**b**) The picture displays a PDMS microfluidic chip for RT-FDC experiments that include connections for sheath (sheath inlet) and cell suspension fluids (sample inlet) as well as an outlet. The picture was done by Sylvi Graupner, kindly provided by Zellmechanik Dresden. (**c**) The sketch shows a representative image of cells entering the channel and reaching the region of interest (dashed rectangle) were the cells are imaged and the contour is detected. (**d**) Scatterplot of peripheral blood measured by RT-FDC. Distribution of cells is represented by the demonstrative density plot obtained from the measurement showing the cross-sectional area (cell size) and the deformation of cells. The gray isoelasticity lines indicate same mechanical properties over different cell sizes. Four distinct populations including red blood cells (RBCs), platelets, lymphocytes, and mono-/granulocytes can be distinguished. Images show representative pictures of all populations obtained from real-time measurement. It is possible to differentiate between different RBC populations: single RBCs, RBC doublets, RBC triplets, and RBC stacks. The isoelasticity lines end at limits of the applicability of the numerical model [21]

complemented with 0.6% (w/v) methyl cellulose. The MB is calibrated to a viscosity of 25 ± 0.6 mPas at 24 °C with a falling ball viscometer (Haake Typ C, Thermo Fisher Scientific Inc.) (*see* **Note 1**).

3 Methods

3.1 Experimental Setup and Principle

RT-FDC combines fluorescence flow cytometry with real-time image analysis for cell mechanical phenotyping (Fig. 1a) [22]. Within the system a laminar flow (0.01–1 μl/s) of sheath and sample medium (MB) is created by a computer-controlled syringe pump at a ratio of 3:1, proven to present the best results for hydrodynamic focusing of cells within the channel (Fig. 1b, c). The microfluidic chip contains a constriction with a typical cross-sectional dimension between 10 and 40 μm. By passing through the constriction in the chip, cells are deformed in a contact-free manner by hydrodynamic shear and normal stresses.

High-speed bright-field microscopy, in tandem with real-time image processing, allows for image capturing and immediate evaluation of the cell deformation. Once a cell is detected in the bright-field image, its cross-sectional area (size in μm^2) and deformation are evaluated based on the assigned cell contour, and the peak maximum, width, and area of the respective fluorescence signal are recorded. The forces that are generated in the channel by the laminar flow of the viscous fluid deform the cells from an isotropic spherical shape to a bullet-like shape. The deformation, *D*, is the main parameter of interest for the mechanical characterization of HSPCs. It is based on the circularity (c) of the cell contour and is defined by the following equation.

$$D = 1 - c = 1 - \frac{2\sqrt{\pi A}}{P},$$

where A is the area enclosed by the convex hull of the detected contour and P the perimeter of this convex hull.

RT-FDC analysis demonstrates that mechanical cell characteristics and cell size are potent discriminators of the major blood cell types (Fig. 1d). When performing RT-FDC on a whole blood sample, along with red blood cells at large deformation, three further cell subpopulations can be distinguished: small and stiff platelets, peripheral blood mononucleated cells, and softer granulocytes [20]. By means of image cell brightness (influenced by intracellular structures), size, and deformation, it is possible to further separate lymphocytes, monocytes, eosinophil, basophil, and neutrophil granulocytes in a single RT-DC measurement of 50 μl blood [24].

In addition to the mechanical phenotype, cell size, cell brightness, and the fluorescence signal recorded during RT-FDC measurements can provide further information about cell type-specific surface protein expression. These combined functionalities allow for unprecedented mechanical characterization of CD34 positive HSPCs in a heterogeneous sample.

Deformation is not the only parameter obtained with RT-DC to describe mechanical properties of the cell. An analytical model developed by Mietke et al. [19] and numerical simulations performed by Mokbel et al. [21] led to a theoretical framework that enables cell Young's modulus assignment based on cell deformation and cell size. This is of particular value, as the deformation of cells measured in RT-FDC is size-dependent and not sufficient for comparison of mechanical properties of cells with different sizes. Together with an image of every single cell, the acquired data are available for multiparametric offline analysis that can further improve the discrimination between different cell types.

3.2 Filter Settings for HSPC Measurement

For PB measurements it is helpful to use filter settings and gating strategies to discriminate between different cell types [21]. RT-FDC provides the following determinants from fluorescence and bright-field images: mean fluorescence intensity, area, deformation, area ratio, and aspect ratio.

1. *Area and deformation filter.*

   For PB a linear range of area and deformation filters can be applied. It is also possible to filter specifically the population of interest with polygon filters. These different types of filters can be created in the open-source analysis software *ShapeOut*. By setting this filter, it is possible to exclude debris, dead or degenerated cells, and stacked cells, similar to gating strategies in conventional flow cytometry.

2. *Area ratio filter.*

   For the correct calculation of deformation or the derived Young's modulus, the detected contour should have a smooth boundary. The *area ratio* filter excludes cells that are not fully convex, by dividing the area of the convex hull of the detected contour by the area of the initially detected contour. According to our experience, setting the *area ratio* filter in the range between 1 and 1.05 or 1.07 is suggested.

3. *Aspect ratio filter.*

   The *aspect ratio* describes the ratio between x length and y length of a detected contour. The *aspect ratio* filter can help to exclude stacked or strongly elongated cells. The acquisition software *ShapeIn II* (Zellmechanik Dresden) and *ShapeOut* allow filtering by *aspect ratio*. For HSPC measurements from PB, values between 0.5 and 2 effectively exclude RBCs, because RBCs are not of interest for later analysis.

3.3 Sample Preparation for HSPC Isolation

G-CSF-mobilized peripheral blood (PB) should be obtained from healthy donors after informed consent (for shown data: ethical approval no. EK221102004) (*see* **Note 2**). Blood sample will be transferred to centrifuge tubes and mixed with PBS.

1. This PBS-diluted blood sample will be carefully coated 2:1 onto a Ficoll layer and centrifuged at 1000 × *g* for 20 min (without brake).
2. Thereafter, the mononuclear cells on top of the Ficoll layer will be separated and washed twice with PBS for further experiments.
3. CD34-positive HSPCs were purified from the mononuclear cell preparations of the leukapheresis samples using CD34 antibody-conjugated magnetic beads according to the manufacturer's instructions.

3.4 RT-FDC Sample Preparation for HSPC Analysis

To prepare HSPCs for the RT-FDC measurement, perform the following steps:

1. Stain HSPCs from PB for 30 min with anti-CD34–APC.
2. Wash cells with PBS and pellet by centrifugation (200 × *g*, 5 min, 23 °C).
3. Resuspend the cell pellet carefully without producing air bubbles in MB.

MB is used to increase the shear stress in the channel constriction allowing for cell deformation under moderate flow rates and to slow down the sedimentation of the cells during the experiment. The final cell concentration for RT-FDC measurements of HSPCs amounts to 1 to 2 × 10^6 cells/ml in MB. 100–1000 μl sample volume is drawn into a 1 ml syringe (*see* **Note 3**).

3.5 Preparation Procedure

1. Switch on the operating equipment: detectors, lasers, microscope, LED lamp, camera, syringe pump, and computer.
2. Start the *ShapeIn II* software (Zellmechanik Dresden) for viewing the current camera image, for controlling the syringe pump, and for recording the measurement. Start a reference move of the syringe pump to calibrate the pump.
3. Before performing the measurement, create a measurement folder to which the data will be saved in *ShapeIn II*.
4. Mount the measurement chip with a selected channel size to the dedicated holder on the microscope stage.
5. Prepare two tubings that are long enough to connect the syringes and the microfluidic chip.
6. Clean sheath and sample flow tubing by flushing them with 5 ml ethanol and subsequently 5 ml deionized water. Blow the tubing dry with compressed nitrogen (*see* **Note 4**).

7. Aliquot 5 ml of 0.6% MB, and filter it through a 0.22 μm syringe filter unit to remove particles and air bubbles from the solution.
8. Fill two 1 ml syringes with the filtered 0.6% MB and remove air bubbles from the syringe.
9. Connect each syringe to a tubing by using a Luer-Lok connector, and fill the tubing with 0.6% MB from the syringe. Make sure that there are no air bubbles present in the syringes and the tubing.
10. Place both syringes in the syringe pump, and connect the tubing of the sheath flow to the sheath inlet to fill the chip. Start the sheath flow within the software with a flow rate of 0.1 μl/s to fill the channel. Immediately after starting the syringe pump, plug an outlet tubing to the fixture.
11. Observe the filling process of the chip within *ShapeIn II*. The chip is filled and bubble-free when a droplet appears at the sample inlet (*see* **Note 5**).
12. Start the sample syringe until a drop emerges at the end of the tubing.
13. Insert the tubing into the sample tube and draw the sample into the tubing by using a negative flow rate of −1 μl/s.
14. Start the sample flow with 0.1 μl/s, and insert the free end of the sample tubing into the middle opening of the chip.
15. When the cells arrive at the channel constriction, set the particular flow rate with the sheath and sample flow ratio of 3:1 (Table 2), and move the region of interest (ROI) to the end of the channel constriction (Fig. 1c). Allow the flow to equilibrate for at least 1 min before starting the measurement.

3.6 Measurement

Classically, two kinds of measurements are performed: the first taken with ROI covers the constriction of the channel to measure cells in the deformed state (bullet-like shape). In the second

Table 2
Measurement flow rate recommendations for 20 μm and 30 μm RT-DC channels and a MB viscosity of 25 mPas

Channel width	Total flow rate	Sample flow rate	Sheath flow rate
20 μm	0.02 μl/s	0.005 μl/s	0.015 μl/s
	0.06 μl/s	0.015 μl/s	0.045 μl/s
	0.18 μl/s	0.045 μl/s	0.135 μl/s
30 μm	0.16 μl/s	0.04 μl/s	0.12 μl/s
	0.24 μl/s	0.06 μl/s	0.18 μl/s
	0.32 μl/s	0.08 μl/s	0.24 μl/s

measurement type, referred to as "reservoir," the ROI is positioned before the channel constriction where negligible hydrodynamic forces affect the cells (isotropic spherical shape).

Setting gates before starting blood measurements is usually performed to exclude invalid events such as cell doublets or larger aggregates, cell debris, or non-intact cells, which were harmed due to handling or other reasons. Two different gating settings can be considered to exclude the acquisition of cellular debris and to narrow down the events of interest.

1. RBC acquisition—Filtering for aspect ratio (*see* Subheading 3.2) values between 2 and 100 effectively includes RBCs. RBCs have in general high aspect ratio values due to their large compliance and their unique shape.
2. Leukocyte acquisition—A standard setting for a 20 μm channel is a minimum height and length of 3 μm, a maximum height of 20 μm, and a maximum length of 80 μm. Furthermore, the aspect ratio is gated to values between 1 and 2 to prevent acquisition of RBCs.

After the cells arrive at the channel constriction and the particular flow rate (Table 2) is adjusted and equilibrated for at least 1 min, the ROI should be set up, and the focus should be adjusted in the way that around the cells a thin bright halo is visible [23]. The acquisition of the cells begins after starting the measurement (*see* **Note 6**). Stop conditions can be preselected before the measurement (manually, time dependent or after specified number of events). Because of their low density (2–3%), to record HSPCs in PB samples, more than 100,000 cells should be recorded. The measurements should be repeated at different flow rates (Table 2) exerting different mechanical stresses. One measurement should be done in the "reservoir" to exclude that the cells are pre-deformed before entering the constriction region. Only when the cells are spherical without any forces, and thus the cross section is circular, the isoelasticity lines can be applied to compare different samples. After the experiments the syringes can be dismounted, and the tubing can be cleaned with ethanol and deionized water [23].

3.7 Data Analysis

RT-FDC data can be analyzed using the freeware *ShapeOut*, provided by Zellmechanik Dresden (https://github.com/ZELLMECHANIK-DRESDEN/ShapeOut). This open-source software enables the analysis of individual cell populations by adding specific filter settings or single-cell inspection by looking at the particular image which is saved for every documented event. The software further allows to derive parameters such as Young's modulus and to apply different statistical models to compare different samples. Moreover, it is possible to export raw or processed data files into the conventional FCS or CSV file format for further comparison or evaluation of the acquired data.

3.7.1 Comparing Mechanical Properties of PB Cell Populations

Cell mechanics is stringently regulated and acts as a quantitative readout for the state of the cytoskeleton [23, 25]. The morpho-rheological phenotype of blood cell populations can be illustrated in a deformation vs. cell size scatter plot that has become a common way to present and compare RT-FDC-generated data (Fig. 1d). Mechanical cell characteristics obtained with this method are potent to discriminate between the major blood cell types; apart from red blood cells at large deformation, three distinct subpopulations were detected: small and stiff platelets, lymphocytes, and granulocytes/monocytes. In the measurement channel, the hydrodynamic shear stress of higher magnitude acts on larger cells as compared to small cells [19, 20, 26]; thus, the cell deformation is size-dependent. With the help of the above-described isoelasticity lines, cells with similar elasticity can be optically identified within the density plots.

To address the question whether morpho-rheology can be a new biomarker for HSPCs, cells were stained with antibodies against $CD34^+$ and measured with RT-FDC. As shown in Fig. 2a, the PB sample includes three subpopulations when CD34-fluorescence is plotted against cell area: a single $CD34^+$ population and two $CD34^-$ populations. The $CD34^-$ population with cell area around 35 μm^2 are lymphocytes, and the $CD34^-$ population with cell area around 70 μm^2 are granulocytes and monocytes (Fig. 2a) according to the populations found in blood measurements (Fig. 1d). Interestingly, the $CD34^+$ cells seem to localize between both $CD34^-$ populations regarding cell area. Rosendahl et al. were already able to show by means of multicolor analysis that $CD34^+$-positive HSPCs clearly distinguish from T cells and monocytes using specific lineage markers [22]. Furthermore they could show that the fingerprint is more than 90% specific for $CD34^+$ cells. After classifying the cells into $CD34^+$ and $CD34^-$ subpopulations on the basis of fluorescence intensity by RT-FDC (Fig. 2b), we obtained deformation versus area plots (Fig. 2c, d), which illustrated $CD34^+$ cells have a mean size of 61.1 ± 1.3 μm^2 and low deformation (0.02–0.05), corresponding to an elastic modulus of around 0.94 ± 0.24 kPa. We confirmed these findings with cells from three different donors. For outliers in the $CD34^+$ population, there was no correlation between CD34 expression levels (fluorescence intensity) and deformation or projected cell area. Compared to the $CD34^+$ fraction (Fig. 2c), $CD34^-$ cells (Fig. 2d) have either a higher elastic modulus like lymphocytes with 1.07 ± 0.38 kPa [20, 21] or a lower elastic modulus like granulocytes/monocytes with 0.908 ± 0.14 kPa and show a more heterogeneous distribution in size.

3.7.2 Data Interpretation

Using morpho-rheological HSPC characterization based on RT-FDC and classical functional readouts will establish the fundamental connection between HSPC cell mechanics and function which is not understood so far. Indeed Gonzalez-Cruz et al. as

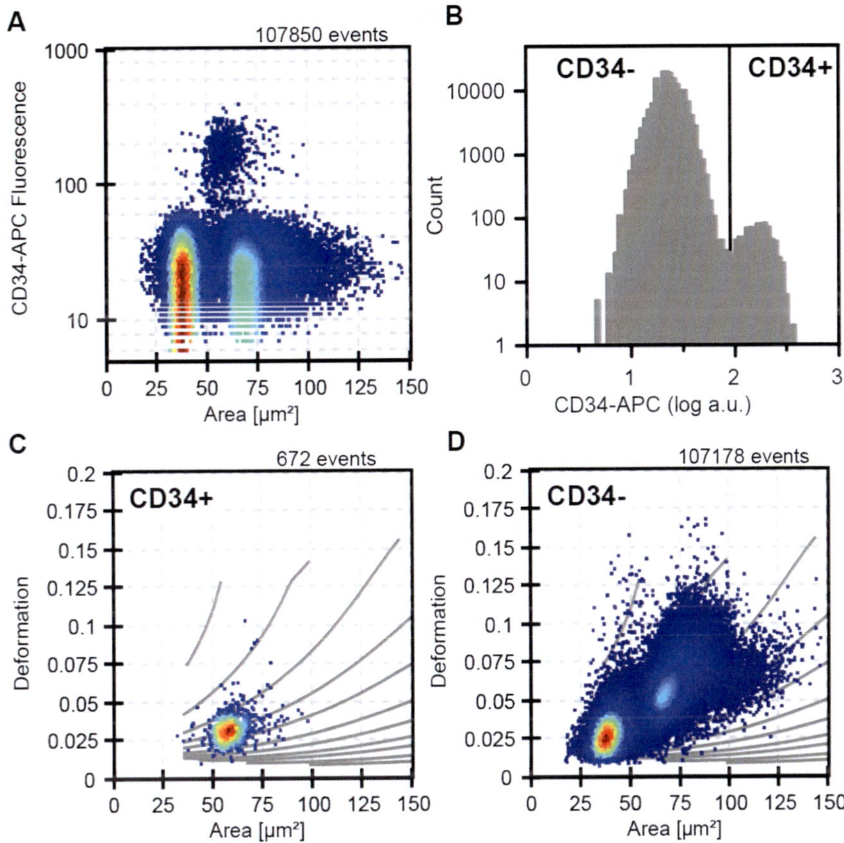

Fig. 2 Morpho-rheological fingerprint of CD34$^+$ HSPCs. HSPCs were stained with anti-CD34 conjugated to allophycocyanin (APC; FL-3 maximum) antibody and measured using RT-FDC. Representative plots of one experiment are shown. Overall three individual experiments were performed. (**a**) CD34-APC fluorescence vs. area scatter plot of PB show three distinct populations. (**b**) Histogram for CD34-APC fluorescence (**c**) CD34$^+$ cells form a homogeneous population regarding deformation and cell area. (**d**) The deformation versus area plot of CD34$^-$ cells shows two populations, lymphocytes and granulo-/monocytes, which are more heterogeneous in deformation and area compared to CD34$^+$ cells

well as Guzniczak et al. showed that mechanical properties are predictive for the differentiation potential of certain stem cells [13, 27]. Furthermore, Ekpenyong et al. pointed out that the mechanical properties of human myeloid precursor cells are essential for timescale relevant processes such as cell migration or circulation, which in turn verifies cell mechanics as a functional marker for this cells [12].

As shown in Fig. 2, there is a specific mechanical phenotype of HSPCs, but it is still unclear if cell mechanics has an impact to the function of HSPCs. RT-FDC is a comprehensive method that firstly combines mechanical characterization of HSPCs and fluorescence as a specific molecular parameter [22]. When analyzing PB with RT-FDC, CD34$^+$ HSPCs emerged as a small and stiff cell population with a unique fingerprint compared to other cells within the sample (Fig. 2c, d). Nevertheless, it is not proven which cells exactly reside in the gate of deformation and area where $CD34^+$ cells are located, as there are $CD34^-$ cells within that region. It is conceivable that these cells are $CD34^-$ stem and progenitor cells, since differentiated granulocytes/monocytes represent a more deformable and larger phenotype and differentiated lymphocytes are smaller. Interestingly, $CD34^+$ HSPCs are exactly in between both populations. It is conceivable to identify these cells without any surface staining or to sort them by means of RT-FDC. This identifies these cells with a unique mechanical phenotype, highlighting cell mechanics as a promising stem cell marker. Cell mechanics is an inherent property of cells, which will allow for label-free isolation of subpopulations from heterogeneous samples. The possibility for sorting in similar microfluidic systems as in RT-FDC, albeit still relying on fluorescence only, has already been demonstrated [28, 29]. Moreover, if the mechanical phenotype of a cell represents its function, it should be possible to identify and sort all functional equal HSPCs within a sample. This will massively increase the success of stem cell transplantation and will reduce severe side effects, due to precise separation of donor material and dosed administration of fully functional characterized cells.

4 Notes

The guidelines written in this chapter illustrate the best procedure to generate comparable results from different donors and measurements. Possible complications that can occur are:

1. A controlled osmolarity (310–330 mOsm/kg) and pH (e.g., 7.4) are recommendable for reproducible measurements.
2. Taking the right anticoagulant: it is advisable to use sodium citrate as the anticoagulant for blood collection. Preliminary tests with EDTA have shown that the cell mechanical properties can already be changed by the anticoagulant.
3. Sedimentation: Especially blood samples tend to sediment quickly despite the high density of the methyl cellulose buffer (MB). Hence, the measurement should be started directly after resuspending the cells. The use of smaller FEP Tubing can be an option to reduce sedimentation in the tubing.
4. During the measurement, cell fragments, cell stacks, dead cells, or dirt may block the channel within the PDMS chip, which can cause a decrease of the measurement frequency. It is absolutely necessary to thoroughly clean the tubing with ethanol and deionized water before the first measurement and to clean them from all liquids with compressed air. Furthermore it is

helpful to filter the cells through a cell strainer before resuspending.

5. Leaking chips should be discarded because the flow rate cannot be controlled. A simple indication for a not well-sealed chip is when liquid leaks out between the glass slide and PDMS. It is absolutely necessary that the chip is completely filled with MB and that there are no air bubbles in the system.

6. No focus and illumination intensity changes while measurement. The focus setting with a slight under focus, which results in a slim bright halo around the cell, has to be found and adjusted before initiation of a measurement (*see* ref. 24).

Acknowledgments

The authors would like to thank Prof. Martin Bornhäuser from the University Hospital Dresden for providing the patients material, Zellmechanik Dresden for providing materials for graphics, and the Microstructure Facility at the Center for Molecular and Cellular Bioengineering (CMCB) at Technische Universität Dresden (in part funded by the State of Saxony and the European Regional Development Fund) and Alejandro Riviera Prieto for help with the production of RT-DC chips. This work was financially supported by the Alexander von Humboldt-Stiftung (Alexander von Humboldt Professorship to J.G.) and the DKMS Mechthild Harf Research Grant (DKMS-SLS-MHG-2016-02 to A.J.).

References

1. Dicke KA, van Noord MJ, Maat B et al (1973) Identification of cells in primate bone marrow resembling the hemopoietic stem cell in the mouse. Blood 42:195–208
2. Abramson S (1977) The identification in adult bone marrow of pluripotent and restricted stem cells of the myeloid and lymphoid systems. J Exp Med 145:1567–1579. https://doi.org/10.1084/jem.145.6.1567
3. Visser JW, Bauman JG, Mulder AH et al (1984) Isolation of murine pluripotent hemopoietic stem cells. J Exp Med 159:1576–1590
4. Bhatia M, Wang JC, Kapp U et al (1997) Purification of primitive human hematopoietic cells capable of repopulating immune-deficient mice. Proc Natl Acad Sci U S A 94:5320–5325. https://doi.org/10.1073/pnas.94.10.5320
5. Bhatia M, Bonnet D, Murdoch B et al (1998) A newly discovered class of human hematopoietic cells with SCID-repopulating activity. Nat Med 4:1038–1045. https://doi.org/10.1038/2023
6. Burt RK, Loh Y, Pearce W et al (2008) Clinical applications of blood-derived and marrow-derived stem cells for nonmalignant diseases. JAMA 299:925–936. https://doi.org/10.1001/jama.299.8.925
7. Cutler C, Antin JH (2001) Peripheral blood stem cells for allogeneic transplantation: a review. Stem Cells 19:108–117. https://doi.org/10.1634/stemcells.19-2-108
8. Arndt K, Grinenko T, Mende N et al (2013) CD133 is a modifier of hematopoietic progenitor frequencies but is dispensable for the maintenance of mouse hematopoietic stem cells. Proc Natl Acad Sci U S A 110:5582–5587. 10.1073/pnas.1215438110/-/DCSupplemental.www.pnas.org/cgi/doi/10.1073/pnas.1215438110
9. Sharma S, Gurudutta GU, Satija NK et al (2016) Stem cell c-KIT and HOXB4 genes: critical roles and mechanisms in self-renewal,

proliferation, and differentiation. Stem Cells Dev 778:755–778. https://doi.org/10.1089/scd.2006.15.755

10. Paschke S, Weidner AF, Paust T et al (2013) Technical advance: inhibition of neutrophil chemotaxis by colchicine is modulated through viscoelastic properties of subcellular compartments. J Leukoc Biol 94:1091–1096. https://doi.org/10.1189/jlb.1012510
11. Lautenschlager F, Paschke S, Schinkinger S et al (2009) The regulatory role of cell mechanics for migration of differentiating myeloid cells. Proc Natl Acad Sci U S A 106:15696–15701. https://doi.org/10.1073/pnas.0811261106
12. Ekpenyong AE, Whyte G, Chalut K et al (2012) Viscoelastic properties of differentiating blood cells are fate- and function-dependent. PLoS One 7(9):e45237. https://doi.org/10.1371/journal.pone.0045237
13. González-Cruz RD, Fonseca VC, Darling EM (2012) Cellular mechanical properties reflect the differentiation potential of adipose-derived mesenchymal stem cells. Proc Natl Acad Sci U S A 109:E1523–E1529. https://doi.org/10.1073/pnas.1120349109
14. Maloney JM, Nikova D, Lautenschläger F et al (2010) Mesenchymal stem cell mechanics from the attached to the suspended state. Biophys J 99:2479–2487. https://doi.org/10.1016/j.bpj.2010.08.052
15. Radmacher M (2007) Studying the mechanics of cellular processes by atomic force microscopy. Methods Cell Biol 83:347–372. https://doi.org/10.1016/S0091-679X(07)83015-9
16. Hochmuth RM (2000) Micropipette aspiration of living cells. J Biomech 33:15–22. https://doi.org/10.1016/S0021-9290(99)00175-X
17. Guck J, Ananthakrishnan R, Mahmood H et al (2001) The optical stretcher: a novel laser tool to micromanipulate cells. Biophys J 81:767–784. https://doi.org/10.1016/S0006-3495(01)75740-2
18. Lincoln B, Wottawah F, Schinkinger S et al (2007) High-throughput rheological measurements with an optical stretcher. Methods Cell Biol 83:397–423. https://doi.org/10.1016/S0091-679X(07)83017-2
19. Mietke A, Otto O, Girardo S et al (2015) Extracting cell stiffness from real-time deformability cytometry: theory and experiment. Biophys J 109:2023–2036. https://doi.org/10.1016/j.bpj.2015.09.006
20. Otto O, Rosendahl P, Golfier S et al (2015) Real-time deformability cytometry as a label-free indicator of cell function. Conf Proc IEEE Eng Med Biol Soc 2015:1861–1864
21. Mokbel M, Mokbel D, Mietke A et al (2017) Numerical simulation of real-time deformability cytometry to extract cell mechanical properties. ACS Biomater Sci Eng 3:2962–2973. https://doi.org/10.1021/acsbiomaterials.6b00558
22. Rosendahl P, Plak K, Jacobi A et al (2018) Real-time fluorescence and deformability cytometry. Nat Methods 15(5):355–358. https://doi.org/10.1038/nmeth.4639
23. Herbig M, Kräter M, Plak K et al (2018) Real-time deformability cytometry: label-free functional characterization of cells. In: Hawley TS, Hawley RG (eds) Flow cytometry protocols. Springer New York, New York, NY, pp 347–369
24. Toepfner N, Herold C, Otto O et al (2018) Detection of human disease conditions by single-cell morpho-rheological phenotyping of blood. elife 7:1–22. https://doi.org/10.7554/eLife.29213
25. Elson EL (1988) Cellular mechanics as an indicator of cytoskeletal structure and function. Annu Rev Biophys Biophys Chem 17:397–430. https://doi.org/10.1146/annurev.bb.17.060188.002145
26. Fletcher DA, Mullins RD (2010) Cell mechanics and the cytoskeleton. Nature 463:485–492. https://doi.org/10.1038/nature08908
27. Guzniczak E, Mohammad Zadeh M, Dempsey F et al (2017) High-throughput assessment of mechanical properties of stem cell derived red blood cells, toward cellular downstream processing. Sci Rep 7:1–11. https://doi.org/10.1038/s41598-017-14958-w
28. Zhu G, Trung Nguyen N (2010) Particle sorting in microfluidic systems. Micro Nanosyst 2:202–216. https://doi.org/10.2174/1876402911002030202
29. Nawaz AA, Chen Y, Nama N et al (2015) Acoustofluidic fluorescence activated cell sorter. Anal Chem 87:12051–12058. https://doi.org/10.1021/acs.analchem.5b02398

Chapter 12

Assessment of Proteolytic Activities in the Bone Marrow Microenvironment

Andreas Maurer, Gerd Klein, and Nicole D. Staudt

Abstract

During cytokine- or chemotherapy-induced hematopoietic stem cell (HSC) mobilization, a highly proteolytic microenvironment can be observed in the bone marrow that has a strong influence on adhesive and chemotactic interactions of HSC with their niches. The increase of proteases during mobilization goes along with a decrease of endogenous protease inhibitors. Prominent members of the proteases involved in HSC mobilization belong to the families of matrix metalloproteinases and cathepsins, which are able to degrade chemokines/cytokines, extracellular matrix components, and membrane-bound adhesion receptors. To determine the functional activity of different proteolytic enzymes, zymographic analyses with different substrates and pH conditions can be employed. An involvement of cysteine cathepsins can be determined by the "active site labeling" technique using a modified inhibitor irreversibly binding to the active center of the enzymes. Intact or degraded chemokines and cytokines, which fall into the range between 1000 and 20,000 Da, can readily be detected by MALDI-TOF analysis. These three methods can help to detect proteolytic activities directly involved in the mobilization process.

Key words Matrix metalloproteinases, Cathepsins, Gelatin zymography, Collagen zymography, Active site labeling, Matrix-assisted laser desorption and ionization time-of-flight mass spectrometry

1 Introduction

In the bone marrow, hematopoietic stem and progenitor cells (HSPC) are located in defined, protecting niches [1–3]. Here HSPC interact with non-hematopoietic niche cells, secreted or membrane-bound cytokines, and a complex extracellular matrix [4, 5]. During the induced mobilization process, the anchorage of HSPC to their niches must be abolished [6]. Key structures for the adhesion of the HSPC to their microenvironment are the vascular cell adhesion molecule-1 (VCAM-1) and the chemokine CXCL12 which bind to the receptors integrin $\alpha4\beta1$ and CXCR4, respectively, on HSPC [7, 8], but certainly additional adhesion receptors are also involved in the retention of the HSPC to their niches. The degradation of these adhesive structures can be

Gerd Klein and Patrick Wuchter (eds.), *Stem Cell Mobilization: Methods and Protocols*, Methods in Molecular Biology, vol. 2017, https://doi.org/10.1007/978-1-4939-9574-5_12,

achieved by a highly proteolytic milieu in the bone marrow which can interfere with the adhesive cell-cell and cell-matrix interactions [9, 10]. High contents of proteolytic enzymes are stored in neutrophil granulocytes, but also osteoblasts, mesenchymal stromal cells and endothelial cells contain different proteases [11, 12] which can be released.

Matrix metalloproteinases (MMPs) are a large family of proteolytic enzymes capable of digesting extracellular matrix components but also different cytokines [13]. The MMP family can be subdivided into collagenases, gelatinases, stromelysins, matrilysins, and membrane-bound MMPs that differ in their substrate specificities [14]. During stem cell mobilization, an involvement of the gelatinases MMP-9 and MMP-2, the collagenase MMP-8, and the membrane-bound MMP-14 (also known as MT1-MMP) has been elucidated [15–18]. MMP-9-deficient mice, however, do not show an impairment of stem cell mobilization indicating that there must be some redundancy in the system [19].

Another large family of proteolytic enzymes are the cysteine cathepsins predominantly found intracellularly in endolysosomal compartments, but some members are also secreted into the extracellular milieu [20, 21]. In the bone marrow, the secretion of cathepsin B, K, L, and X has been detected in osteoblastic cells [22]. These proteases are able to digest the cytokine CXCL12, thereby modulating the important adhesive CXCR4-CXCL12 axis [12, 22].

Two other secreted proteases with a functional involvement in stem cell mobilization are the serine protease cathepsin G and neutrophil elastase that are both released upon stimulation with G-CSF [23].

The functional activities of proteases can be assessed by zymography, a technique based on SDS-polyacrylamide gel electrophoresis (SDS-PAGE) [24]. Depending on the specificity of the protease under study, different substrate zymographies can be employed. Collagen type I, gelatin, and casein are common substrates that are copolymerized with the polyacrylamide matrix [25, 26]. After protein separation by SDS-PAGE, often both the proenzymes and the activated enzymes can be detected since SDS in the gel induces a non-proteolytic activation of the proenzymes. The proteases can degrade the substrate in the gel matrix, and after staining of the gel, their presence becomes visible as cleared zones. Collagen zymography is the preferred method for the collagenases MMP-1, MMP-8, and MMP-13, while gelatin zymography is used for the gelatinases MMP-2 and MMP-9 and also for the acidic cathepsins L and K, and by casein zymography, one can easily detect the matrilysin MMP-7 and the stromelysin MMP-11.

An established method for the detection of cysteine cathepsins in conditioned media or cell lysates is active site labeling with a biotinylated probe that can irreversibly bind to the active center of the enzymes [27]. The proteases are then detected after SDS-PAGE, blotting and labeling with a streptavidin-coupled reagent.

For the detection of proteolytic activities on different chemokines and growth factors, matrix-assisted laser desorption and ionization time-of-flight (MALDI-TOF) mass spectrometry can be employed [28].

This chapter describes in detail protocols used for zymography, active site labeling, and MALDI-TOF analysis as a comprehensive tool for the analysis of different proteolytic enzymes which can be of functional importance in the mobilization process of HSPC out of their niches.

2 Materials

2.1 Materials and Reagents for the Performance of Zymographic Gels

1. SDS electrophoresis chamber with cassettes or glass slides.
2. DC power supply for electrophoresis.
3. 30% acrylamide/0.8% bisacrylamide, store at +4 °C in the dark up to 1 year.
4. Separation gel solution: 1.5 M Tris–HCl pH 8.8/0.4% SDS buffer. Dissolve 181.7 g Tris base in 900 mL of water, then adjust the pH value to 8.8 using 4 M HCl, and then add 4 g sodium dodecyl sulfate (SDS). Adjust the volume with dd H_2O to 1000 mL. Store at +4 °C for up to 1 year.
5. Gelatin solution: dissolve 10 mg/mL in dd H_2O.
6. Collagen type I solution: dissolve 15 mg of rat tail collagen type I in 5 mL 10 mM acetic acid by stirring at 4 °C. This process may take up to 48 h. Store aliquots at −20 °C.
7. Casein solution: dissolve 10 mg/mL in dd H_2O. Eventually you have to add minimal amounts of 0.1 M NaOH to get a clear solution. Store aliquots at −20 °C.
8. 10% APS solution: dissolve 1 g of ammonium persulfate in 10 mL dd H_2O, and store at +4 °C for 1 month.
9. TEMED: *N,N,N′,N′*-tetramethylethylenediamine. Store at +4 °C in the dark.
10. Isopropanol (>97%).
11. Stacking gel solution: 0.5 M Tris–HCl pH 6.8/0.4% SDS buffer. Dissolve 60 g Tris base in 900 mL of water, adjust the pH value with 4 M HCl to 6.8, add 4 g SDS, and fill up to 1000 mL. Store at +4 °C.
12. SDS running buffer: 50 mM Tris base, 380 mM glycine, and 0.1% SDS.
13. 10 mM 4-aminophenylmercuric acetate (APMA) in dimethylsulfoxide.

14. Activation buffer for MMP zymogens: 50 mM Tris–HCl pH 7.5, 200 mM NaCl, 10 mM $CaCl_2$, and 50 μM $ZnCl_2$. For activation add 1 mM APMA to the activation buffer.
15. Nonreducing loading buffer (6× concentrate): 1.25 mL 0.5 M Tris–HCl pH 6.8, 1.2 g SDS, 6 mL glycerol (v/v), 6 mg bromophenol blue, and add H_2O to a final volume of 10 mL. Do not add any reducing agents such as dithiothreitol (DTT) or mercaptoethanol.
16. Molecular weight size marker for nonreducing gels.
17. Renaturation buffer for MMPs: 50 mM Tris–HCl pH 7.5, 2.5% Triton X-100 (v/v).
18. Renaturation buffer for acidic activities: 65 mM Tris–HCl pH 7.4, 20% glycerol.
19. Developing buffer for MMPs (neutral proteolysis): 50 mM Tris–HCl pH 7.5, 150 mM NaCl, 10 mM $CaCl_2$, 1 μM $ZnCl_2$, 0.015% Brij35.
20. Developing buffer for acidic activities (sodium phosphate buffer pH 6.0): prepare 500 mL of 0.1 M NaH_2PO_4 solution, adjust the pH to 6.0 using Na_2HPO_4, and then add 19 mg EDTA per 500 mL buffer (1 mM EDTA). Before usage freshly add 60 mg DTT per 200 mL buffer (1.5 mM DTT).
21. 0.25% Coomassie brilliant blue (CBB) staining solution: Dissolve 2.5 g CBB R250 powder in 45% MeOH/10% acetic acid (combine 450 mL MeOH, 450 mL H_2O, and add 100 mL glacial acetic acid). Can be stored at 20–25 °C (RT).
22. CBB destaining solution: 45% MeOH/10% acetic acid. Store at RT.

2.2 Materials and Reagents for Active-Site Labeling

1. Serum-free conditioned cell culture media and cell lysates in RIPA lysis buffer without protease inhibitors: 50 mM Tris–HCl pH 7.2, 150 mM NaCl, 0.1% SDS, 1% sodium deoxycholate, and 1% Triton X-100.
2. Ultrafiltration spin columns with a 10 kDa cutoff.
3. DCG-04, 5 mM stock solution in DMSO. Store aliquots at −20 °C.
4. E-64 protease inhibitor and 5 mM stock solution in DMSO. Store at −20 °C.
5. 2-(*N*-morpholino) ethanesulfonic acid (MES) buffer.
6. Complete active site labeling buffer (I) consisting of 50 mM citrate buffer pH 5.0/5 mM DTT/2.5 mM EDTA.
7. Complete active site labeling buffer (II) consisting of 25 mM sodium acetate buffer pH 3.5/5 mM DTT/2.5 mM EDTA.
8. 10× concentrated 50 mM dithiothreitol (DTT) stock solution, diluted in 25 mM sodium acetate or 50 mM citrate buffer, and aliquots stored frozen at −20 °C.

9. Loading buffer containing DTT (6× concentrate): Use nonreducing loading buffer (*see* Subheading 2.1, **item 13**) and freshly add 10 mM DTT.
10. SDS-polyacrylamide gels (8–12% separation gels).
11. Running buffer: 50 mM Tris pH 8.6, 380 mM glycine, 0.1% SDS.
12. PVDF (polyvinylidenfluoride) membranes.
13. Transfer buffer for blotting: 25 mM Tris base, 142 mM glycine, 20% methanol (pH 8.3).
14. Protein- and biotin-free blocking solution for Western blot [e.g., Roti® Block (Carl Roth, Karlsruhe, Germany) diluted according to the manufacturer's instructions].
15. 10× TBS: dissolve 24.2 g Tris base and 80 g NaCl in 950 mL of dd H_2O, and adjust the pH to 7.0 using 4 M HCl. Fill up to 1000 mL.
16. 1× TBST20: dilute 10× TBS 1:10 in water and add 10 mL of 10% Tween-20 solution.
17. Streptavidin-horseradish peroxidase conjugate, 1:1000.
18. Enhanced chemiluminescence (ECL) reagents.

2.3 Materials and Reagents for MALDI-TOF Analysis

1. MALDI-TOF instrument (e.g., Reflex IV or Autoflex, Bruker Daltonics, Bremen, Germany).
2. Target plate (e.g., Anchorchip target, Bruker Daltonics).
3. DHB matrix solution: 10 mg 2,5-dihydroxybenzoic acid dissolved in 333 μL acetonitrile and 666 μL of 0.1% aqueous trifluoroacetic acid.
4. Protein standard for calibration (8–25 kDa range).
5. Recombinant human cathepsins CatX, CatB, CatK, and CatL.
6. SDF-1α and SDF-1ß: stock solutions of 100 ng/μL in dd H_2O. Store aliquots at −80 °C.
7. Box with crushed ice.
8. Incubation buffers: 50 mM sodium citrate buffer pH 5.0 for CatB, MES buffer/5 mM DTT for CatK and CatL, and 25 mM sodium acetate buffer pH 3.5/5 mM DTT for CatX according to the manufacturer's instructions for best activity of the cathepsins (*see* **Note 1**).
9. Reaction tubes for incubation (e.g., PCR tubes).
10. Tabletop centrifuge suitable for the reaction tubes.
11. Incubator/thermo cycler (*see* **Note 2**).
12. 10% trifluoroacetic acid solution.

3 Methods

3.1 Detection of Matrix Metalloproteinases or Cathepsins by Different Zymographies

Assembling gels for zymography analysis, you start with pouring the separation gel. After its polymerization the stacking gel is poured on top of the separation gel. Immediately insert a suitable sample comb between the glass plates without creating bubbles. After polymerization, the protease-containing nonreduced samples can be run on the gel.

1. For the separation gel, use an 8–12% polyacrylamide gel depending on the molecular weight of the investigated proteases. For MMPs of the gelatinase subfamily, we recommend using a 12% gel consisting of 11 mL of 30% acrylamide/0.8% bisacrylamide, 6.9 mL 1.5 M Tris–HCl pH 8.8/0.4% SDS, 2.75 mL gelatin (0.1% in total), 6.8 mL dd H_2O, 91.5 μL 10% APS, and 19 μL TEMED. When preparing the mixture, make sure to add the APS and TEMED as the last reagents as this step starts the polymerization of the gels.
2. For collagen and casein zymography, all reagents for the separation gel are identical with gelatin zymography, except the gelatin solution. This solution is replaced by collagen type I or casein solutions, respectively (*see* **Note 3**).
3. Directly add 2 mL of isopropanol on top of the non-polymerized separation gel to achieve an even gel surface.
4. After polymerization of the separation gel (which takes about 20–30 min), you can remove all isopropanol, and carefully wash the separation gel surface with dd H_2O to get rid of any residual isopropanol. Then pour off all H_2O before adding the stacking gel solution. The stacking gel consists of 1.65 mL of 30% acrylamide/0.8% bisacrylamide, 3.5 mL 0.5 M Tris–HCl pH 6.8/0.4% SDS, 7.65 mL dd H_2O, 62.5 μL APS, and 12.5 μL TEMED.
5. Zymogens of MMPs are activated by incubation of the proteases (or conditioned media or cell lysates) with 1 mM APMA in activation buffer for 3 h at 37 °C (*see* Fig. 1a).
6. The samples containing the proteases are run under nonreducing conditions on the SDS-polyacrylamide gels containing the copolymerized substrates (*see* **Note 4**). Do not boil your samples. Instead, unfold the enzymes by incubating them at 55 °C for 5 min prior to loading onto the gels. Make sure to use a suitable molecular weight size marker side-by-side to your samples in nonreducing loading buffer to identify your proteases of interest by their specific molecular weights.
7. After running the gels at 120 V for approximately 1 h, carefully take them out from in-between the glass or plastic plates, and wash them three times in renaturation buffer for 10 min. This

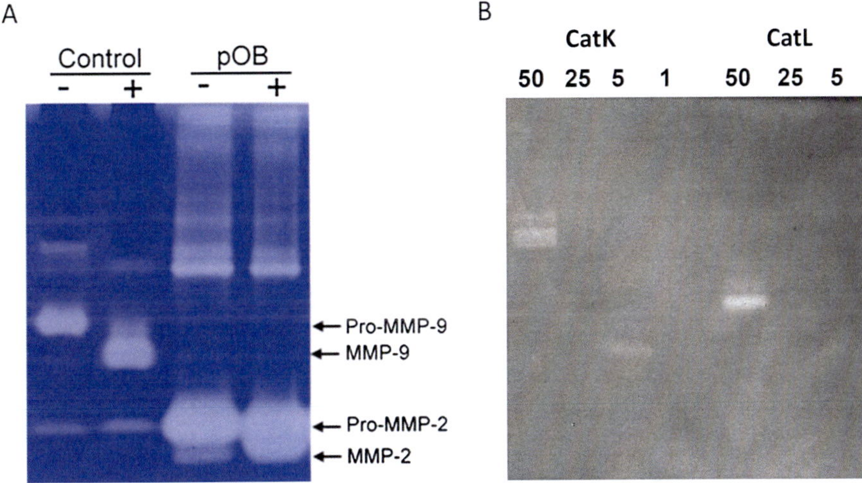

Fig. 1 Gelatin zymography at neutral pH and acidic pH. (**a**) Clear zones in the CBB-stained gelatin gel indicate the presence and activity of gelatinolytic matrix metalloproteinases. In this example, conditioned media from human primary osteoblasts (pOB) reveal pro-MMP-2 and MMP-2 expression. In the control, blood serum samples of G-CSF-mobilized patients were used to indicate positions of the pro- and active forms of MMP-2 and MMP-9. Samples that were developed at neutral pH were directly analyzed (−) or after activation with 1 mM APMA (+). (**b**) For the detection of cysteine cathepsins, different amounts (50 ng, 25 ng, 5 ng, and 1 ng) of the recombinant proteases cathepsin K (CatK) and cathepsin L (catL) were used with (50 ng) or without (25 ng, 5 ng, 1 ng) the leupeptin inhibitor. Pro-cathepsin K and cathepsin L are affected by autoproteolysis, and their gelatinolytic activity can only be seen in the presence of leupeptin

step is crucial to get rid of the SDS in the gel (*see* **Note 5**). Then incubate the gels from overnight to 3 days in the developing buffer with soft agitation.

8. Gels are stained in CBB staining solution for 1 h. Afterward the gels are transferred to the destaining solution for 10 min (or more, if needed) until the desired staining, and the best contrast is reached. The proteolytic bands should appear as clear light zones according to the molecular weight of your protease of interest (*see* Fig. 1). Based on the proteolytic degradation of the substrate in this area, the zones appear clear, whereas your background with the intact substrate appears dark blue in color (*see* **Note 6**).
9. Analysis: If you already know what protease you are investigating, we strongly recommend performing a side-by-side analysis next to your recombinant protease (if available). It might be extremely helpful to try different buffers according to the pH and buffer optima of the different proteolytic families. Using buffers of different pH values might also help if you are getting too many bands per sample (*see* **Note 7**). To verify your results, perform Western blotting side-by-side using specific antibodies to your protease of interest.

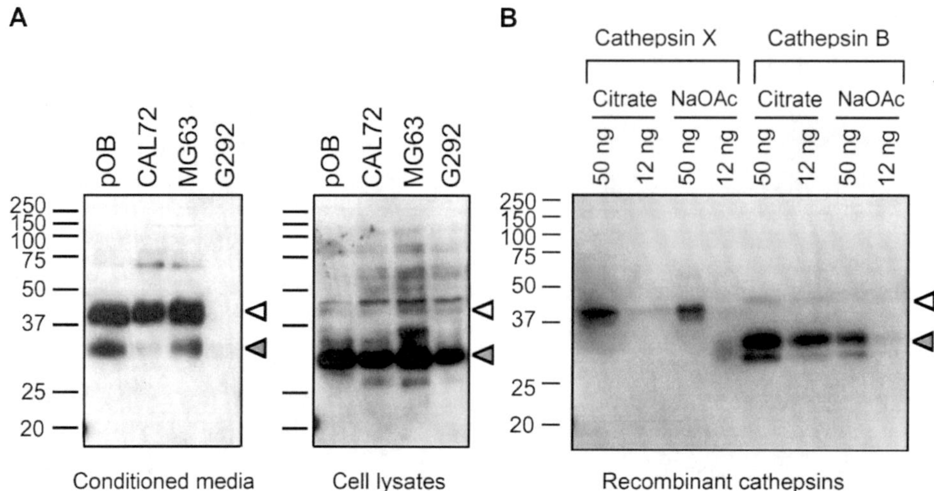

Fig. 2 Active site labeling of recombinant cathepsins and cathepsins in cell culture-conditioned media versus cell lysates. (**a**) In the lysates of different human osteoblastic cells (pOB) and cell lines (CAL72, MG63, G292) or in their cell culture supernatants, different cysteine cathepsins can be detected in each individual sample. (**b**) Labeling of human cathepsin X or cathepsin B with DCG-04 was performed both in sodium acetate buffer pH 3.5 and in citric acid buffer pH 5.0 using either 50 ng or 12 ng of the purified recombinant protease. Using different labeling buffers, one can easily distinguish between the different cathepsin proteases. In the example shown, pro-cathepsin B at acidic pH is auto-proteolytically processed to its active form (filled triangles), and only minimal amounts of the pro-form are still present. In contrast, pro-cathepsin X and the unfolded cathepsin X will still appear at the same molecular weight (open triangles)

3.2 Active-Site Labeling (ASL): Detection of Several Members of a Protease Family by Their Active Sites in Cell Lysates or Cell-Conditioned Media

The biotinylated activity-based probe DCG-04 as an analog of the E-64 broad-spectrum inhibitor of cysteine cathepsins, which irreversibly binds to the active center of the proteases, can be used to detect several cysteine cathepsins at once in a sample. We used this method to characterize the proteolytic microenvironment in the osteoblastic bone marrow niche [22] (*see* Fig. 2a).

For a strong detection of the proteases, the sample volume to be labeled has to be adjusted for different sample types (e.g., conditioned media, cell lysates). Then the sample is incubated with the labeling reagent DCG-04. The labeled enzymes are separated by SDS-PAGE, transferred onto PVDF membranes, and detected with streptavidin-peroxidase. The protease bands are visualized by chemiluminescence staining.

3.2.1 Preparation of Samples (Proteases in Conditioned Media or Recombinant Proteases) and Labeling of Cysteine Cathepsins with DCG-04

1. For the preparation of conditioned media, cultivate 3.5×10^6 cells in 10 mL of serum-containing medium in regular cell culture flasks or 10 cm cell culture dishes. The next day adherent cells are carefully washed twice with 15–20 mL serum-free medium before adding 10 mL fresh serum-free medium for conditioning. After 2 days, analyze the cells for viability by microscopy to make sure they can tolerate serum starvation. Collect the media and centrifuge at $10{,}000 \times g$ to get rid of

dead or floating cells. Concentrate the conditioned media by a factor of ten using ultrafiltration spin columns with a cutoff of 10 kDa.

2. For the preparation of cell lysates, carefully wash cells twice in PBS, then add 1 mL of RIPA lysis buffer per 3.5×10^6 cells, and incubate for at least 1 h at +4 °C. Use a cell scraper to remove adherent cells, and then pipet the cell lysate into a fresh microcentrifuge tube. Spin down cell lysate at $10{,}000 \times g$ for 10 min at +4 °C to get rid of cell debris. Put the cleared lysate into a fresh tube. If storage for further assays is desired, we recommend to divide your lysates in smaller aliquots, snap-freeze them on dry ice, and keep the samples at −80 °C.
3. For active site labeling of cysteine cathepsins in the concentrated culture media and cell lysates, we recommend a volume of 10 μL of 10× conditioned medium and 2–5 μL of the lysates. However, optimal volumes and protein concentrations have to be adjusted by each user and depend on the cell type under study (*see* Fig. 2a).
4. Label your sample in a total volume of 23 μL using 50 μM DCG-04 in 25 mM sodium acetate/2 mM EDTA buffer containing 20 μM DTT at +37 °C for 30 min.
5. When using purified recombinant cathepsin proteases, we recommend to label them under the same buffer conditions as the cell lysates or cell culture media (*see* **Note 8**). Several of the used cysteine proteases have similar molecular weights and running characteristics in the SDS-PAGE as zymogens or active proteases. It therefore might be hard to distinguish them by molecular weight. Using recombinant cathepsins an amount of 50 ng protease per labeling reaction seems to be an appropriate amount leading to good signal intensities (*see* Fig. 2b).

3.2.2 Separation of Active-Site Labeled Cysteine Cathepsins in Samples on SDS-PAGE, Blotting, and Detection of the Cysteine Proteases

1. Run 10 μL of each sample in an SDS-PAGE gel at 120 V.
2. Soak a PVDF membrane in methanol for 30 s, transfer it to dd H_2O for 5 min, and incubate it in transfer buffer.
3. Transfer the separated proteins from the gel onto a PVDF membrane by semidry blotting at 150 mA for 90 min.
4. Block nonspecific binding to the PVDF membrane using a blocking reagent for at least 1 h.
5. Incubate the membrane for 1 h with streptavidin-HRP while shaking.
6. Wash your blots thoroughly changing the TBST-20 solution at least five times. We recommend an overnight incubation with shaking in the TBST-20 washing solution (*see* **Note 9**). The next day, development with enhanced chemiluminescence reagents can be performed (*see* **Note 10**).

3.3 Characterization of Proteolytic Activities by MALDI-TOF Analysis

To test if a cytokine might be a substrate of a given protease, the recombinant cytokine can be incubated with the recombinant protease under suitable conditions and analyzed for degradation using MALDI-TOF mass spectrometry (*see* Fig. 3a, b).

3.3.1 Degradation of Cytokines

1. Prepare a sufficient number of reaction tubes on ice to set up the incubation reactions with a total volume of 6 μL. Label your tubes properly, and always make sure to have your undigested control without protease side-by-side to your protease-containing samples for comparison and to test stability of your protein in the buffer used.
2. For kinetic observations prepare a master mix of your cytokine in reaction buffer to make sure the concentrations are even in the tubes. In the example shown in Fig. 3a, 3 μM SDF-1β was

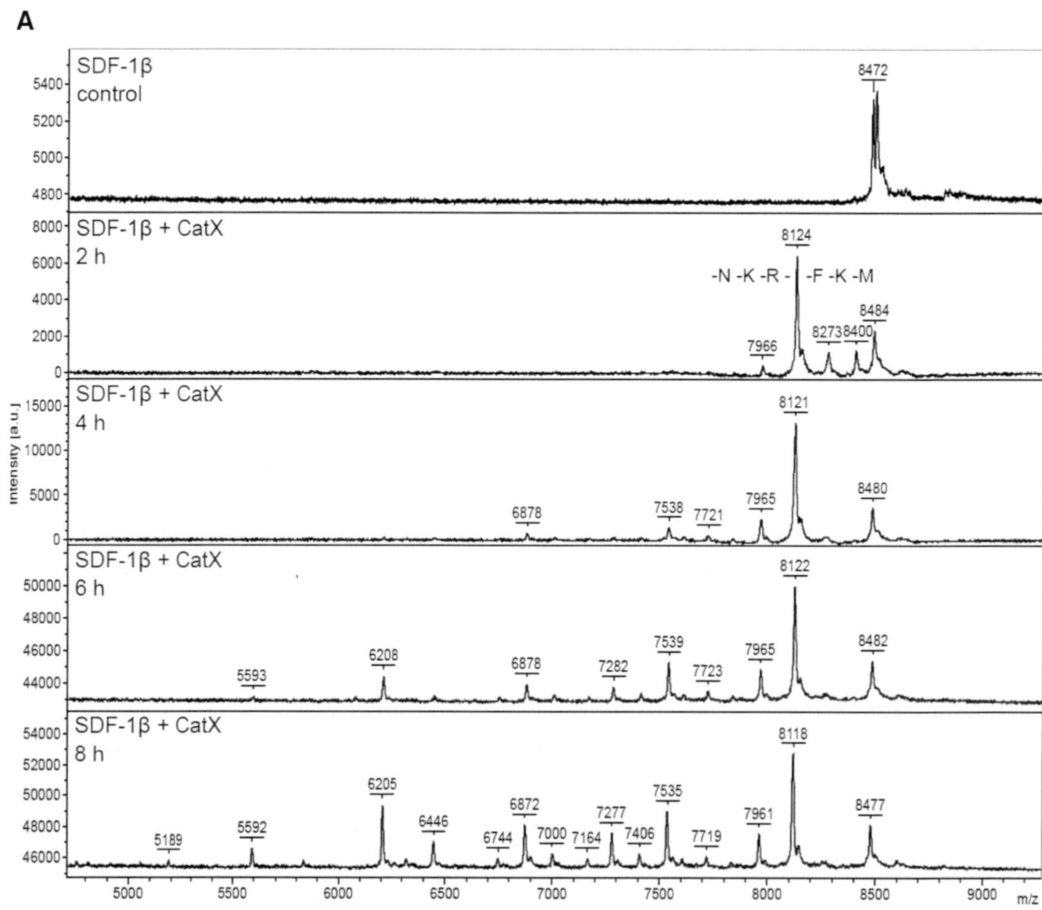

Fig. 3 (**a**) MALDI-TOF analysis of the full-length SDF-1β and its cleavage products. Over time, the carboxy-monopeptidase cathepsin X cleaves single amino acids at the carboxy-terminal end in SDF-1β. (**b**) MALDI-TOF analysis of SDF-1α with different cathepsins. The cytokine SDF-1α was digested with cathepsin K, B, and X. MALDI-TOF analysis clearly revealed different digestion mechanisms

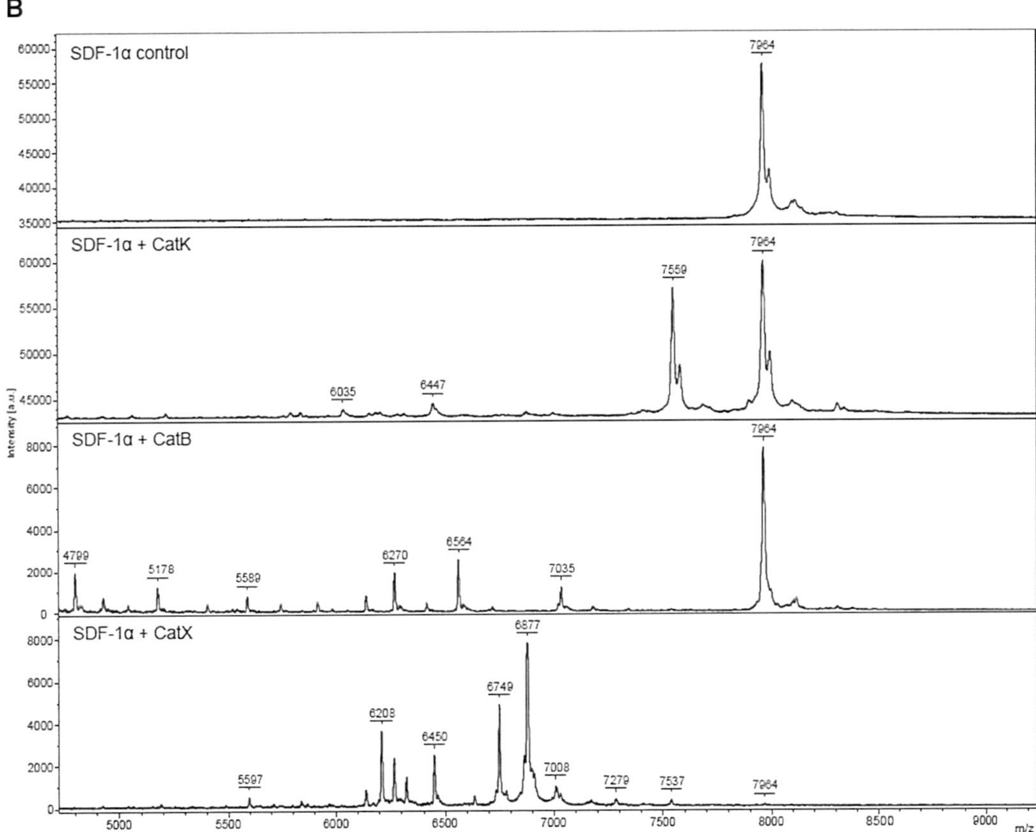

Fig. 3 (continued)

incubated with 0.38 μM CatX. For each sample a volume of 1.56 μL SDF-1β (100 ng/μL) was diluted with 1.93 μL sodium acetate reaction buffer pH 3.5, and 2.5 μL pre-diluted (1:12) recombinant CatX was added for each reaction. Make sure that your cytokine in the control sample has the same concentration.

3. After mixing distribute the master mix to different reaction tubes.
4. Spin your reaction tubes.
5. Transfer the tubes to the thermocycler and incubate at a temperature fitting the optimal temperature of your enzyme. For CatX, in the example, a temperature of 30 °C was used (*see* **Note 2**).
6. After different time points (e.g., 2 h, 4 h, 6 h, and 8 h), stop the reaction in a single tube by adding trifluoroacetic acid to a total of 1%, and store the sample at −20 °C until MALDI-TOF measurement is performed.

3.3.2 Analysis by Mass Spectrometry

1. Apply 2 μL of 2,5-dihydroxybenzoic acid (DHB) matrix solution onto a target spot for each sample. Also prepare a spot for the protein calibration standard, as well as a matrix-only spot to detect artifacts or contamination.
2. Add 1 μL of sample to the spot, mix by pipetting, and let it dry (*see* **Note 1**).
3. Load the target into the mass spectrometer.
4. Select an adequate protocol for the expected protein size. The precise settings (laser energy, voltages, delays, etc.) depend on your individual mass spectrometer. Linear time-of-flight mode is recommended to avoid background from post-source decay.
5. Measure the protein standard and calibrate the mass spectrometer (m/z to time-of-flight).
6. Measure the samples using the selected method. DHB forms needle-like crystals and a heterogeneous spot. Look for sites within your sample that give high-quality spectra (so-called sweet spots) and average (sum) a good amount of such spectra to get a representative overview.
7. Save the summed spectrum, and open it within the analysis application.
8. Manually pick the visible protein peaks to analyze their corresponding masses (average m/z) (*see* **Note 11**).
9. Make a table of detected masses and identify the undigested cytokine and the cleaved forms.
10. Use the amino acid table (*see* Table 1) and the protein sequence from the datasheet to identify the cleaved (terminal) amino acids (*see* **Note 12**).

4 Notes

1. For setting-up your degradation reaction for MALDI-TOF, the best choice is to use volatile buffer substances such as NH_4OAc or NH_4HCO_3, since salts interfere with crystallization of the matrix. If buffers with high content of nonvolatile salts are used, additional desalting (e.g., using ZipTips) might be required to achieve good crystallization, desorption, and ionization, depending on your mass spectrometry setup.
2. Using small reaction tubes and a heated incubator lid prevents drying of the sample and thus suboptimal reaction conditions.
3. For collagen zymography a concentration of 0.3 mg/mL rat tail collagen type I in the separation gel is recommended. 0.2 mg/mL casein in the separation gel is used for casein zymography.

Table 1
Expected mass loss for individual amino acids

Amino acid	*m/z* difference	Amino acid	*m/z* difference	Amino acid	*m/z* difference	Amino acid	*m/z* difference
-A-	71	-C-	103	-D-	115	-E-	129
-F-	147	-G-	57	-H-	137	-I-	113
-K-	128	-L-	113	-M-	131	-N-	114
-P-	97	-Q-	128	-R-	156	-S-	87
-T-	101	-V-	99	-W-	186		

4. Before you run your zymography gel, make sure to use the right assay conditions considering the pH optima of your proteases in the buffers. It is advantageous knowing the characteristics of your proteases of interest when it comes to activation, stability, and pH/temperature optima to choose the right buffers. It is also advantageous knowing the detection limit. If the proteases of interest are available as recombinant proteins, test them first in your assay, and vary the loading to see how sensitive the assay is in your hands.
5. Make sure your renaturation time is long enough to get rid of the SDS in your gel. It is very important to reach nonreducing conditions for proper activities.
6. If you do not get proper activity, consider that you may have to add additional reagents (e.g., matrix metalloproteinases need metal ions as cofactors, acidic pH is important for cysteine cathepsins). If the gels are run at low temperature, the gel performance is improved.
7. If you get too many signals in your zymogram and you cannot identify all the proteases in your data set, it is recommended to use more stringent buffers.

   If you, for example, want to exclude matrix metalloproteinase activity, make sure to use EDTA in your activity buffer. You can also add protease-specific inhibitors to the buffer. If the inhibitor binds covalently to the protease you are intending to suppress, add the inhibitor directly to the sample before running the gel. Make sure always having the non-blocked sample side-by-side to your blocked sample.
8. For the identification of a specific protease activity, we recommend running the sample under investigation next to the recombinant protease, if available.
9. If you have a lot of background on your blots after active site labeling, prolong and intensify your shaking by increasing the shaking speed and duration. Optionally you can extend your

washing steps of the very same experiment even after you have already added ECL reagents. Wash then again, and add fresh ECL substrate for a second time for higher clarity of chemiluminescent bands corresponding to your proteases of interest.

10. To verify that the signals obtained are protease-specific, you can run replicates of the samples that you pre-incubated (blocked) with the non-biotinylated E-64 inhibitor before labeling. Any signal also obtained in the E-64-pretreated samples can be considered nonspecific, e.g., due to endogenously biotinylated proteins such as pyruvate carboxylase interfering with the streptavidin detection system.
11. In the example shown in Fig. 3, m/z ratios were measured. Typically small proteins mainly acquire one charge and are present as $[M + H]^+$ ion under the conditions used, so m/z directly corresponds with molecular weight. Double-charged molecules only appear with low intensity.
12. First measure the undigested cytokine and the matrix-only spot to identify the cytokine peak. Verify the correct mass from the datasheet, and compare it to the calculated mass from online tools, e.g., ExPASyProtParam. Oxidation of methionine (+16 Da) might occur to variable extent under storage or incubation conditions.

References

1. Morrison SJ, Scadden DT (2014) The bone marrow niche for haematopoietic stem cells. Nature 505:327–334
2. Crane GM, Jeffery E, Morrison SJ (2017) Adult haematopoietic stem cell niches. Nat Rev Immunol 17(9):573–590
3. Wei Q, Frenette PS (2018) Niches for hematopoietic stem cells and their progeny. Immunity 48(4):632–648
4. Gattazzo F, Urciuolo A, Bonaldo P (2014) Extracellular matrix: a dynamic microenvironment for stem cell niche. Biochim Biophys Acta 1840(8):2506–2519
5. Klamer S, Voermans C (2014) The role of novel and known extracellular matrix and adhesion molecules in the homeostatic and regenerative bone marrow microenvironment. Cell Adhes Migr 8(6):563–577
6. Greenbaum AM, Link DC (2011) Mechanisms of G-CSF-mediated hematopoietic stem and progenitor mobilization. Leukemia 25 (2):211–217
7. Papayannopoulou T, Craddock C, Nakamoto B, Priestley GV, Wolf NS (1995) The VLA4/VCAM-1 adhesion pathway defines contrasting mechanisms of lodgement of transplanted murine hemopoietic progenitors between bone marrow and spleen. Proc Natl Acad Sci U S A 92(21):9647–9651
8. Greenbaum A, Hsu YM, Day RB, Schuettpelz LG, Christopher MJ, Borgerding JN, Nagasawa T, Link DC (2013) CXCL12 in early mesenchymal progenitors is required for haematopoietic stem-cell maintenance. Nature 495(7440):227–230
9. Levesque J-P, Hendy J, Takamatsu Y, Williams B, Winkler IG, Simmons PJ (2002) Mobilization by either cyclophosphamide or granulocyte colony-stimulating factor transforms the bone marrow into a highly proteolytic environment. Exp Hematol 30 (5):440–449
10. Marquez-Curtis L, Jalili A, Deiteren K, Shirvaikar N, Lambeir AM, Janowska-Wieczorek A (2008) Carboxypeptidase M expressed by human bone marrow cells cleaves the C-terminal lysine of stromal cell-derived factor-1alpha: another player in hematopoietic stem/progenitor cell mobilization? Stem Cells 26:1211–1220
11. Levesque J-P, Takamatsu Y, Nilsson SK, Haylock DN, Simmons P (2001) Vascular cell

adhesion molecule-1 (CD106) is cleaved by neutrophil proteases in the bone marrow following hematopoietic progenitor cell mobilization by granulocyte colony-stimulating factor. Blood 98(5):1289–1297

12. Staudt ND, Maurer A, Spring B, Kalbacher H, Aicher WK, Klein G (2012) Processing of CXCL12 by different osteoblast-secreted cathepsins. Stem Cells Dev 21(11):1924–1935
13. Overall CM (2002) Molecular determinants of metalloproteinase substrate specificity: matrix metalloproteinase substrate binding domains, modules, and exosites. Mol Biotechnol 22 (1):51–86
14. Page-McCaw A, Ewald AJ, Werb Z (2007) Matrix metalloproteinases and the regulation of tissue remodelling. Nat Rev Mol Cell Biol 8(3):221–233
15. Janowska-Wieczorek A, Marquez LA, Dobrowsky A, Ratajczak MZ, Cabuhat ML (2000) Differential MMP and TIMP production by human marrow and peripheral blood CD34(+) cells in response to chemokines. Exp Hematol 28(11):1274–1285
16. Steinl C, Essl M, Schreiber TD, Geiger K, Prokop L, Stevanovic S, Pötz O, Abele H, Wessels JT, Aicher WK, Klein G (2013) Release of matrix metalloproteinase-8 during physiological trafficking and induced mobilization of human hematopoietic stem cells. Stem Cells Dev 22(9):1307–1318
17. Shirvaikar N, Marquez-Curtis LA, Shaw AR, Turner AR, Janowska-Wieczorek A (2010) MT1-MMP association with membrane lipid rafts facilitates G-CSF—induced hematopoietic stem/progenitor cell mobilization. Exp Hematol 38(9):823–835
18. Golan K, Vagima Y, Goichberg P, Gur-Cohen S, Lapidot T (2011) MT1-MMP and RECK: opposite and essential roles in hematopoietic stem and progenitor cell retention and migration. J Mol Med (Berl) 89 (12):1167–1174
19. Levesque JP, Liu F, Simmons PJ, Betsuyaku T, Senior RM, Pham C, Link D (2004) Characterization of hematopoietic progenitor mobilization in protease-deficient mice. Blood 104:65–72
20. Brix K, Dunkhorst A, Mayer K, Jordans S (2008) Cysteine cathepsins: cellular roadmap to different functions. Biochimie 90 (2):194–207
21. Reiser J, Adair B, Reinheckel T (2010) Specialized roles for cysteine cathepsins in health and disease. J Clin Invest 120 (10):3421–3431
22. Staudt ND, Aicher WK, Kalbacher H, Stevanovic S, Carmona AK, Bogyo M, Klein G (2010) Cathepsin X is secreted by human osteoblasts, digests CXCL-12 and impairs adhesion of hematopoietic stem and progenitor cells to osteoblasts. Haematologica 95:1452–1460
23. Lapidot T, Petit I (2002) Current understanding of stem cell mobilization: the roles of chemokines, proteolytic enzymes, adhesion molecules, cytokines, and stromal cells. Exp Hematol 30:973–981
24. Vandooren J, Geurts N, Martens E, Van den Steen PE, Opdenakker G (2013) Zymography methods for visualizing hydrolytic enzymes. Nat Methods 10:211–220
25. Inanc S, Keles D, Oktay G (2017) An improved collagen zymography approach for evaluating the collagenases MMP-1, MMP-8, and MMP-13. BioTechniques 63(4):174–180
26. Yasumitsu H (2017) Serine protease zymography: low-cost, rapid, and highly sensitive RAMA casein zymography. Methods Mol Biol 1626:13–24
27. Fonović M, Bogyo M (2007) Activity based probes for proteases: applications to biomarker discovery, molecular imaging and drug screening. Curr Pharm Des 13(3):253–261
28. Cho YT, Su H, Wu WJ, Wu DC, Hou MF, Kuo CH, Shiea J (2015) Biomarker characterization by MALDI-TOF/MS. Adv Clin Chem 69:209–254

Chapter 13

Analysis of the Complement Cascade Activation During Mobilization of Hematopoietic Stem/Progenitor Cells

Anna I. Grabowska and Jakub M. Hawryluk

Abstract

It has been shown that the complement cascade is involved in the process of mobilization of hematopoietic stem cells, from their niche in the bone marrow to the peripheral blood. Based on this knowledge modulation of complement, cascade activation may enable the development of better mobilization strategies for poorly mobilizing patients. Herein we present a mobilization protocol in mice model, useful for studying the effect of the complement activation in the mobilization process.

Key words Mobilization, Complement cascade, Hematopoietic stem cells, G-CSF, Innate immunity

1 Introduction

Hematopoietic stem/progenitor cells (HSPCs) are retained in bone marrow (BM) niches, and under steady-state conditions, only a small number of these cells are constitutively released into the peripheral blood (PB) [1]. During stress situations like infection, tissue injury, strenuous exercise, or pharmacologically induced mobilization, HSPCs egress in higher numbers from the BM microenvironment into PB [2]. It has been shown that the mobilization of HSPCs involves a granulocyte and monocyte burst to induce an activation of the complement cascade (ComC) that results in sterile inflammation in BM [1, 3] and consequently activates the ComC. The complement system is a part of the innate immune response which is characterized by an enzymatic proteolytic cascade that releases cleaved C fragments (Fig. 1) [4]. Mariusz Ratajczak's group could demonstrate that from the classical, alternative, and mannan-binding lectin (MBL) pathways, the latter two are the most important in inducing mobilization of HPSCs from BM into PB [5].

Anna I. Grabowska and Jakub M. Hawryluk contributed equally to this work.

Gerd Klein and Patrick Wuchter (eds.), *Stem Cell Mobilization: Methods and Protocols*, Methods in Molecular Biology, vol. 2017, https://doi.org/10.1007/978-1-4939-9574-5_13, © Springer Science+Business Media, LLC, part of Springer Nature 2019

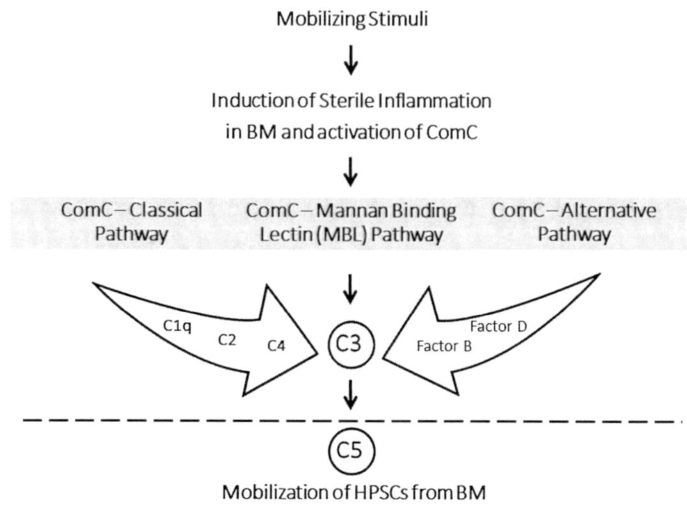

Fig. 1 ComC system contribution to mobilization of HPSCs from BM, reproduced and modified from [1]. ComC activation leads to cleavage of several bioactive fragments like C1q, C2–C5. Mobilization of HSPCs is triggered by activation of the ComC, which leads to a sequence of events which releases cells from the BM involving the generation of ComC cleavage fragments

Further studies have demonstrated that while ComC induces HPSC mobilization, the proximal and distal parts of this cascade show opposite effects. Analysis of murine models with C3 or C5 deficiency has shown that while C3 cleavage fragments increase retention of HSPCs in BM, C5 fragments have the opposite effect [5–7]. On the basis of these results, one can state that C3-deficient mice are easy mobilizers and C5-deficient mice, in contrast, are poor mobilizers. Additionally, the Cq1-knockout mice (with disruption of the classical pathway of ComC) exhibited normal mobilization of HSPCs. Clinical mobilization is performed in order to harvest stem cells for hematopoietic autologous or allogenic transplantation [2, 8], and efficiency of this process is critical for a successful therapy. At the molecular level, the disruption of the signaling ligand-receptor SDF-1-CXCR4 and/or VCAM-1-VLA-4 axes is responsible for the release of HSPCs from BM niches into PB [1]. The most widely applied and safe mobilizing agent is the cytokine granulocyte colony-stimulating factor (G-CSF) that induces sterile inflammation in BM microenvironment [9]. Other agents such as AMD3100 [10, 11], which blocks CXCR4 or zymosan in mice [4], are also employed in this process.

In this chapter we will provide standardized protocols to induce and assess the mobilization of HPSCs in mice. In brief, following mobilization, we measure in PB (1) the total number of white blood cells (WBC), (2) clonogenic colony-forming unit granulocyte/macrophage (CFU-GM) progenitors, and (3) the number of mobilized HPSCs (Sca-1$^+$ CD45$^+$ Lin$^-$ cells) and SKL cells (Sca-1$^+$ c-Kit$^+$ Lin$^-$ cells). We also show typical examples of images of flow cytometry analyses and CFU-GM colony assays.

2 Materials

2.1 Animals

The following protocol should be performed in compliance with all institutional, national, and international guidelines for animal care.

Depending on the species used in the experiment, animals should be free from pathogens, with matched sexes. Transgenic strains of mice with knocked-out genes associated with complement cascade are useful tools for analysis of the ComC system (*see* **Note 1**). Loss of function of specific genes could provide valuable information about the involvement of particular proteins in complement cascade activation.

Some examples of transgenic animals with complement defects (all these mice are available from the Jackson Laboratory, Bar Harbor, Maine, USA) are shown below:

C1 knockout mice.

B6N(Cg)-$C1qa^{tm1b(EUCOMM)Wtsi}$/3J.

MBL knockout mice.

B6.129S4-$Mbl1^{tm1Kata}$ $Mbl2^{tm1Kata}$/J.

C3 knockout mice.

B6;129S4-$C3^{tm1Crr}$/J.

2.2 Reagents and Equipment

1. G-CSF—granulocyte colony-stimulating factor (Neupogen 960 μg/ml). Store at +4 °C in a glass container.
2. 70% ethanol.
3. FBS—fetal bovine serum, store at −20 °C.
4. PBS—phosphate-buffered saline. Store at room temperature (RT).
5. Penicillin/streptomycin solution 100× concentration.
6. Human methylcellulose complete medium (components in 100 ml medium): 1.4% methylcellulose (1500 cps) in Iscove's Modified Dulbecco's Medium; 25% fetal bovine serum, 2% bovine serum albumin, 2 mM L-glutamine, 5×10^{-5} M 2-mercaptoethanol, 50 ng/ml recombinant human stem cell factor (SCF), 10 ng/ml recombinant human GM-CSF, 10 ng/ml recombinant IL-3, and 3 IU/ml recombinant human erythropoietin (Epo).
7. CFU-GM colony formation medium: human methylcellulose complete medium with 1% penicillin/streptomycin and L-glutamine. To prepare the medium for CFU-GM colony

formation, add 0.5 ml penicillin/streptomycin solution (100×) and 1 ml of 200 mM L-glutamine solution to 100 ml human methylcellulose complete medium (listed above). Store at +4 °C. Before using warm up to +37 °C.

8. RPMI-1640 medium with 2 mM L-glutamine.
9. Cell culture medium: RPMI +2% FBS—add 1 ml FBS to 49 ml RPMI-1640; store at +4 °C. Before usage warm up to RT.
10. IL-3—interleukin-3. Stock 5 ng/μl. Dissolved in PBS. Store in −80 °C.
11. hGM-CSF—human granulocyte-macrophage colony-stimulating factor. Stock 5 ng/μl. Dissolved in PBS. Store in −80 °C.
12. L-Glutamine 200 mM solution in dd H_2O. Store at −20 °C.
13. Red blood cell lysing buffer 10× (containing 1.5 M ammonium chloride, 0.1 M potassium carbonate, and 1 mM EDTA). Store at +4 °C. To prepare a working concentration of lysing buffer, add 5 ml red blood cell lysing buffer to 45 ml nanopure water, and store at +4 °C. Before using warm up to RT.
14. 0.5 M EDTA (ethylene-diamine-tetra-acetate) diluted in dd H_2O, filtered. Store in 4 °C.
15. Turk's solution.
16. Appropriate antibodies for flow cytometry, and store in +4 °C (*see* **Note 2**).
17. Microvette tubes.
18. Insulin syringes.
19. Tissue culture 12-well plates.
20. Refrigerated centrifuge.

3 Methods

3.1 Mobilization of HSPCs in Mice

Mice were subcutaneously injected with 200 μg/kg human G-CSF once daily for 3 or 6 consecutive days (*see* **Note 3**).

1. Transfer G-CSF to Eppendorf tube.
2. Prepare the appropriate amount of G-CSF, and dilute with PBS buffer to a final volume of 100 μl solution. Transfer the solution into 1 ml syringe.
3. Remove air bubbles from the syringe before subcutaneous injection.
4. 6 h after the last G-CSF administration, withdraw blood from the retro-orbital plexus of the mouse for hematology analysis (*see* Subheading 3.2), or bleed mouse from main vein (*vena cava*) for flow cytometry and colony formation analyses.

3.2 Blood Sampling

For an estimated weight (approx. 25 g) of a mouse, it is recommended to withdraw not more than 200 μl of blood at a single bleeding.

1. Bleed the mouse from retro-orbital plexus (for the analysis of white blood cells—WBC).
2. Place 50 μl of blood in EDTA-coated Microvette tubes (*see* **Note 4**).
3. Vortex tubes gently.
4. Perform mice euthanasia (*see* **Note 5**).
5. Disinfect the mouse with 70% ethanol.
6. Place the mouse on the pad and fix the front paws with pins.
7. Cut the mouse from the crotch area to the chest.
8. Unhook the *vena cava*.
9. Gently tap an insulin syringe with 70 μl EDTA into vein, and collect the blood (*see* **Note 6**).
10. Transfer blood into a 5 ml tube, add 1 ml of red blood cell lysing solution,and mix well.

3.3 Hematological Analysis

After HSPCs' mobilization, the level of white blood cells needs to be determined. Mobilizing agents should elevate WBC level; therefore WBC level should be higher than in a control group treated with PBS (*see* Fig. 2). Samples need to be analyzed within 1 h after collection.

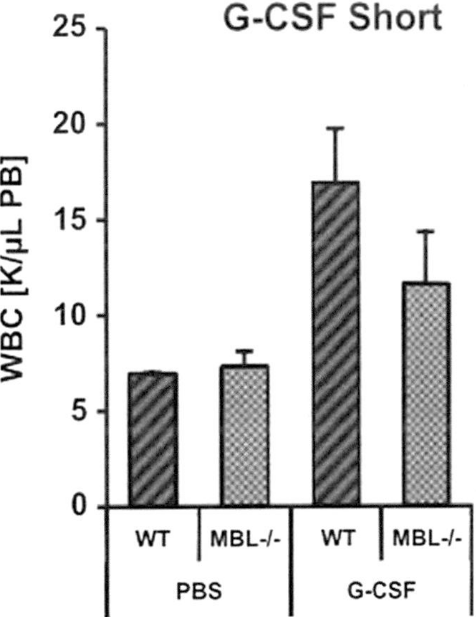

Fig. 2 Effect of inhibition of the coagulation cascade during mobilization of HSPCs in mannan-binding lectin (MBL)-deficient mice. WBC level after short mobilization with 100 μg/kg G-CSF per 3 days. Control mice were injected with PBS. MBL-deficient mice were compared to WT mice

1. Place a Microvette tube, which contains blood and EDTA, in veterinary hematology analyzer Hemavet.
2. Set analysis of WBC.

3.4 Blood Lysis

1. Fill the tube with lysing solution for lysis of red blood cells.
2. Incubate for 10 min at RT.
3. Centrifuge blood for 10 min with 600 × *g* at +4 °C.
4. Decant the supernatant.
5. Add 3 ml lysing solution to the tube and the precipitated cell pellets.
6. Incubate for 10 min at RT.
7. Add 2 ml RPMI +2% FBS.
8. Centrifuge for 10 min with 600 × *g* at +4 °C.
9. Decant the supernatant.
10. Resuspend the cell pellets in 1 ml RPMI +2% FBS.
11. Take 5 μl of cell suspension and add to 95 μl of Turk's solution (*see* **Note 7**).
12. Fill the cytometer tube with RPMI +2% FBS.
13. Centrifuge for 10 min with 600 × *g* at +4 °C.
14. While samples are being centrifuged, you can count the cells (*see* **Note 8**).
15. Decant the supernatant.
16. Suspend cells in 300 μl RPMI +2% FBS.
17. Take 1–2 × 10^6 cells for CFU-GM colony analysis, transfer to 5 ml tube (*see* **Note 9**), and use the rest of cells for flow cytometry analysis.

3.5 Flow Cytometry Analysis

1. In order to check the efficiency of performed mobilization process, identify the presence of HSCs and SKL cells. Listed antibodies (*see* **Note 2**) are designed for the detection of cell surface markers and clusters of characteristic differentiation antigens. HSCs are commonly characterized by the absence of lineage-specific marker expression. Mouse hematopoietic stem cells are considered Sca-1^+, $CD45^+$, Lin^- and SKL cells Sca-1^+, c-Kit^+, and Lin^- (*see* Fig. 3). Take the rest of cells, and with RPMI+2% FBS, bring the volume up to 300 μl.
2. Add mixture of antibodies to the tube with cells and mix well.
3. Incubate for 30 min at +4 °C in the dark.
4. Centrifuge cells for 10 min with 600 × *g* at +4 °C.
5. Decant the supernatant.
6. Wash pellet in 1 ml PBS, and fill the tube with PBS.

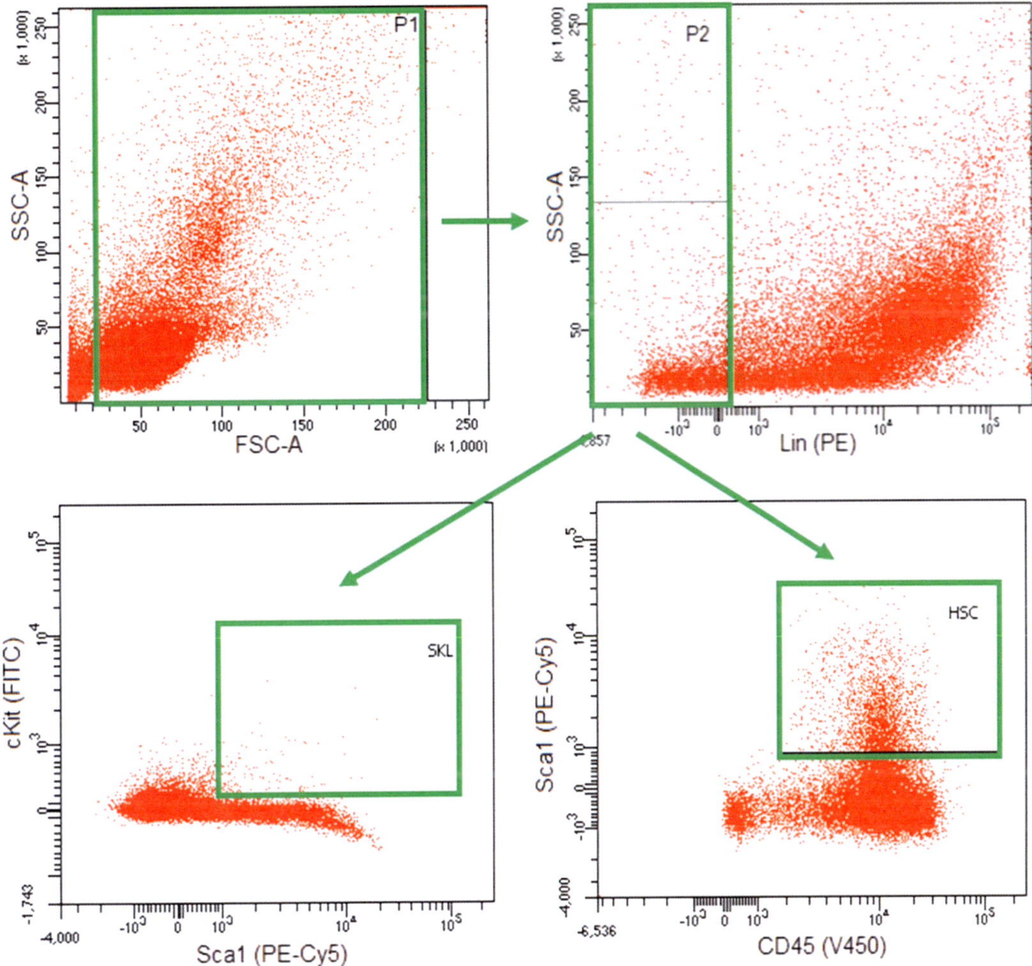

Fig. 3 Gate strategy for SKL and HSC cells from mobilized peripheral blood. All blood cells were gated through the FSC-A vs SSC-A plot (**a**). Plots were further interrogated by the ratios of height to width in forward scatter and side scatter to gate lineage (Lin) negative markers (**b**). Double-positive population for c-Kit (CD117) and Sca-1 was selected as SKL population of cells (**c**). Double-positive population for CD45 and Sca-1 was selected as HSCs population of cells (**d**)

7. Centrifuge for 10 min with 600 × *g* at +4 °C.
8. Resuspend cells in 350 μl RPMI +2% FBS.
9. The point of interest is the subpopulation of Sca-1$^+$ c-Kit$^+$ lineage $^-$ cells as markers of mobilized hematopoietic cells.

3.6 Colony-Forming Assay

Granulocytes and macrophages are an important part of innate immunological responses strictly involved in HSPCs' mobilization. CFU-GM assay is another indicator of HSPCs' generation. CFU-GM colonies are plated on 12 well-plates. 1–2 × 10^6 cells/well are required.

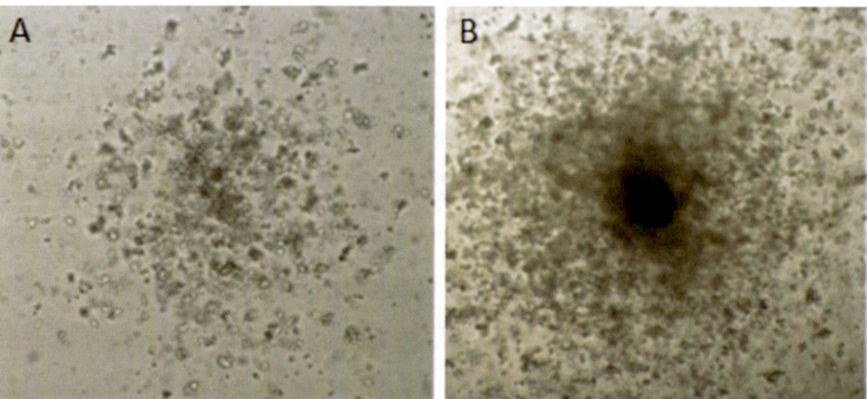

Fig. 4 An optimized mobilization protocol allows for CFU-GM colony formation. Representative CFU-GM-derived colony after 7 days of culture

1. Resuspend 1 or 2 × 10^6 cells in 300 μl RPMI+2% FBS.
2. Add growth factors (make a mixture: 2 μl IL-3 and 5 μl GM-CSF for each sample; distribute 7 μl from the mixture to test tubes with cells).
3. Add 700 μl human methylcellulose medium with antibiotics and L-glutamine to the tube with cells and mix well, avoiding the formation of air bubbles.
4. Plate the cells, and try to avoid air bubbles (*see* **Note 10**).
5. Add PBS between wells to keep plate microenvironment humid.
6. Incubate for a week.
7. Count CFU-GM colonies (*see* **Note 11**) (Fig. 4).

4 Notes

1. Transgenic lines of mice with defective ComC elements (such as C3-, C5-, Cq1-knockout mice) could be used in the experiment to exclude the involvement of a particular pathway in the process of mobilization. However, in that case it is important to exclude defects in hematopoiesis, in particular in animals under steady-state conditions. It is recommended to check (1) PB cell counts, (2) red blood cell parameters, (3) numbers of BM-residing HSPCs, and (4) numbers of clonogenic progenitors and compare the results to the WT mice.
2. Antibodies are sensitive to light. When you are preparing them, work in the dark, and turn off the lamp in the laminar chamber. Store all antibodies at 4 °C in a container which will protect them from the light. All antibodies are mixed in one tube as a cocktail. The following panels allow for separation of specific

cell populations: Lin^- population, Sca-1^+ population, c-Kit $^+$ population, and $CD45^+$ population.

Anti-mouse lineage marker (Lin) antibodies:

(a) PE rat anti-mouse CD45R/B220.

(b) PE rat anti-mouse Ter119/erythroid cells.

(c) PE rat anti-mouse CD11b.

(d) PE hamster anti-mouse TCR β chain.

(e) PE hamster anti-mouse γδ T-cell receptor.

(f) PE rat anti-mouse Ly-6G and Ly-6C.

Antibodies for hematopoietic SKL cells:

(a) Biotin rat anti-mouse Ly6A/E + PE-Cy5 Streptavidin (detects Sca-1^+ cells).

(b) FITC rat anti-mouse CD117 (detects c-Kit $^+$ cells).

(c) Antibodies for hematopoietic stem cells (HSCs):

(d) All anti-mouse lineage marker (Lin) antibodies.

(e) All SKL antibodies.

(f) V450 rat anti-mouse CD45 (HSCs are $CD45^+$).

3. We conduct mobilization procedure for a short time period (3 days) and for a long time period (6 days) in parallel. In case, if there are no significant differences in mobilization results, we reduce the dosage to 100 μg/kg G-CSF. We use filgrastim (commercial name: Neupogen), which is an analogue of the human granulocyte colony-stimulating factor. Prepare appropriate volume of G-CSF (Neupogen) at a concentration of 960 μg/ml. For example, if you perform mobilization of a mouse with 200 μg/kg G-CSF, then for a 25 g body weight mouse, 5.21 μl G-CSF (Neupogen) is needed in 100 μl of PBS solution. If you use six mice, make a mix for seven individuals, in case of any losses.
4. The blood for hematological analysis in Hemavet should be handled no longer than 2 h after the withdrawal.
5. Euthanasia of mice can be performed by cervical dislocation, however, only by personnel with documented training. Euthanasia can also be performed by intraperitoneal injection of ketamine-xylazine mixtures with dosages of 100 mg/kg ketamine and 10 mg/kg xylazine delivered in 1 ml syringe. It is important to remember that the ketamine/xylazine mixture is not stable and fresh mixtures should be prepared on a weekly basis.
6. Collect whole blood from the main vein and heart. Take 25-gauge needle, and collect the blood from the main vein; however, be careful and do not tear this vessel. Additionally you can take the blood from the heart.

7. Turk's *solution* destroys red blood cells and stains the nuclei of the white blood cells.
8. Count the cells in three to four squares in counting chamber such as a Burker hemocytometer according to the formula:

$$\text{Average number of cells from squares} \times 10{,}000 \times \text{dilution of the sample.}$$

For example, in four squares you have 34, 32, 35, and 37 cells; the average number is 34.5. If you put 5 μl of cells from 1 ml into 95 μl Turk's solution, your dilution is 20 times. $34.5 \times 10{,}000 \times 20$ is 6.9×10^6 cells in 1 ml.

9. If you have 6.9×10^6 cells and resuspend them in 300 μl, and you take 43.47 μl from this cell suspension, the final number of the cells will be 1×10^6 cells to seed them for CFU-GM colony formation assay.
10. You can get air bubbles by precipitation with a pipette tip. Be careful and avoid it.
11. Observe and count colonies under a microscope at magnification of 40×.

Acknowledgments

We thank Mateusz Adamiak (Department of Regenerative Medicine and Center for Preclinical Studies and Technology, Warsaw Medical University, Warsaw, Poland) for his helpful suggestions and critical reading of the manuscript and Kamila Bujko (Stem Cell Institute at James Graham Brown Cancer Center, University of Louisville, Louisville, KY, USA) for her kind providing of figure representing flow cytometry analysis. We thank Tomasz Wilanowski (Nencki Institute of Experimental Biology, Polish Academy of Sciences, Warsaw, Poland) for reviewing the protocol as a native speaker.

References

1. Ratajczak MZ, Adamiak M, Plonka M, Abdel-Latif A, Ratajczak J (2018) Mobilization of hematopoietic stem cells as a result of innate immunity-mediated sterile inflammation in the bone marrow microenvironment-the involvement of extracellular nucleotides and purinergic signaling. Leukemia 32 (5):1116–1123. https://doi.org/10.1038/s41375-018-0087-z
2. Ratajczak MZ, Borkowska S, Ratajczak J (2013) An emerging link in stem cell mobilization between activation of the complement cascade and the chemotactic gradient of sphingosine-1-phosphate. Prostaglandins Other Lipid Mediat 104-105:122–129
3. Ratajczak MZ, Kim CH, Wojakowski W, Janowska-Wieczorek A, Kucia M, Ratajczak J (2010) Innate immunity as orchestrator of stem cell mobilization. Leukemia 24 (10):1667–1675
4. Reca R, Cramer D, Yan J, Laughlin MJ, Janowska-Wieczorek A, Ratajczak J et al (2007) A novel role of complement in mobilization: immunodeficient mice are poor

granulocyte-colony stimulating factor mobilizers because they lack complement-activating immunoglobulins. Stem Cells 25 (12):3093–3100

5. Bujko K, Rzeszotek S, Hoehlig K, Yan J, Vater A, Ratajczak MZ (2017) Signaling of the complement cleavage product anaphylatoxin C5a through C5aR (CD88) contributes to pharmacological hematopoietic stem cell mobilization. Stem Cell Rev 13(6):793–800
6. Ratajczak J, Reca R, Kucia M, Majka M, Allendorf DJ, Baran JT et al (2004) Mobilization studies in mice deficient in either C3 or C3a receptor (C3aR) reveal a novel role for complement in retention of hematopoietic stem/progenitor cells in bone marrow. Blood 103 (6):2071–2078
7. Lee HM, Wu W, Wysoczynski M, Liu R, Zuba-Surma EK, Kucia M et al (2009) Impaired mobilization of hematopoietic stem/progenitor cells in C5-deficient mice supports the pivotal involvement of innate immunity in this process and reveals novel promobilization effects of granulocytes. Leukemia 23 (11):2052–2062
8. Itkin T, Kumari A, Schneider E, Gur-Cohen S, Ludwig C, Brooks R et al (2017) MicroRNA-155 promotes G-CSF-induced mobilization of murine hematopoietic stem and progenitor cells via propagation of CXCL12 signaling. Leukemia 31(5):1247–1250
9. Petit I, Szyper-Kravitz M, Nagler A, Lahav M, Peled A, Habler L et al (2002) G-CSF induces stem cell mobilization by decreasing bone marrow SDF-1 and up-regulating CXCR4. Nat Immunol 3(7):687–694
10. Lee HM, Wysoczynski M, Liu R, Shin DM, Kucia M, Botto M et al (2010) Mobilization studies in complement-deficient mice reveal that optimal AMD3100 mobilization of hematopoietic stem cells depends on complement cascade activation by AMD3100-stimulated granulocytes. Leukemia 24(3):573–582
11. Devine SM, Vij R, Rettig M, Todt L, McGlauchlen K, Fisher N et al (2008) Rapid mobilization of functional donor hematopoietic cells without G-CSF using AMD3100, an antagonist of the CXCR4/SDF-1 interaction. Blood 112(4):990–998

Chapter 14

A Freezing Protocol for Hematopoietic Stem Cells

Petra Pavel and Sascha Laier

Abstract

Especially in the field of autologous transplantation, it has been found necessary to develop methods that ensure long-term storage with maintenance of functionality of the cells to bridge the therapy-related temporal separation of collection and application.

Based on the experiences of more than 40 years, some practical considerations, especially regarding the cell concentration, final volume, and possibly other exogenous substances, should be considered when establishing a protocol for the routine cryopreservation of peripheral blood stem cells. In the following chapter, we describe a freezing protocol for cryopreservation of peripheral blood stem cells which was used and optimized over the past 8 years and was applied to the cryopreservation of more than 2000 peripheral stem cell transplants.

Key words Cryopreservation, Cryoprotectants, Autologous peripheral blood stem cells, Autologous stem cell transplantation

1 Introduction

For a widespread use of autologous stem cell transplantation, two factors were crucial. The use of G-CSF (granulocyte-colony stimulating factor) to mobilize hematopoietic stem cells into the peripheral blood and the establishment of appropriate collection procedures (cell separation by apheresis) have enabled collection of therapeutically sufficient doses of stem cells as a routine procedure. However, especially in the field of autologous transplantation, it has also been found necessary to develop methods that ensure long-term storage with maintenance of functionality of the cells to bridge the therapy-related temporal separation of collection and application. This was achieved by developing a process which enables stem cells to be frozen and stored at low temperature. When optimizing these predominantly empirically developed procedures, a number of problems had to be solved in order to avoid cell lysis, cell damage, and loss of function due to freezing of heterologous cell structure. Because of heterogeneous nucleation, which dominates ice crystal formation in biological systems above

Gerd Klein and Patrick Wuchter (eds.), *Stem Cell Mobilization: Methods and Protocols*, Methods in Molecular Biology, vol. 2017, https://doi.org/10.1007/978-1-4939-9574-5_14, © Springer Science+Business Media, LLC, part of Springer Nature 2019

temperatures ranging from −5 °C to about −30 °C, freezing causes osmotic dehydration, irreversible membrane damages, and subsequent cell lysis while thawing. The cells are invariably excluded from the growing ice matrix. This inevitable consequence of increased solutes concentration was conceptualized as a major damaging factor during the freezing of cells.

The first freeze-survival demonstration was probably reported by Gonzales and Luyet in 1950 [1] with tissue derived from an embryonic four-chambered heart (from chick embryos). The scientists incubated beating hearts from chick embryos in 30% ethylene glycol for 6 min and then placed them directly in liquid nitrogen for at least a minute and sometimes up to an hour. More than half of the treated embryos survived the solidification; however most of these surviving embryos showed some sign of injury. Schopf-Ebner et al. [2] were able to show that 5–20% of cardiac cells from 8-day chicken embryos survived and resumed contracting after addition of dimethyl sulfoxide (Me_2SO) and controlled cooling (1 °C per min) down to −78 °C.

In the following years, a large number of more or less theoretically substantiated experiments to optimize the freezing and thawing process were performed. The limiting factors for survival and functionality of cells and tissue turned out to be the type and quantity of cryoprotective agents; the cooling rate during freezing, including the greatest possible control of ice formation; and the heating rate while thawing.

1.1 Cooling and Heating Rates

Detailed experiments to optimize cooling rates while deep freezing and while thawing of heterogeneous cell suspensions were performed by Mazur et al. [3, 4]. They found out that for cells to resist the freezing process, an optimal cooling rate can be defined, with increased cell injury both at very low cooling rates (equated to osmotic stress and other associated mechanisms during freezing) or very high cooling rates (where intracellular ice is produced). The balance between these two factors results in a maximum chance of survival for a particular cell type across a limited cooling rate profile, producing what came to be known as *Mazur's bell-shaped survival curves*.

1.2 Controlled Ice Nucleation in Cryopreservation

Apart from the cooling rate, minimization of cell damage during the freezing process also involves the control of ice nucleation (crystallization) and the avoidance of an intermediate temperature rise due to the heat of crystallization (temperature range between −5 °C and −30 °C).

Cell damage can be minimized, if the ice nucleation and crystallization triggers both extracellularly and intracellularly and, at the same time, to different cell compartments and therefore proceeds as homogeneously as possible. There are different methods to control ice nucleation in cryopreservation [5].

1. *Seeding*: The trigger for ice nucleation is the physical introduction of a small ice crystal into an undercooled sample. However, this has the potential risk of contaminations. To remove the potential contamination, it is common practice nowadays to induce ice nucleation by manually *generating a cold spot on the outside of the closed container*.
2. *Chemical nucleants:* The inclusion of specific ice-nucleating catalysts in the suspending medium assures homogenous ice nucleation as much as possible. This method is not common practice as it is usually not GMP-compliant or not biocompatible.
3. *Electro-freezing:* The homogenous ice nucleation will be caused by a high voltage, applied to a metal electrode. There are however a number of practical problems when inducing and using high voltage.
4. *Mechanical methods:* Methods such as shaking, taping, and application of ultrasound are known; these methods are difficult to standardize.
5. *Shock cooling:* Following initial slow cooling, the sample is cooled rapidly and then further exposed to a complex set of temperature ramps. It is likely that a "cold spot" is formed at the wall of the sample leading to local ice nucleation (seeding).
6. *Pressure shift:* In a pressure chamber, the samples are pressurized and then cooled to the desired nucleation temperature. The pressure is then reduced to induce nucleation. As with electro-freezing, there are also a number of problems in practical use with this method.

Comparative studies have shown that freezing of different cell types by controlled ice nucleation reduces cell damage. However, no relevant difference was found between the different methods of ice nucleation control. Both the optimized cooling rates and the controlled ice nucleation can be standardized and implemented into practice with reasonable effort using a freezer (controlled-rate freezing = CRF) which can provide variable cooling rates and controlled ice nucleation by seeding due to a cold spot outside the container.

1.3 Cryoprotectants

Cryoprotective agents (CPA) invariably play the key role in allowing cells to be processed for storage at deep cryogenic temperatures and be recovered with high levels of functionality. A cryoprotectant is per definition any solute which, when added to cells in their medium, allows for higher post-thaw recovery rates than if it were not present [10].

In the 1960s Karow et al. studied the pharmacology of a wide spectrum of CPAs [6–8]. He clearly defined the two major classes of CPAs: small-molecular-weight penetrating agents and high-molecular-weight non-penetrating agents.

He listed some 56 chemicals (comprising both CPA classes) which demonstrated efficacy defined as post-thaw survival of 40% or greater in different cell types [7]. From the 1960s to the 1990s, activity in cryopreservation increased tenfold which led to a pragmatic consolidation of the CPAs applied in the majority of the reported studies.

Karow's "list of 56" was reduced down to some 20 agents by Ashwood-Smith [9]. Of these 20 agents, only half were identified as potentially effective. Elliot et al. have updated Ashwood-Smith's list based on ongoing evidence since his publication [10]. Contemporary in a similar vein to Karow, Hubalek [11] reviewed the CPAs used for cryopreservation of microorganisms and identified some 55 agents. However only eight of these agents ensured a survival of over 40% after freezing. These eight CPAs were also represented by Ashwood-Smith.

CPAs are multimodal agents, which may be loosely described as nonspecific drugs. They can influence a variety of normal biochemical processes. In his studies of pharmacology of these solutes responsible for penetrating CPAs, Karow highlighted a resulting conflict between impact of protection and toxicity resulting from the use of these solutes. He indicated that these toxic effects could be mitigated by developing CPA protocols where the final concentration, temperature, and time of exposure before cryopreservation were optimized.

Dimethyl sulfoxide (Me_2SO) was identified as a very effective cryoprotectant for all classes of cells. Me_2SO is an intracellular cryoprotectant as it can move across the cell membrane by displacing the water within the cell, thus preventing the formation of ice crystals in the cell and protecting the cell from rupture [12, 13]. The toxic effects of Me_2SO on cells depend on the temperature and the exposure time of both the pre-freeze and the post-thaw periods [12–14].

Since Me_2SO is also used therapeutically, there is relatively detailed data available on pharmacology and toxicology. Me_2SO has been used as the standard cryoprotectant for the cryopreservation of hematopoietic stem cells since the end of the 1990s [15]. To minimize Me_2SO toxicity to the cells (ex vivo), as well as during the use of the cell preparations after thawing (in vivo), several options have been discussed and investigated [16–21]:

- Avoid using toxic concentrations.
- Increase of the cryoprotective effect by combining with non-penetrating cryoprotectants.
- Membrane stabilization by adding high-molecular-weight additives (HES, HSA, Dextran).

- Rapid removal of the cryoprotectant after thawing (partly significant cell loss).
- Add "toxicity neutralizers" (additional potential side effects).

Cryopreservation at the lowest possible concentration of Me_2SO in combination with membrane stabilization by high-molecular-weight solutions has been proven to be effective for cryopreservation of cell suspension, particularly of stem cell preparations from bone marrow, umbilical cord blood, and apheresis.

A series of optimization studies of cryopreservation protocols showed that neither different concentrations of Me_2SO (1–10%, possibly in combination with up to 10% ethylene glycol) nor type and amount of additives or their combination (HES, HSA, dextran, AB serum, Plasma-Lyte A, glucose) caused relevant differences in recovery of CD34+ cells or viability of nucleated cells [22–24].

Exposure of stem cells from umbilical cord blood or peripheral blood stem cells at 22 °C with Me_2SO in a concentration of up to 10% is relatively well tolerated (10.9% loss of viability of the CD34+ cells for PBSC and 12.6% for umbilical cord blood) [25].

Typically, cryoprotocols are currently used in which 5–10% of Me_2SO is added. If available (peripheral blood stem cells), this is pre-diluted prior to adding to the cells by donor plasma. If donor plasma is not available (bone marrow or umbilical cord blood), isotonic solutions, respectively, isomolar or colloidal, such as HES, HSA, or dextran, are used to dilute the Me_2SO prior to adding to the cells or even diluting the cells themselves.

Based on the above thoughts, some practical considerations, especially regarding cell concentration, final volume, and possibly other exogenous substances, should be considered when establishing a protocol for the routine cryopreservation of peripheral blood stem cells.

Goals:

- Minimizing the cytotoxic effect of Me_2SO before and during cryopreservation and while thawing (ex vivo).
- Minimizing the cytotoxic effect of Me_2SO during and after the infusion of the cells (in vivo).
- Maximizing the cryoprotective effect of Me_2SO and possibly other additives to maintain viability and functionality of the cells.
- Prevention of cell damage, cell lysis, or cell loss due to interim storage, processing, and cryopreservation.
- If possible exclusion of other exogenous substances.
- If necessary, the use of exogenous substances (resuspension media) whose suitability for infusion has been proven.

2 Materials

1. Safety cabinet (class A in class B, *see* **Note 1**).
2. Refrigerator.
3. Centrifuge (for blood bags), e.g., Rotixa 50 RS/Hettich.
4. Plasma extractor.
5. Sterile connecting device (e.g., TSCD-II).
6. Blades for TSCD.
7. Tube sealer.
8. Controlled-rate freezer.
9. Scales.
10. Particle monitor.
11. Sterile cryobags (filling volume 100 ml).
12. Sterile cryotubes (1.8 ml).
13. Sterile transfer bags with coupler (600 ml/400 ml).
14. Benja-Mix non-vented spike.
15. Sterile coupler with septum.
16. Sterile three-folder connector.
17. Sterile four-folder connector.
18. Sterile closing caps (Luer Lock).
19. Sterile syringes (50 ml (Luer Lock), 5 ml (Luer Lock), 2 ml).
20. Sterile 18-gauge needle.
21. Composol® PS sterile solution or, e.g., 5% HSA.
22. CryoSure-DMSO (Me_2SO, e.g., Wak-Chemie Medical GmbH).
23. Culture vial (e.g., BACTEC Plus aerobic F, BACTEC Plus anaerobic F).
24. Aluminum container for cryobags.
25. Settle plates/contact plates.
26. Sterile disinfectants (wipes, spray).
27. Coolpacks or cooling plates.
28. Labels for containers and tubes.

3 Methods

Carry out all procedures at room temperature (18–22 °C) unless otherwise specified. The sampling to detect the content of the stem cell preparation is done after the end of collection (*see* **Note 2**).

Begin the cryopreservation procedure as soon as possible but latest 68 h and complete at the least 72 h after the end of collection (*see* **Note 3**).

3.1 Calculations

1. Calculate the target volume and the number of cryopreserved aliquots dependent on the content of CD34+ cells, the concentration of nucleated cells, and the number of required doses (transplants).
2. Goals (*see* **Note 4**):

Volume:	100 ml/bag
Number of aliquots:	1–3
Concentration of NC:	$\leq 5 \times 10^8$ NC/ml
Volume of cell suspension:	70 ml/bag
Concentration of Me_2SO:	10% in the final product ≤33.3% in the cryomedium
Cryomedium volume:	30 ml/bag
Composition:	Me_2SO (33.3%): autologous plasma or medium (66.7%) = 1:2

3.2 Preliminary Operations

1. Print out all labels for all secondary containers, tubes, settle plates, contact plates, protocols, and cryobags.
2. Prepare all protocols for processing in the cleanroom area.
3. Introduce all materials into the cleanroom area according to local requirements.
4. Enter the cleanroom area according to the local gowning procedure.
5. Start all devices necessary for processing according to local requirements.
6. Prepare and declare all needed bags, containers, tubes, settle plates, contact plates, and forms.
7. Start particle monitoring and open the settle plates.

3.3 Concentration (See Fig. 1)

If the volume of the cell suspension is >70% of the final volume, the cell suspension should be concentrated.

1. Connect a transfer bag (600 ml) to the bag with the stem cell preparation by using a sterile docking device (the tubes of the bags should still be as long as possible).
2. Transfer the cell suspension into the transfer bag.
3. Empty the tubes and shut them, using clamps to close the bag.

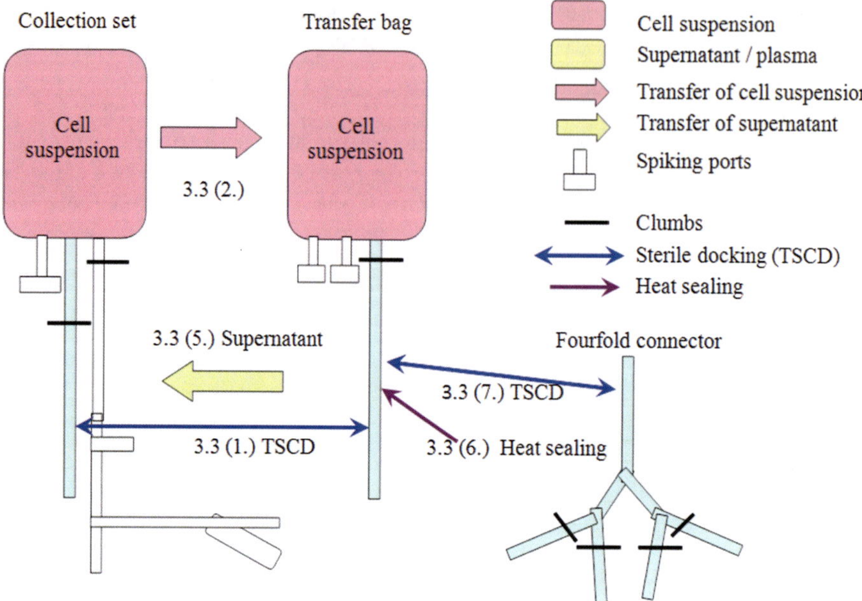

Fig. 1 Concentration of peripheral blood stem cells

4. Centrifuge the bag at room temperature or cooled (450 × *g*, 10 min, medium brake).
5. Transfer the bags into a plasma extractor, and press off the supernatant into the empty bag. Target 70 ml/per bag after concentration (*see* **Note 4**).
6. Separate the bags by heat sealing (3 welds—the tube to the transfer bag with cells should still be as long as possible).
7. Couple a four-folder connector to the transfer bag with cells by sterile docking close to the bag.
8. Store in a refrigerator at a temperature between +2 °C and +6 °C.

3.4 Dilution (See Fig. 2)

Alternatively to Subheading 3.3, if the volume of the cell suspension is <70% of the final volume, the cell suspension should be diluted. Preferably use donor plasma of the same collection for dilution (*see* **Note 5**).

1. Connect the bag with the stem cell preparation by sterile docking to the plasma bag (the tubes to the bags should still be as long as possible).
2. Transfer the plasma into the bag with the cell suspension (control by scale) until the target volume is reached.
3. Separate the bags by heat sealing (3 welds—the tube to the transfer bag with cells should still be as long as possible).

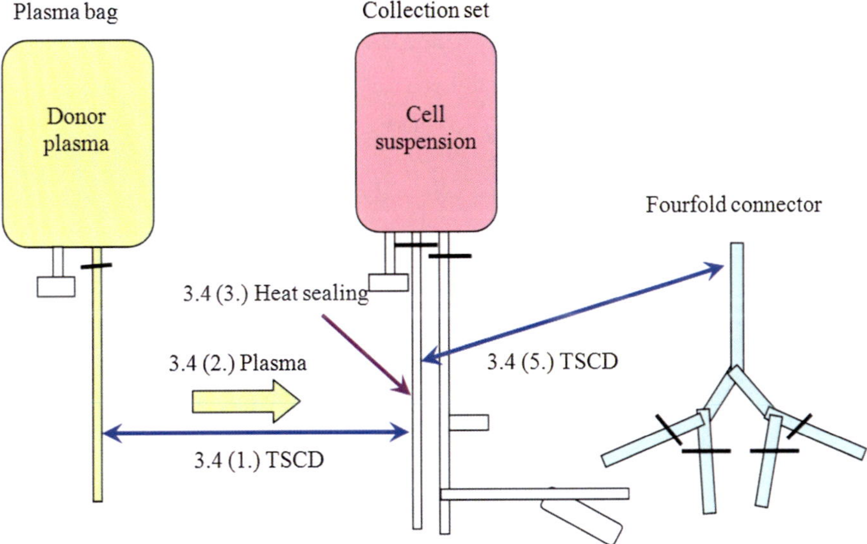

Fig. 2 Dilution of peripheral blood stem cells with donor plasma

4. Empty the tubes and shut them by using clamps to close the bag.
5. Couple a four-folder connector to the transfer bag with cells by sterile docking close to the bag.
6. Store in a refrigerator at a temperature between +2 and +6 °C.

3.5 Preparation of the Cryomedium (See Fig. 3)

For the preparation of the cryomedium, preferably use donor plasma of the same collection (*see* **Note 5**).

1. Connect the plasma bag with a three-folder connector by spiking to one of the ports.
2. Connect the plasma bag to a transfer bag (400 ml) by spiking, and transfer the calculated volume of plasma into the 400-ml transfer bag (control by scale).
3. Shut the tube of the transfer bag (plasma) close to the bag with a clamp.
4. Separate the bags by heat sealing (3 welds—close to the transfer bag).
5. Spike a coupler with septum to the second port of the transfer bag.
6. Open the vial with Me_2SO, and take the calculated volume of Me_2SO with an 18-gauge needle into a 50-ml syringe.
7. Transfer the Me_2SO through the coupler by spiking it into the transfer bag with plasma.
8. Mix thoroughly and store in a refrigerator between +26 °C and +6 °C.

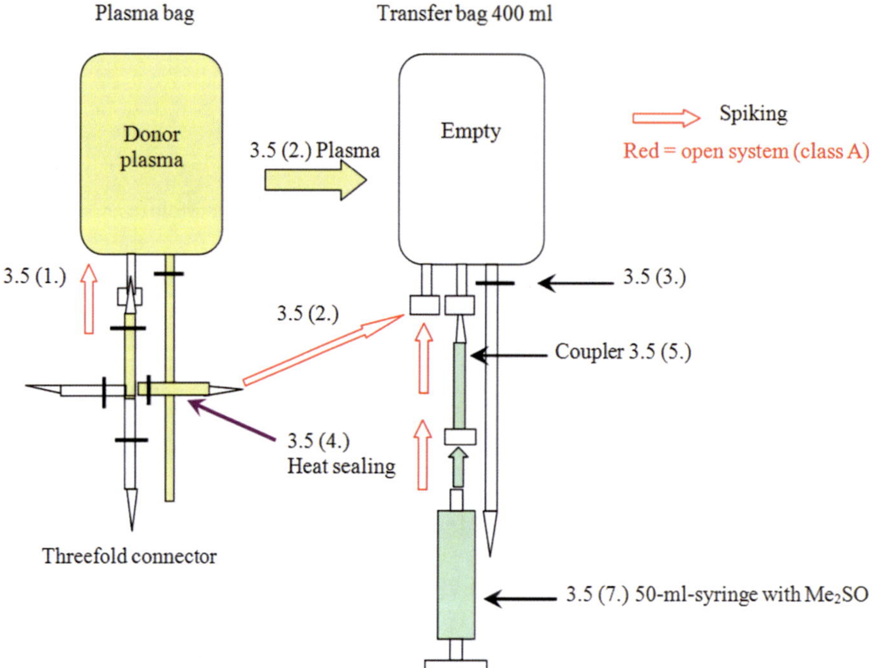

Fig. 3 Preparation of cryomedium with donor plasma

3.6 Addition of the Cryomedium (See Fig. 4)

At this time the controlled-rate freezer has to be started and precooled to starting temperature (=0 °C). All components should be precooled, and the processing steps should be done on coolpacks or coolplates (*see* **Note 6**).

All clamps of tubes which are not needed for transfer cells or the medium should be closed. Open only the clamp of tubes, which are necessary for the current processing step. Use the same Luer Lock connection only once if possible, but not more than twice.

1. Transfer the bag with the prepared cell suspension, including the four-folder connector, into the sterile cabinet as well as the bag with cryomedium, with coolpacks placed between them.
2. Prepare the intended number of cryobags (unpack and label), and place coolpacks in between.
3. Connect the bags with cell suspension and cryomedium by sterile docking or spiking.
4. *Log the time (start of exposure of cells to Me_2SO—critical step).*
5. Transfer the cryomedium into the bag with cell suspension—the first third slowly, then speedily—and mix continuously and thoroughly (*see* **Note 7**).
6. Empty the tubes and shut them with clamps to close the bag.
7. Separate the bags by heat sealing (3 welds).

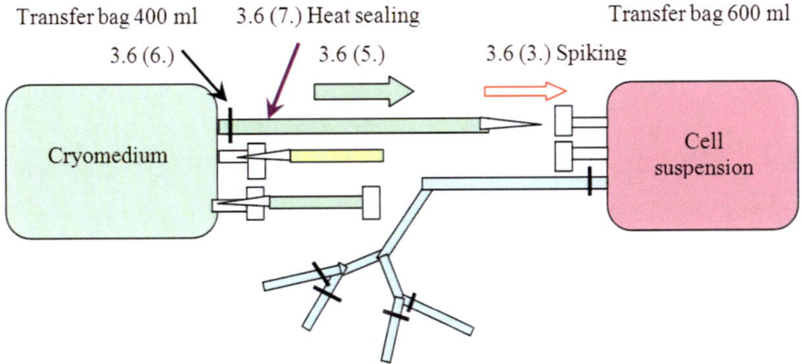

Fig. 4 Addition of cryomedium to the cell suspension

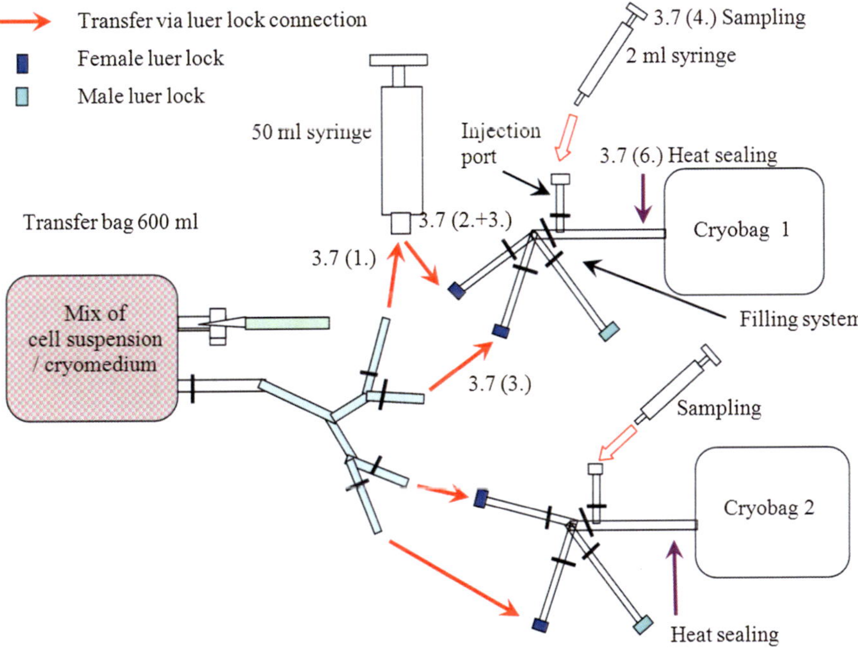

Fig. 5 Transfer/splitting of the cell suspension-cryomedium mix into the cryobags

3.7 Splitting and Transfer to the Cryobags (See Fig. 5)

1. Connect a 50 ml syringe to the first connection of the four-folder connector (Luer Lock).
2. Separate the syringe from the connector (close the Luer Lock), and couple it to the filling system of the cryobag (Luer Lock).
3. Transfer 50 ml of the cell suspension-cryomedium mix into the cryobag, and repeat this for the next 50 ml.
4. With a 2 ml syringe and an 18-gauge needle, pierce through the membrane port of the filling set, and take 2 ml cell suspension as sample for microbiological control.

5. Inoculate the sample into the culture vials (1 ml aerobic/1 ml anaerobic).
6. De-aerate the cryobag, and disconnect the filling set close to the bag (heat sealing—3 welds).
7. Repeat—if necessary—to fill the next cryobag, but use another Luer Lock connection.
8. Take the remaining cell suspension-cryomedium mix (about 5 ml) with a 5 ml syringe, and transfer it into the cryotubes as samples for quality control and retention (*see* **Note 8**).
9. Place the cryobags and cryotubes between coolpacks or on coolplates, and immediately transfer them out of the clean-room area to the controlled-rate freezer.
10. Place the cryobags in precooled aluminum container and together with the cryotubes (pilot tubes) into the controlled-rate freezer.
11. Start the freezing procedure (*see* **Note 9** and Fig. 6).
12. *Log the time (end of exposure of cells to* Me_2SO*—critical step).*

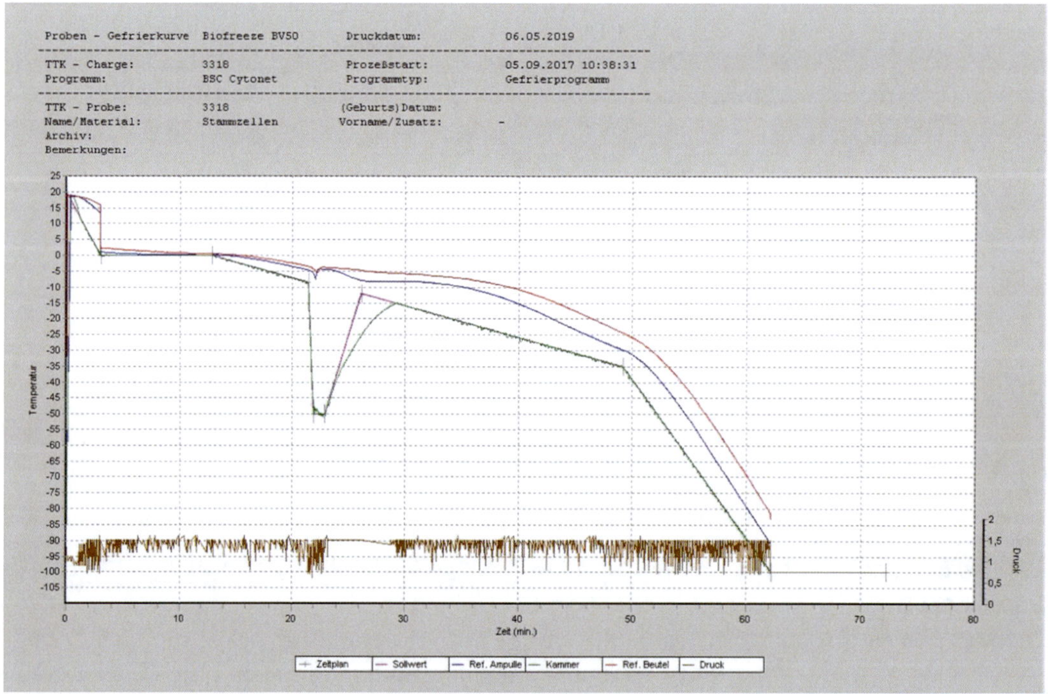

Fig. 6 Exemplary freezing rate in a controlled-rate freezer

3.8 Freezing, Storage, and Quality Control

1. After finalizing of the programmed freezing procedure, transfer the cryobags and cryotubes soon (not later than 1 h) into the storage container (storage in the gas phase of liquid nitrogen at <-140 °C).
2. Print out the freezing protocol.
3. Test not earlier as 24 h after cryopreservation at least the recovery (target $\geq$80%) and the vitality (target $\geq$70% but at least 50%) of nucleated cells (*see* **Note 10**).

4 Notes

1. Processing (concentration, dilution, preparation of cryomedium, addition of cryomedium to the cell suspension, splitting, and sampling) should be performed within a cleanroom. All processing steps which required the opening of closed containers have to be performed in a safety cabinet (grade A) surrounded by a cleanroom grade B area. All processing steps with opened containers should be monitored by counting of air particles and test of environmental contaminants (settle plates and contact plates).
2. Critical to the result of the cryopreservation are, on the basis of literature data and own experience, the following goals of collection:
   (a) Concentration of nucleated cells (NC): $\leq 2 \times 10^8$ NC/ml, but max. 5×10^8 NC/ml.
   (b) Volume: 100–400 ml.
   (c) Hematocrit $\leq$ 5% but max. 10%.
3. Critical to the result of the cryopreservation are, on the basis of literature data and own experience, the following conditions for intermediate storage until cryopreservation:
   (a) Temperature: +2 °C to + 6 °C.
   (b) Concentration of NC: $\leq 2 \times 10^8$ NC/ml.
   (c) Duration: $\leq$48 h after the end of collection (strong recommendation) but max. $\leq$68 h after the end of collection.
4. Based on our experience, cryopreservation of nucleated cells is possible up to a concentration of 5×10^8 NC/ml without any relevant loss of viability when using 10% Me_2SO. Thus, in 80% of cases, the volume of a transplant can be reduced to 100 ml and correspondingly to a maximum of 10 ml of Me_2SO per transplantation. The prerequisite in most cases is that the stem cell preparations are reduced in volume after collection and, if necessary, after dilution for intermediate storage. For this purpose, process parameters must be set so that the residual

volume is large enough to avoid cell losses when pressing off the supernatant. On the other hand, there must be sufficient reserve for the addition of the cryoprotective agent. In order to minimize the osmotic stress for the cells, it is recommended not to add the Me_2SO undiluted to the cell suspension, but to dilute in advance, preferably with donor plasma.

This results in the following procedure specifications for a standard volume of 100 ml per bag: Volume after concentration = 70 ml

Cryoprotective solution (10 ml Me_2SO + 20 ml donor plasma or medium) = 30 ml

If several portions are produced (usually a maximum of 3 portions), the ratio is multiplied accordingly. This also ensures a good standardization of the process without new calculations in each case.

5. For the dilution that may be required, donor plasma should preferably be used. If this is not available or not available in sufficient quantity, other suitable additives can be used (isotonic, possibly colloidal). For regulatory reasons, infusion solutions should be chosen which have drug approval or have proven to be suitable for infusion in humans (e.g., various storage solutions for blood components for transfusion). For practical reasons, it is advisable to choose solutions that are offered in bags with a tube system for sterile docking, since hereby a dilution can be carried out even in the unclassified area outside the cleanroom. The addition of colloidal solutions, e.g., human albumin, seems to be dispensable according to our experience (tested with up to 50% content of dilution medium).

6. The critical points for cell protection in cryopreservation have already been mentioned above. In addition to that, it is advisable to reduce the toxic effect of Me_2SO on the cells ex vivo by further measures.

   To reduce the osmotic stress on the cells, the addition of the Me_2SO should not be undiluted but diluted in a 1:1 or 1:2 ratio with a suitable medium, preferably donor plasma. When selecting the medium, the same considerations should be considered as for dilution for temporary storage.

   Duration and temperature of exposure of cells to Me_2SO should be reduced to a minimum.

   Cell suspension and cryoprotective solution should be cooled down to approximately 4–6 °C before mixing; further handling up to the beginning of the freezing process should be carried out on cold plates or cool packs. The time before the start of the freezing process in which the cells are exposed to Me_2SO should not exceed 30 min. In individual cases, no quality losses were noted even when exceeded for up to 3 min (own data).

7. To ensure that all aliquots are homogeneous, the cell suspension and cryoprotectant solution are mixed prior to the separation on the cryobags.
8. It is recommended to prepare a sufficient number of aliquots (pilot tubes) for quality control and retention. We calculate one pilot tube for quality control before release, one pilot tube as retention sample and for every cryobag an additional pilot tube, which can be used for retest after expiry or in case of special requests.
9. The freezing kinetics should have a low cooling rate at least until the crystallization point. After seeding and compensation of the heat of crystallization, the cooling rate can be increased. As a general rule, this should be completed after one hour at -100 °C. Subsequently, the preparations should be transferred promptly to storage containers for long-term storage with ≤ -140 °C. The best standardization and control of the freezing process is possible by using a so-called controlled-rate freezer.

   The programmed and controlled release of liquid nitrogen into the chamber of the freezer ensures variable cooling rates, seeding due to temperature shock and compensation of the heat of crystallization during ice formation.

   As an example Fig. 6 shows the freezing rate including information on the chamber temperature and the temperature in a reference bag (100 ml) and a pilot tube (2 ml) filled with cryoprotective solution.
10. In case of relevant process changes (i.e., changes of exogenous substances), the functionality (clonogenicity) has to be tested additionally.

References

1. Gonzales F, Luyet B (1950) Resumption of heart-beat in chick embryo frozen in liquid nitrogen. Biodynamica 7:1–5
2. Schöpf-Ebner E, Gross WO, Bucher OM (1968) Pulsatile activity of isolated heart muscle cells after freezing storage. Cryobiology 4:200–203
3. Mazur P (1965) Causes of injury in frozen and thawed cells. Fed Proc 24:S175–S182
4. Mazur P (1970) Cryobiology: the freezing of biological systems. Science 168:939–949
5. Morris GJ, Acton E (2013) Controlled ice nucleation in cryopreservation—a review. Cryobiology 66:85–92
6. Karow AM Jr (1969) Biological effects of cryoprotectants as related to cardiac cryopreservation. Cryobiology 5:429–443
7. Karow AM Jr (1969) Cryoprotectants—a new class of drugs. J Pharm Pharmacol 21:209–223
8. Karow AM Jr (1974) Cryopreservation: pharmacological considerations. In: Karow AMS, Karow GJM, Humphries AI (eds) Organ preservation for transplantation. Little Brown & Co, Boston, pp 86–107
9. Ashwood-Smith MJ (1987) Mechanisms of cryoprotectant action. Symposia Soc Exp Biol 41:395–406
10. Elliott GD, Wang S, Fuller BJ (2017) A review of the actions and applications of cryoprotective solutes that modulate cell recovery from ultra-low temperatures. Cryobiology 76:74–91
11. Hubalek Z (2003) Protectants used in the cryopreservation of microorganisms. Cryobiology 46:205–229

12. Berz D, McCormack EM, Winer ES, Colvin GA, Quesenberry PJ (2007) Cryopreservation of hematopoietic stem cells. Am J Hematol 82:1–8
13. Choi CW, Kim BS, Shin SW, Kim YH, Kim JS (2007) Long-term engraftment stability of peripheral blood cells cryopreserved using the dump-freezing method in a −80°C mechanical freezer with 10% dimethyl sulfoxide. Int J Hematol 3:245–250
14. Arakawa T, Carpenter JF, Kita YA, Crowe JH (1990) The basis for toxicity of certain cryoprotectants: a hypothesis. Cryobiology 27:401–415
15. Günzel P, Eichler H (2008) Zusammenfassung und Bewertung der toxiko-pharmakologischen Daten und Informationen zu Dimethylsulfoxid (DMSO) im Hinblick auf seine Verwendung in Stammzellzubereitungen als gutachterliche Ergänzung der Gemeinsamen Stellungnahme der Fachgesellschaften DGTI, DGHO und GPOH zu Genehmigungsverfahren von Stammzellzubereitungen (https://dgti.de/service/news/stellungnahmen-der-dgti.html)
16. Kim KM, Huh JY, Kim JJ, Kang MS (2017) Quality comparison of umbilical cord blood cryopreserved with conventional versus automated systems. Cryobiology 78:65–69
17. Li L, Chen Z, Zhang M, Panhwar F, Gao C, Zhao G, Jin B, Ye B (2017) Cell membrane permeability coefficients determined by single-step osmotic shift are not applicable for optimization of multi-step addition of cryoprotective agents: as revealed by HepG2 cells. Cryobiology 79:82–86
18. Freimark D, Sehl C, Weber C, Hudel K, Czermak P, Hofmann N, Spindler R, Glasmacher B (2011) Systematic parameter optimization of a ME_2SO- and serum-free cryopreservation protocol for human mesenchymal stem cells. Cryobiology 63:67–75
19. Son JH, Heo YJ, Park MY, Kim HH, Lee KS (2010) Optimization of cryopreservation condition for hematopoietic stem cells from umbilical cord blood. Cryobiology 60:287–292
20. Hopkins JB, Badeau R, Warkentin M, Thorne RE (2012) Effect of common cryoprotectants on critical warming rates and ice formation in aqueous solutions. Cryobiology 65:169–178
21. Shleback AA, Marley SB, Roberts IA, Davidson RJ, Goldman JM, Gordon MY (1999) Optimal timing for processing and cryopreservation of umbilical cord hematopoietic stem cells for clinical transplantation. Bone Marrow Transplant 23:131–136
22. Johnson LN, Winter KM, Reid S, Hartkopf-Theis T, Marks DC (2011) Cryopreservation of buffy-coat-derived platelet concentrates in dimethyl sulfoxide and platelet additive solution. Cryobiology 62:100–106
23. Solves P, Mirabet V, Planelles D, Carbonell-Uberos F, Roig R (2008) Influence of volume reduction and cryopreservation methodologies on quality of thawed umbilical cord blood units for transplantation. Cryobiology 56:152–158
24. Best A, Hidalgo G, Mitchell K, Yanelli JR (2007) Issues concerning the large scale cryopreservation of peripheral blood mononuclear cells (PBMC) for immunotherapy trials. Cryobiology 54:294–297
25. Yang H, Zhao H, Acker JP, Liu JZ, Akabutu J, McGann LE (2005) Effect of dimethyl sulfoxide on post-thaw viability assessment of CD45+ and CD34+ cells of umbilical cord blood and mobilized peripheral blood. Cryobiology 51:165–175

Chapter 15

Assessment of Young and Aged Hematopoietic Stem Cell Activity by Competitive Serial Transplantation Assays

Yu Wei Zhang and Nina Cabezas-Wallscheid

Abstract

Healthy hematopoietic stem cells (HSCs) are capable to self-renew and reconstitute the complete hematopoietic system. Upon aging, there is an increased incidence of blood-related diseases. Age-related phenotypes have been widely studied by bone marrow transplantation experiments, where reconstitution of the transplanted cells is a direct measure of HSC activity. In this protocol we describe a competitive bone marrow transplantation assay to functionally test young and old HSCs.

Key words Hematopoietic stem cells, Transplants, Young, Aged, Bone marrow, Competitive

1 Introduction

Hematopoietic stem cells (HSCs) hold the capacity to generate a series of multipotent progenitors that differentiate into lineage-committed progenitors and subsequently mature cells [1–5]. HSCs remain quiescent for most of their life span [6–9, 18]. By contrast, downstream committed progenitors show very high proliferative capacities [10–13]. The maintenance of the hematopoietic lineage is critical for homeostasis and enables the rapid and robust response to physiological stresses, such as cytokines, infection, and injury [14–17]. It remains to be elucidated how a stem cell determines whether to maintain its multipotency or to differentiate to generate mature cells.

Upon aging an increased frequency of anemias [19], myeloproliferative diseases, leukemia [20], as well as decreased lymphopoiesis [21, 22] have been observed. The manifestation of age-related defects in the HSC population can alter the functional dynamics of self-renewal and differentiation [23, 24]. Several studies suggest that aged HSCs exhibit a myeloid bias, impaired DNA repair, and shortening of their telomeres which contribute to the decline in homing and reconstitution capacities [25–28]. Cell extrinsic

Gerd Klein and Patrick Wuchter (eds.), *Stem Cell Mobilization: Methods and Protocols*, Methods in Molecular Biology, vol. 2017, https://doi.org/10.1007/978-1-4939-9574-5_15,

alterations due to aging are associated with decreased bone marrow cellularity, enhanced adipogenesis, and changes in extracellular matrix components [29]. All these factors are inversely correlated with a decline in HSC function. Young HSCs transplanted to aged niches exhibit deficient homing and a decreased potential for differentiation as well as a bias to differentiate toward the myeloid lineage [30]. It is important to remember that some of these aging phenotypes are still debated and additional research is required to elucidate the biology of aged HSCs.

Transplantation assays is the golden standard method to address the functional potential of HSCs [31]. HSC potency is quantified by the frequency of their progeny over long periods of time. To distinguish cells derived from wild-type (WT) and genetically modified HSCs, researchers have utilized congenic C57BL/6 (CD45.2) and B6.SJL (CD45.1) mouse strains [32], as well as their F1 hybrids (CD45.1/2). Bone marrow cells are isolated from CD45.2 and CD45.1/2 donor mouse strains and transplanted into lethally irradiated CD45.1 recipient mice. Mouse monoclonal antibodies targeting specific epitopes in the CD45 regions are used to distinguish the origin of the hematopoietic cells [33, 34]. In combination with blood lineage markers, the CD45.1-CD45.2 system allows experimenters to detect myeloid and lymphoid outcome, but not the erythroid differentiation potential of the two competitors.

Compared to WT congenic C57BL/6 (CD45.2) cells, WT B6. SJL (CD45.1) cells have been reported to show decreased homing efficiency, a reduced number of transplantable HSCs, and a cell intrinsic engraftment disadvantage [35]. WT C57BL/6 (CD45.2) cells show higher myeloid, B and T cell engraftment. The molecular cause of the inherent bias that causes an unfair advantage to CD45.2 is not well understood. Recently, a novel mouse strain has been developed to abolish this advantage [36]. This mouse model named CD45.1^{STEM} was created de novo from the WT C57BL/6N (CD45.2) strain by a single amino acid change in the *Ptprc* gene encoding for the CD45 antigen, making it detectable with the CD45.1 antibody instead. CD45.1^{STEM} has been proven to be functionally equivalent to its CD45.2 counterpart [36]. Using CD45.2 and CD45.1^{STEM} in competitive transplantation assays serves as a more accurate and fair comparison of HSC biology.

Here we present a transplantation protocol that will guide you through the necessary steps needed to plan and execute a successful competitive transplantation experiment for young and aged hematopoietic cells. Young hematopoetic cells of a genetically modified organism (GMO) (i.e., knockout) will compete against its young wild-type (WT) counterpart. Similarly, aged hematopoietic cells with the same genetic modifcation will compete with their aged WT counterpart. This competitive bone marrow transplantation scheme will delineate the importance of particular genes during aging.

2 Materials

2.1 Selection of Mice (See Notes 1)

- For competitive assay CD45.2 and CD45.1/2 donor mice are transplanted into CD45.1 recipient mice (Fig. 1).

Mice	Strain	Young	Old
CD45.2 donor (GMO or WT)	C57BL/6	8–12 weeks	18–24 months[a]
CD45.1/2 competitor (WT)	F1 hybrid of C57BL/6 × B6.SJL	8–12 weeks	18–24 months[a]
CD45.1 recipient (WT)	B6.SJL	8–12 weeks	18–20 months or 8–12 weeks[b]

[a]Mice that are 18–24 months old are equivalent to the human age of 56–69 years old. Therefore, aged HSC mice used for the experiment should be older than 18 months. Of note, mice that are 10–14 months old are equivalent to the human age of 38–47 years old and thereby categorized as middle-aged [37].
[b]The availability of 18-month-old mice is usually limited. Therefore, some groups opt to use young mice as recipients. Take into consideration that the niche influences HSC behavior and therefore proper controls are needed.

2.2 Dissection and Isolation of Bone Marrow

- CO_2 chamber (optional).
- 70% ethanol.
- Dissection tools (dissecting scissors, dissecting tray, iris forceps, steel dissecting pins, and scalpel).

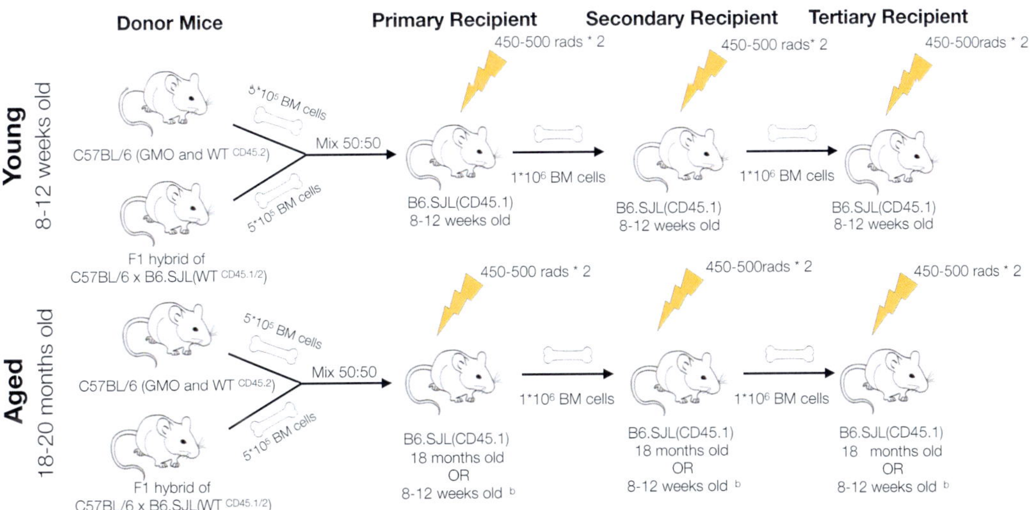

Fig. 1 Experimental strategy comparing HSC function of young or old mice. Bone marrow cells from donor WT, GMO CD45.2 and competitor WT CD45.1/2 mice are isolated. A 50:50 mix of CD45.2:CD45.1/2 bone marrow cells are transplanted into lethally irradiated CD45.1 recipients (450–500 rad × 2 times). At 16 weeks posttransplantation, the primary recipients are sacrificed, bone marrow cells are isolated and transplanted into secondary recipients. This process might be repeated if tertiary transplantations are required

- Phosphate-buffered saline (PBS).
- 6-well plate filled with 5 ml PBS.
- Mortar and pestle.
- 40 μm cell strainer.
- 50 ml Falcon tube.
- Cell counter (e.g., hemocytometer).

2.3 For Irradiation

- Supplemental antibiotic water: 5 ml Trimethosel in 1 l of tap water.
- Irradiator.
- Container.

2.4 For Injections

1. 29-gauge needles (*see* **Note 2**).
2. 1 ml syringe.
3. Gauze sponges.
4. Heating lamp or warm water bath (maximal 40°C).
5. Restrainer.
6. 70% ethanol.

2.5 FACS Analysis

1. Blood lancet.
2. 1.5 ml EDTA polypropylene microcentrifuge tube.
3. Erythrocyte-lysing buffer (150 mM NH_4Cl, 10 mM $KHCO_3$, 0.1 mM Na_2EDTA; dissolve in H_2O; pH 7.2–7.4; store at room temperature for up to 6 months).
4. PBS.
5. Fetal bovine serum (FBS).
6. Antibodies (*see* **Note 3**).

Antibody	Fluorochrome	Population	Clone
CD45.1	Alexa Fluor 700	CD45.1 mouse cells	A20
CD45.2	APC-Cy7	CD45.2 mouse cells	104
CD11b	PE	Myeloid cells	M1/70
CD3	FITC	T cells	UCHT1
B220	APC	B-cells	RA3-6B2
Gr-1	PE-Cy7	Granulocytes	RB6-8C5

7. FACS tubes conical 4.5 ml.
8. Flow cytometry device.

3 Methods

3.1 Preparation of Bone Marrow Cells for Transplantation

Before performing the animal experiment, consult with the local animal welfare authorities or in-house animal specialists regarding the rules and regulations governing animal experiments in your region.

1. Euthanize donor mice through either cervical dislocation or CO_2 asphyxiation.
2. Place mouse on its back, secure animal onto dissecting tray by pinning front palms so that they are raised diagonally from its body, and disinfect by spraying down with 70% ethanol.
3. Isolate the tibias, femurs, and hipbones and place into 6-well plate filled with ice-cold PBS. Excessive muscle tissue will loosen after 5 min in the PBS, and bones can be further cleaned using the scalpel to remove excess muscle tissue.
4. To isolate bone marrow cells, transfer bones to the mortar. Add 5 ml of PBS and crash the bones thoroughly with the pestle. Grind bones until they lose their pink color and become pale white/pink.
5. Collect the cell suspension, and pipette up and down few times to break apart the clumps of cells.
6. Place a 40 μm cell strainer over a 50 ml Falcon tube, and filter the cell suspension from **step 5** through it.
7. Wash the mortar with additional 5 ml of PBS, and crush the bones for 20 sec. Again, filter the cell suspension through the cell strainer into the 50 ml Falcon tube.
8. For a competitive transplantation, ideally 2–5 × 10^5 total bone marrow cells from CD45.2 donor mice are supplemented with the same number of bone marrow cells from the CD45.1/2 competitor. The transplantation scheme is depicted below and shown in Fig. 1.

Age	Competition	Ratio of Injected Cells
Young	(GMO or WT)$^{CD45.2}$: WT$^{CD45.1/2}$	5 × 10^5 CD45.2: 5 × 10^5 CD45.1/2
Old	(GMO or WT)$^{CD45.2}$: WT$^{CD45.1/2}$	5 × 10^5 CD45.2: 5 × 10^5 CD45.1/2

9. Before transplantation, to confirm 50:50 ratio of CD45.2 to CD45.1/2 bone marrow cells, prepare an aliquot of the cells from **step 8**. Stain aliquot with the antibodies against CD45.1 and CD45.2 and analyze by flow cytometry
10. Adjust the 50:50 mix of CD45.1 and CD45.1/2 cells from **step 8** in a volume appropriate for intravenous injection (100 μl per recipient).

3.2 Irradiation of Recipient Mice

1. 1–5 days prior to irradiation, place CD45.1 recipient mice on supplemental antibiotic water.
2. Place the recipient mouse into the acrylic container (*see* **Note 4**). Multiple mice can be irradiated simultaneously, depending on the size of the irradiator.
3. One day prior to injection, give the recipient mice a lethal dose of total body irradiation (TBI). Ideally, split the entire 900-950 rad lethal dose of TBI into two sublethal doses: 450 rad in the morning and 4–6 h later another dose of 450–500 rad.
4. Give antibiotic water to irradiated mice for at least 14–28 days postirradiation and carefully monitor.

3.3 Intravenous Injection of Total Bone Marrow Cell Suspension into the Lateral Tail Vein of Irradiated Recipient Mice

1. Prior to injection, warm up the tail of the recipient under the heat lamp or in warm water to dilate the lateral veins. Researchers using warm water need to maintain the temperature at 40 °C.
2. For injection, place each mouse into a mouse restrainer (*see* **Note 5**).
3. Dampen a gauze sponge with 70% ethanol, and disinfect the injection site.
4. Using the non-dominant hand, hold the mouse at the most distal end of the tail (approximately 1/3 from the tip). Bend the tail over the index finger, and use the thumb to secure it to this position.
5. Locate the tail veins that run on the lateral sides of the tail.
6. Holding the syringe strictly parallel to the tail with the lumen of the needle facing upwards, insert the needle about 5 mm into the tail vein.
7. Inject slowly. In case of resistance, the needle should be removed and reinserted above the first site.
8. After injection, remove the needle without rotating or moving the needle while in the vein. Apply gentle compression to the tail until bleeding has stopped.
9. Mice need to be monitored for several days after transplantation since insufficient cell transplantation can lead to death within 2 weeks.

3.4 Collection of Peripheral Blood

1. Restrain the mouse with the lateral surface of the head facing upwards. You can alternate between the two sides of the head every 4 weeks.
2. At this position, the superficial temporal vein runs from the orbit to the base of the ear diagonally. Use the lancet to stab halfway between the ear and the mandible using enough pressure to produce a small incision.

3. The drops of blood should flow into the lancet and be collected into the EDTA-coated tube. Approximately, 100 μl of blood can be collected with this method.
4. To stop the bleeding, place a light pressure on the cheek using gauze for a second.

3.5 Analysis of Peripheral Blood by FACS

1. To deplete red blood cells, add 500 μl of erythrocyte-lysing buffer to each tube, and incubate at room temperature for 5–10 min. For convenience, the cell suspension can be transferred into a 4.5 ml FACS tube.
2. Add 2 ml of PBS with 2% FBS to quench the red blood cell lysis buffer.
3. Centrifuge at 400 × *g* for 5 min. Aspirate the supernatant and be careful not to touch the bottom of the tube.
4. If red blood cells still remain, repeat the red cell lysis one to two additional times (**steps 1–3**) until there is a clear white/pink pellet.
5. Prepare complete antibody cocktail mix in PBS. In addition, prepare compensation tubes: single color controls and unstained control.
6. Add 50 μl of antibody cocktail mix per sample; vortex gently; incubate at 4 °C for 20 min.
7. Add 2 ml of PBS to wash cells; vortex gently
8. Centrifuge at 400 × *g* for 5 min, and aspirate supernatant leaving approximately 100–200 μl at the bottom. The tubes are now ready for FACS analysis.
9. The analysis of the different lineages of the peripheral blood can be performed using the gates described below (Fig. 2).

Peripheral blood analysis is performed starting at week 4 posttransplantation and then followed by intervals of 4 weeks between bleedings. At 16 weeks posttransplantation, mice are sacrificed, and bone marrow cells from the tibia, femur, and hipbones are harvested. Reconstitution analysis is performed on bone marrow cells. Approximately 20 million bone marrow cells are sufficient to analyze all types of reconstituted hematopoietic cells, including HSC cell cycle analysis. In case of impaired HSC self-renewal, this phenotype might be observed only upon serial transplantation [38]. Thus, secondary or even tertiary transplantation assays are recommended to assess long-term HSC activity. To perform secondary transplantation, 1–3 × 10^6 total bone marrow cells from the sacrificed recipients are injected into new cohort of irradiated CD45.1 mice.

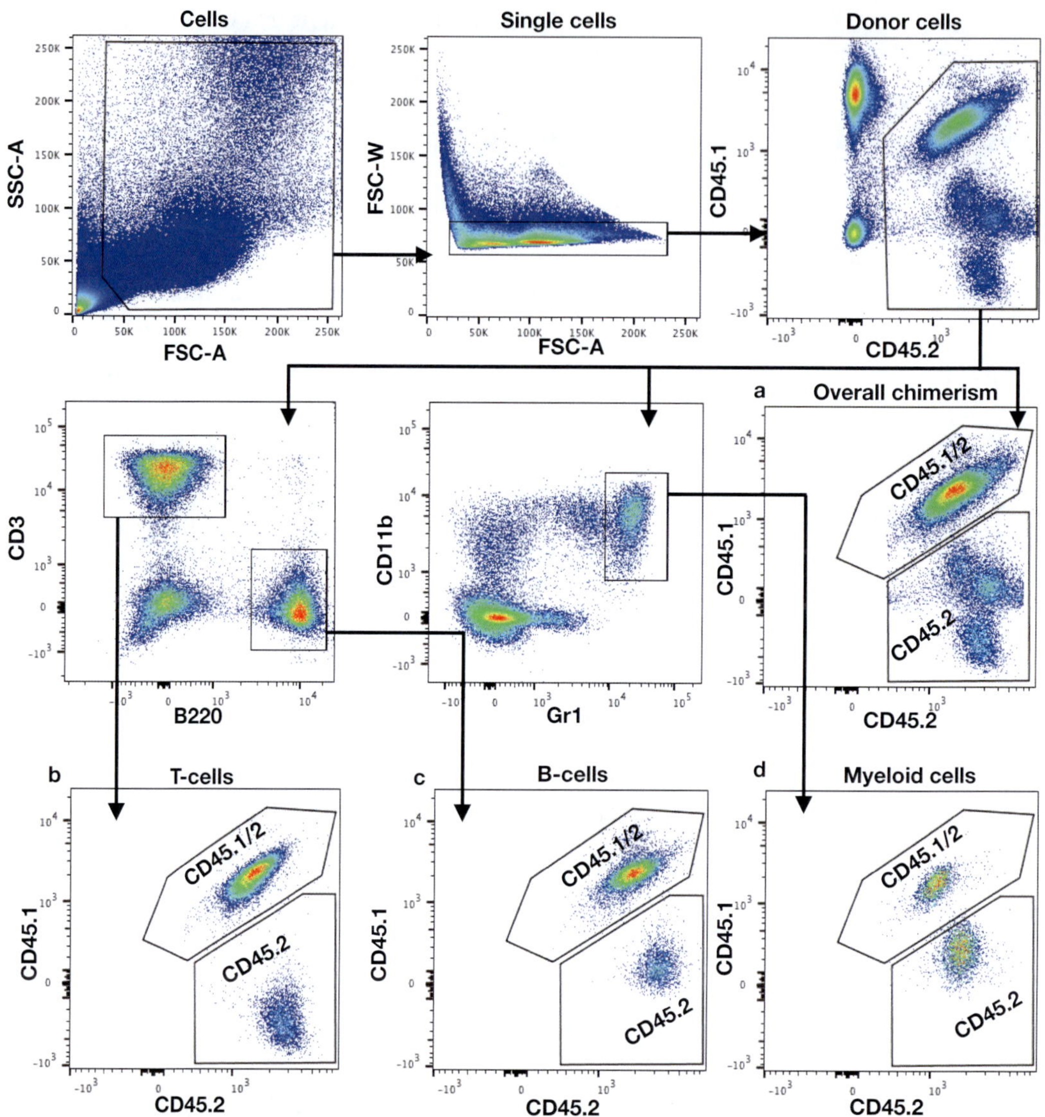

Fig. 2 FACS analysis of peripheral blood from transplanted mice. Representative gating scheme for competitive bone marrow transplantation. B, T, and myeloid cell populations are depicted. Comparison of CD45.1/2 and CD45.2 constitution potential within (**a**) total bone marrow compartment, (**b**) T cells, (**c**) B cells, and (**d**) myeloid cells

4 Notes

1. Ideally, CD45.1^{STEM} and CD45.1^{STEM}/2 mice would be used for competitive transplantation assays. Given that the B6.SJL (CD45.1) mouse strain is more widely available and for practical reasons this protocol will use B6.SJL (CD45.1) as the recipient. We highly recommend using CD45.2 (WT) transplanted with

CD45.1/2 donor mice as controls in parallel to the CD45.2 (GMO) transplantations. However, CD45.1 mice can be replaced with CD45.1^{STEM} if they are available to you.

CD45.1 mice are designated as recipient, since the heterozygous genotype of CD45.1/2 mouse strains make it less frequent compared to the homozygote CD45.1 mouse strains.

2. Avoid using smaller gauge (<29 gauge) needles as this can lead to increase shearing of cells in the inocula.
3. Alternative panel for FACS. When using the FITC channel for other purposes such as a GFP reporter cells, we suggest an alternative antibody cocktail for FACS.

   Other alternative fluorochromes might be equally used.

Antibody	Fluorochrome	Clone
CD45.1	PE-Cy7	A20
CD45.2	Pacific Blue	104
CD11b	APC Cy7	M1/70
CD3	PE	UCHT1
B220	Alexa Fluor 700	RA3-6B2
Gr-1	APC	RB6-8C5

4. To help reduce the stress of mice, place some of the nesting material into the acrylic container.
5. An alternative method to restrain and inject mice.

   Materials

   (a) 1-pound weight.

   (b) 1l beaker.

   (c) 29-gauge needles.

   (d) 1 ml syringe.

   Method

   (a) With tape or another adhesive, secure a 1-pound weight onto the outside bottom of a 1l beaker.

   (b) Bend the tip of the syringe 90° with the lumen of the needle facing upwards. By bending the needle 90°, the injection can then be done parallel to the tail with greater ease.

   (c) Place the mice under the 1l beaker, with their tail exposed from the small spout. The weight should be heavy enough to secure the mice under the beaker; otherwise add more weights.

   (d) Inject with the lumen of the needle facing upwards.

This method gives more stability to inject and mice show less stress.

References

1. Weissman IL (2000) Stem cells: units of development, units of regeneration, and units in evolution. Cell 100(1):157–168
2. Orkin SH, Zon LI (2008) Hematopoiesis: an evolving paradigm for stem cell biology. Cell 132(4):631–644
3. He S, Nakada D, Morrison SJ (2009) Mechanisms of stem cell self-renewal. Annu Rev Cell Dev Biol 25(1):377–406
4. Seita J, Weissman IL (2010) Hematopoietic stem cell: self-renewal versus differentiation. Wiley Interdiscip Rev Syst Biol Med 2 (6):640–653
5. Cabezas-Wallscheid N, Klimmeck D, Hansson J, Lipka Daniel B, Reyes A, Wang Q, Weichenhan D, Lier A, von Paleske L, Renders S, Wünsche P, Zeisberger P, Brocks D, Gu L, Herrmann C, Haas S, Essers MAG, Brors B, Eils R, Huber W, Milsom MD, Plass C, Krijgsveld J, Trumpp A (2014) Identification of regulatory networks in HSCs and their immediate progeny via integrated proteome, transcriptome, and DNA methylome analysis. Cell Stem Cell 15(4):507–522
6. Wilson A, Laurenti E, Oser G, van der Wath RC, Blanco-Bose W, Jaworski M, Offner S, Dunant CF, Eshkind L, Bockamp E, Lió P, MacDonald HR, Trumpp A (2008) Hematopoietic stem cells reversibly switch from dormancy to self-renewal during homeostasis and repair. Cell 135(6):1118–1129
7. van der Wath RC, Wilson A, Laurenti E, Trumpp A, Liò P (2009) Estimating dormant and active hematopoietic stem cell kinetics through extensive modeling of bromodeoxyuridine label-retaining cell dynamics. PLoS One 4(9):e6972
8. Qiu J, Papatsenko D, Niu X, Schaniel C, Moore K (2014) Divisional history and hematopoietic stem cell function during homeostasis. Stem Cell Reports 2(4):473–490
9. Bernitz JM, Kim HS, MacArthur B, Sieburg H, Moore K (2016) Hematopoietic stem cells count and remember self-renewal divisions. Cell 167(5):1296–1309.e1210
10. Cantor AB, Orkin SH (2001) Hematopoietic development: a balancing act. Curr Opin Genet Dev 11(5):513–519
11. Miyamoto T, Iwasaki H, Reizis B, Ye M, Graf T, Weissman IL, Akashi K (2002) Myeloid or lymphoid promiscuity as a critical step in hematopoietic lineage commitment. Dev Cell 3(1):137–147
12. Adolfsson J, Månsson R, Buza-Vidas N, Hultquist A, Liuba K, Jensen CT, Bryder D, Yang L, Borge O-J, Thoren LAM, Anderson K, Sitnicka E, Sasaki Y, Sigvardsson M, Jacobsen SEW (2005) Identification of Flt3+ lympho-myeloid stem cells lacking erythro-megakaryocytic potential: a revised road map for adult blood lineage commitment. Cell 121 (2):295–306
13. Forsberg EC, Serwold T, Kogan S, Weissman IL, Passegué E (2006) New evidence supporting megakaryocyte-erythrocyte potential of Flk2/Flt3+ multipotent hematopoietic progenitors. Cell 126(2):415–426
14. Essers MAG, Offner S, Blanco-Bose WE, Waibler Z, Kalinke U, Duchosal MA, Trumpp A (2009) IFNα activates dormant haematopoietic stem cells in vivo. Nature 458:904–908
15. Baldridge MT, King KY, Boles NC, Weksberg DC, Goodell MA (2010) Quiescent haematopoietic stem cells are activated by IFN-γ in response to chronic infection. Nature 465:793–797
16. Takizawa H, Regoes RR, Boddupalli CS, Bonhoeffer S, Manz MG (2011) Dynamic variation in cycling of hematopoietic stem cells in steady state and inflammation. J Exp Med 208 (2):273–284
17. Walter D, Lier A, Geiselhart A, Thalheimer FB, Huntscha S, Sobotta MC, Moehrle B, Brocks D, Bayindir I, Kaschutnig P, Muedder K, Klein C, Jauch A, Schroeder T, Geiger H, Dick TP, Holland-Letz T, Schmezer P, Lane SW, Rieger MA, Essers MAG, Williams DA, Trumpp A, Milsom MD (2015) Exit from dormancy provokes DNA-damage-induced attrition in haematopoietic stem cells. Nature 520:549–552
18. Cabezas-Wallscheid N, Buettner F, Sommerkamp P, Klimmeck D, Ladel L, Thalheimer FB, Pastor-Flores D, Roma LP, Renders S, Zeisberger P, Przybylla A, Schönberger K, Scognamiglio R, Altamura S, Florian CM, Fawaz M, Vonficht D, Tesio M, Collier P, Pavlinic D, Geiger H, Schroeder T, Benes V, Dick TP, Rieger MA, Stegle O, Trumpp A (2017) Vitamin A-retinoic acid signaling regulates hematopoietic stem cell dormancy. Cell 169(5):807–823.e19
19. Guralnik JM, Eisenstaedt RS, Ferrucci L, Klein HG, Woodman RC (2004) Prevalence of anemia in persons 65 years and older in the United States: evidence for a high rate of unexplained anemia. Blood 104(8):2263–2268
20. Lichtman MA, Rowe JM (2004) The relationship of patient age to the pathobiology of the clonal myeloid diseases. Semin Oncol 31 (2):185–197

21. Linton PJ, Dorshkind K (2004) Age-related changes in lymphocyte development and function. Nat Immunol 5:133–139
22. Miller JP, Allman D (2003) The decline in B lymphopoiesis in aged mice reflects loss of very early B-lineage precursors. J Immunol 171 (5):2326–2330
23. Rossi DJ, Bryder D, Zahn JM, Ahlenius H, Sonu R, Wagers AJ, Weissman IL (2005) Cell intrinsic alterations underlie hematopoietic stem cell aging. Proc Natl Acad Sci U S A 102 (26):9194–9199
24. Geiger H, de Haan G, Florian MC (2013) The ageing haematopoietic stem cell compartment. Nat Rev Immunol 13:376–389
25. Rossi DJ, Bryder D, Seita J, Nussenzweig A, Hoeijmakers J, Weissman IL (2007) Deficiencies in DNA damage repair limit the function of haematopoietic stem cells with age. Nature 447:725–729
26. Cho RH, Sieburg HB, Muller-Sieburg CE (2008) A new mechanism for the aging of hematopoietic stem cells: aging changes the clonal composition of the stem cell compartment but not individual stem cells. Blood 111 (12):5553–5561
27. Dykstra B, Olthof S, Schreuder J, Ritsema M, de Haan G (2011) Clonal analysis reveals multiple functional defects of aged murine hematopoietic stem cells. J Exp Med 208 (13):2691–2703
28. Flach J, Bakker ST, Mohrin M, Conroy PC, Pietras EM, Reynaud D, Alvarez S, Diolaiti ME, Ugarte F, Forsberg EC, Le Beau MM, Stohr BA, Méndez J, Morrison CG, Passegué E (2014) Replication stress is a potent driver of functional decline in ageing haematopoietic stem cells. Nature 512(7513):198–202
29. Tuljapurkar SR, McGuire TR, Brusnahan SK, Jackson JD, Garvin KL, Kessinger MA, Lane JT, O'Kane BJ, Sharp JG (2011) Changes in human bone marrow fat content associated with changes in hematopoietic stem cell numbers and cytokine levels with aging. J Anat 219(5):574–581
30. Ergen AV, Boles NC, Goodell MA (2012) Rantes/Ccl5 influences hematopoietic stem cell subtypes and causes myeloid skewing. Blood 119(11):2500–2509
31. Harrison D (1980) Competitive repopulation: a new assay for long-term stem cell functional capacity. Blood 55(1):77–81
32. Shen FW, Saga Y, Litman G, Freeman G, Tung JS, Cantor H, Boyse EA (1985) Cloning of Ly-5 cDNA. Proc Natl Acad Sci U S A 82 (21):7360–7363
33. Yakura H, Shen FW, Bourcet E, Boyse EA (1983) On the function of Ly-5 in the regulation of antigen-driven B cell differentiation. Comparison and contrast with Lyb-2. J Exp Med 157(4):1077–1088
34. Spangrude G, Heimfeld S, Weissman I (1988) Purification and characterization of mouse hematopoietic stem cells. Science 241 (4861):58–62
35. Waterstrat A, Liang Y, Swiderski CF, Shelton BJ, Van Zant G (2010) Congenic interval of CD45/Ly-5 congenic mice contains multiple genes that may influence hematopoietic stem cell engraftment. Blood 115(2):408–417
36. Mercier FE, Sykes DB, Scadden DT (2016) Single targeted exon mutation creates a true congenic mouse for competitive hematopoietic stem cell transplantation: the C57BL/6-CD45.1 (STEM) mouse. Stem Cell Reports 6 (6):985–992
37. Flurkey KCJ, Harrison DE (2007) The mouse in aging research. In: Fox JG et al (eds) The mouse in biomedical research, 2nd edn. American College Laboratory Animal Medicine, Elsevier, pp 637–672
38. Eaves CJ (2015) Hematopoietic stem cells: concepts, definitions, and the new reality. Blood 125(17):2605–2613

Chapter 16

Zebrafish Xenografts for the In Vivo Analysis of Healthy and Malignant Human Hematopoietic Cells

Martina Konantz, Joëlle S. Müller, and Claudia Lengerke

Abstract

The zebrafish is a powerful vertebrate model for genetic studies on embryonic development and organogenesis. In the last decades, zebrafish were furthermore increasingly used for disease modeling and investigation of cancer biology. Zebrafish are particularly used for mutagenesis and small molecule screens, as well as for live imaging assays that provide unique opportunities to monitor cell behavior, both on a single cell and whole organism level in real time. Zebrafish have been also used for in vivo investigations of human cells transplanted into embryos or adult animals; this zebrafish xenograft model can be considered as an intermediate assay between in vitro techniques and more time-consuming and expensive mammalian models.

Here, we present a protocol for transplantation of healthy and malignant human hematopoietic cells into larval zebrafish; transplantation into adult zebrafish and possible advantages and limitations of the zebrafish compared to murine xenograft models are discussed.

Key words Zebrafish, Xenograft, Yolk sac, Duct of Cuvier, HSC

1 Introduction

Over the past decades, the zebrafish has developed into a powerful vertebrate model for genetic studies of embryonic development and disease. These investigations are extremely feasible in zebrafish due to the high fecundity and short generation times as well as the ease to perform transient and stable genetic modifications [1]. Furthermore, molecular pathways are highly conserved between zebrafish and mammals indicating that discoveries made in fish are in many cases relevant for humans [2]. Chemical carcinogenesis screens providing cancer models resembling human disease and transgenic zebrafish expressing human and mouse oncogenes and tumor suppressors have enabled studies on tumor formation and maintenance in zebrafish [3–6]. Most importantly, however, zebrafish embryos lack adaptive immunity enabling engraftment without the need of immunosuppression [7]. Additionally, the transparency,

Gerd Klein and Patrick Wuchter (eds.), *Stem Cell Mobilization: Methods and Protocols*, Methods in Molecular Biology, vol. 2017, https://doi.org/10.1007/978-1-4939-9574-5_16,

which occurs both naturally during development and through mutations in adults, allows following of cancer cell progression microscopically without the need of invasive methods as it is the case in murine xenograft models [8, 9].

Due to obvious ethical and practical limitations, in vivo studies on human cells are limited to xenografts. While these assays are so far best established in immunocompromised mice [10, 11], several reports document the feasibility of xenografting human cells to zebrafish [12]. Since the zebrafish embryo is mostly transparent, engraftment of human cancer cells can be followed over time even on a single-cell level through, e.g., high-resolution imaging techniques, by flow cytometry analyses or other methods of choice [13–15]. If human cancer cells are transplanted into the vessel-free area of the yolk sac, they are able to proliferate or to migrate toward other tissues [12–14, 16–19]. Furthermore, they are able to induce angiogenesis upon transplantation, to extravasate, and to induce micrometastasis or interact with the host microenvironment [13, 20–26]. Most of these studies have been performed with solid tumor cells, while only a few data exists for healthy and malignant hematopoietic cells [14, 15, 27–32].

Most of the studies concerning healthy hematopoietic cells involved transplantation in adult zebrafish preconditioned with irradiation or dexamethasone [32–37]. Increasing efforts have been made to engineer adult immunocompromised zebrafish, which promise to allow the study of xenografts in the adult microenvironment [32, 34, 37–39]. A recent study by Staal and colleagues shows the feasibility of HSPC (human hematopoietic stem and progenitor cell) transplantation into transgenic zebrafish embryos. They show that fluorescently labeled HSPCs survive up to 6 days posttransplantation (dpt) and that these cells can respond in vivo to the stromal-derived factor CXCL12. Furthermore, the authors demonstrate that the transplanted cells differentiate along the myeloid lineage and also survive if transplanted retro-orbitally into adult zebrafish [31]. However, several data suggest that the cross-reactivity of cytokines and proteins is in general lower between zebrafish and human as compared to mouse and human [40–43]. Next to temperature (29 °C in zebrafish vs. 37 °C in human) and anatomical size differences ("large human cells in narrow fish vessels"), this likely represents the major factor that limits the utility of this model. Not surprisingly thus, human cytokine knock-in zebrafish are underway and may significantly improve the results obtained with human hematopoietic xenografts. Of note, several studies have also shown the usability of nanotherapeutics together with zebrafish xenografts [44–46].

Here, we focus on the transplantation of healthy and malignant hematopoietic cells in the zebrafish embryo, in which blood vessels or immune cells are fluorescently labeled. Analysis involves quantification based on flow cytometry analysis and interaction with the host microenvironment upon transplantation into the vessel-free area of the yolk and transplantation of HSCs into the duct of Cuvier.

2 Materials

2.1 Cell Suspension Preparation

2.1.1 Isolation of Cord Blood-Derived CD34+ HSPCs

1. Dulbecco's Phosphate-Buffered Saline (DPBS).
2. SepMate (STEMCELL Technologies, Vancouver, Canada).
3. Histopaque®-1077.
4. RBC lysis buffer: 8.3 g ammonium chloride (NH_4Cl), 1.0 g potassium hydrogen carbonate ($KHCO_3$), 200 μl 0.5 M EDTA, and 1 l of distilled water.
5. MACS buffer: 500 ml DPBS, 2.5 g bovine serum albumin (BSA), and 5 ml 0.2 M EDTA.
6. MACS purification magnet with an indirect CD34 MicroBead Kit (Miltenyi Biotec, Bergisch Gladbach, Germany).
7. Freezing medium: fetal calf serum (FCS) and 10% dimethyl sulfoxide (DMSO).

2.1.2 Tumor Cell Suspension

1. Cell culture flasks 75 cm^2.
2. Cell culture medium: RPMI 1640, 10% FCS, and 1% Penicillin/Streptomycin.
3. Human tumor cell lines. Here we use the human SKM-1 cell line.
4. Neubauer cell counter and trypan blue.

2.2 Cell Labeling

1. 1 mg/ml CellTracker (CM-Dil, Molecular Probes®, Life technologies) stock: lyophilized pellet, DMSO, and storage at −20 °C. Working concentration 5 μg/ml in DPBS (*see* **Note 1**).

2.3 Preparation of Zebrafish Embryos for Transplantation and Microscopy

1. 2-day-old zebrafish larvae (reporter line with fluorescent blood vessels or fluorescent immune cells). Here we use either *Tg (kdrl:eGFP)*^{la116} or *Tg(mpeg:gal4;UAS:Kaede)*$^{gl24/rk7}$ as well as wild-type zebrafish embryos [47, 48].
2. Larvae must be without chorion. Either use forceps or dechorionate with 0.04% Pronase E overnight.
3. 60× egg water E3 stock solution (pH 7.2): 17.4 g sodium chloride, 0.8 g potassium chloride, 2.9 g calcium chloride dihydrate, 4.89 g magnesium chloride hexahydrate, and 1 l of distilled water.
4. Egg water: 16 ml 60× E3, 100 μl methylene blue, and 984 ml of distilled water.

5. 2% Tricaine (pH 7): 10 g Tricaine (MS-222, Sigma-Aldrich®), 10.5 ml Tris–HCl, and 489.5 ml of distilled water. Store aliquots at −20 °C. Use a 0.04% working solution.
6. Petri dish (100 × 20 mm).
7. Petri dish (100 × 20 mm) coated with 1.5% (w/v) agarose, dissolved in egg water with mold.
8. Pasteur plastic and glass pipettes (3.5 ml).
9. Borosilicate glass capillary needles: OD = 1.2 mm, ID = 0.94 mm; L = 100 mm with filament, tips broken at OD = 20 μm, and length of the tip = 9 mm (Biomedical Instruments, Zöllnitz, Germany).
10. Microloader tips (0.5–20 μl Microloader, Eppendorf).
11. Forceps (Dumont #5) for manual chorion removal and needle cutting.
12. Pneumatic Pico Pump PV820, World Precision Instruments (WPI).
13. Micromanipulator (WPI).
14. Stereo microscope (ZEISS Stemi 508).
15. 35 °C incubator.
16. Stereo fluorescence microscope (ZEISS SteREO Discovery. V20).
17. Confocal microscope (ZEISS LSM 710).
18. Glass-bottomed petri dish 35 × 10 mm.
19. Agarose Low Melt.
20. Digital MC170 HD camera attached to a Leica DM 2000 LED.

2.4 Flow Cytometry Quantification

1. Liberase™ TM Research Grade.
2. Bovine serum albumin (BSA).
3. 35 μm cell strainer on a BD Falcon polystyrene tube (BD Biosciences, Franklin Lakes, NJ, USA).
4. Live/dead staining 7-AAD.
5. Thermocycler.
6. Flow cytometer.

2.5 H&E Staining on Zebrafish Slices

1. 4% paraformaldehyde.
2. Tissue processing machine (Tissue Processor TPC15, MEDITE GmbH, Burgdorf, Germany).
3. Paraffin embedding machine (TBS88 Paraffin Embedding System, MEDITE GmbH).
4. Microtome.

5. SuperFrost® Plus glass slides.
6. Eosin (eosin 1% aqueous solution/hematoxylin (H&E) staining).
7. Tissue stainer (COT20, MEDITE GmbH).

3 Methods

3.1 Zebrafish Embryo Preparation

1. Keep eggs at 28–29 °C in petri dishes with egg water ($n \leq 60$).
2. Remove the chorion at 1 dpf (day postfertilization). This can be achieved manually or enzymatically by adding 10 μl 0.04% Pronase E overnight. Keep the embryos at 28–29 °C until cell injection.

3.2 Isolation of Cord Blood-Derived CD34+ HSPCs

1. Dilute cord blood 1:2 in DPBS.
2. Prepare SepMate with Histopaque®-1077, and continue according to the manufacturer's protocol.
3. Dissolve the obtained cell pellet in 30 ml RBC lysis buffer, and let stand for 2 min.
4. Centrifuge for 10 min at 300 × *g* and remove the supernatant. Repeat if the cell pellet appears still red.
5. Resuspend the pellet in DPBS and count cells.
6. Centrifuge for 10 min at 300 × *g* and remove the supernatant.
7. Resuspend the pellet at 10^8 cells/ml in MACS buffer.
8. Isolate CD34+ HSPCs from mononuclear cells (MNCs) by MACS purification according to the manufacturer's protocol.
9. Store in FCS with 10% dimethyl sulfoxide (DMSO) in liquid nitrogen or use freshly.

3.3 Cell Suspension Preparation

1. Count cells and collect by centrifugation (300 × *g* for 7 min in a benchtop centrifuge).
2. Adjust cell suspension concentration to 10^6 cells/ml in DPBS.
3. Stain cells according the manufacturer's protocol. Incubate for 5 min at 37 °C and then for 20 min at 4 °C.
4. Finally make a cell suspension in 35 μl DPBS, and keep cells at room temperature during transplantation.

3.4 HSPC and AML Cell Transplantation

1. Load the needle with the cell suspension, and place in the micromanipulator (*see* **Note 2**).
2. Cut the end of the needle with forceps under the microscope (*see* **Note 3**).
3. Anesthetize larvae with 0.04% Tricaine prior to transplantation.
4. Load embryos to the mold of the agarose-coated plate.

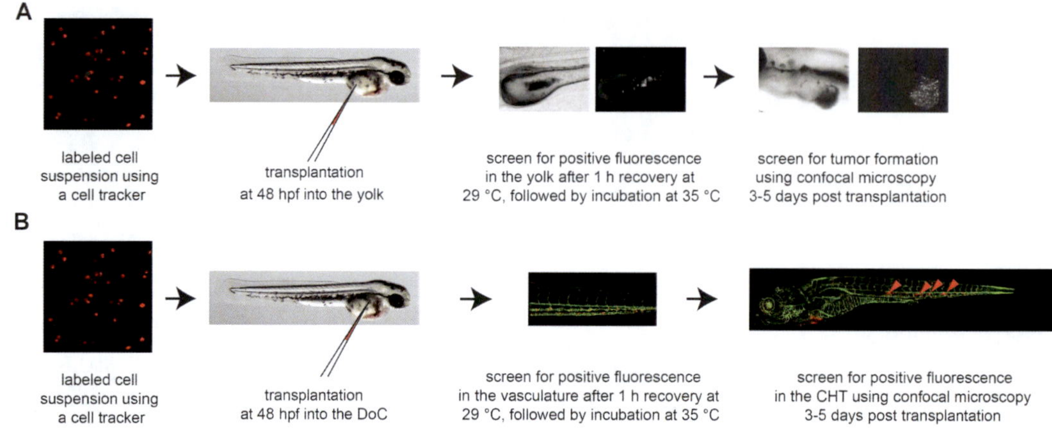

Fig. 1 Tumor cell engraftment in zebrafish embryo. Schematic representation of a 48-hpf-old zebrafish embryo and the implantation sites. Human SKM-1 cancer cells (**a**) or healthy human cord blood-derived HSPCs were labeled using CellTracker CM-DiI and then transplanted either into the yolk (**a**) or the duct of Cuvier (**b**) of 48 hpf zebrafish embryos. After 1 h of recovery at 29 °C, transplanted embryos were screened for positive fluorescence and then transferred to 35 °C for the following days until final analysis

5. Perform test injection onto the agarose plate and count number of cells.
6. Adjust injection pressure and time to be able to transplant ~200–500 cells per larva.
7. In order to transplant into the yolk sac, inject into the vessel-free area of the yolk (Fig. 1a).
8. In order to transplant into the vasculature, reach the distal branch of the DoC close to the opening into the heart cavity (Fig. 1b).
9. Repeat until appropriate numbers of larvae are successfully transplanted.
10. Transfer injected zebrafish in a new petri dish containing fresh egg water without anesthetic, and keep at 29 °C for 1 h (*see* **Note 4**).

3.5 Verification of Correct Cell Transplantation in Zebrafish Larvae

1. Anesthetize larvae with 0.02% Tricaine.
2. Check for successful transplantation (cell number and location of cells).
3. Discard embryos that were not injected properly using a fluorescence microscope.
4. Transfer to 34–35 °C.

3.6 Proliferation Analysis Using Flow Cytometry Methods

1. Transfer five to ten embryos in 50 μl PBS in a 1.5 ml tube, and add 4 μl Liberase™ (*see* **Note 5**).
2. Incubate 15–30 min on a thermocycler with 30 °C and 800 rpm.

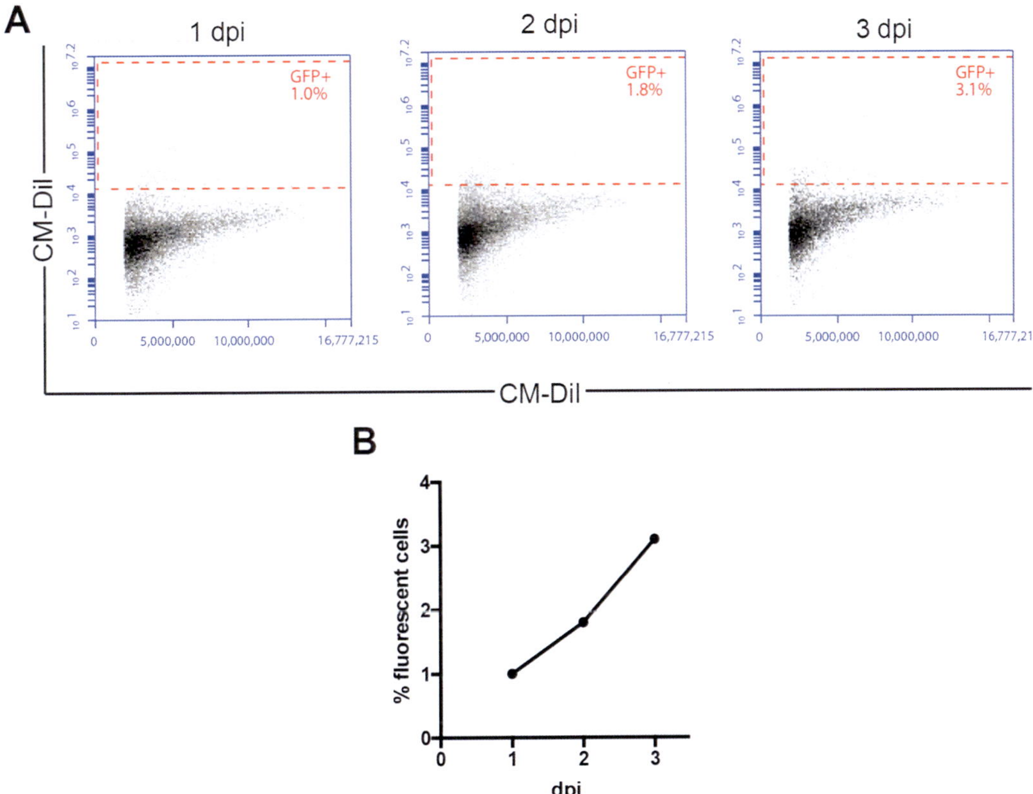

Fig. 2 Flow cytometric analysis of zebrafish xenografts after injection of CM-Dil-labeled cells. (**a**) Shown are representative flow cytometry plots displaying the percentages of cells that are labeled in red at 1, 2, or 3 days postinjection (dpi). CM-Dil-positive cells were monitored by gating for red fluorescent after live/dead cell discrimination using 7-AAD. (**b**) Summarized data from (**a**) over time

3. Dissolve the embryos by pipetting.
4. Add 4 μl of 5% BSA in DPBS.
5. For flow cytometric analysis, add 150 μl DPBS and pass the cell solution through a 35 μm cell strainer on a BD Falcon polystyrene tube.
6. Perform live/dead staining with 7-AAD according to the manufacturer's protocol prior to the measurement (*see* **Note 6**).
7. Analyze for positive fluorescence on single living cells (*see* Fig. 2).

3.7 Analysis of Tumor Formation Using Histopathological Methods

1. Fix xenografts with tumorlike structures at day 3 or 5 after injection using 4% paraformaldehyde in DPBS overnight at 4 °C.
2. Perform three 5 min washing steps in PBS.
3. Prepare the embryos for embedding by using a tissue processor.

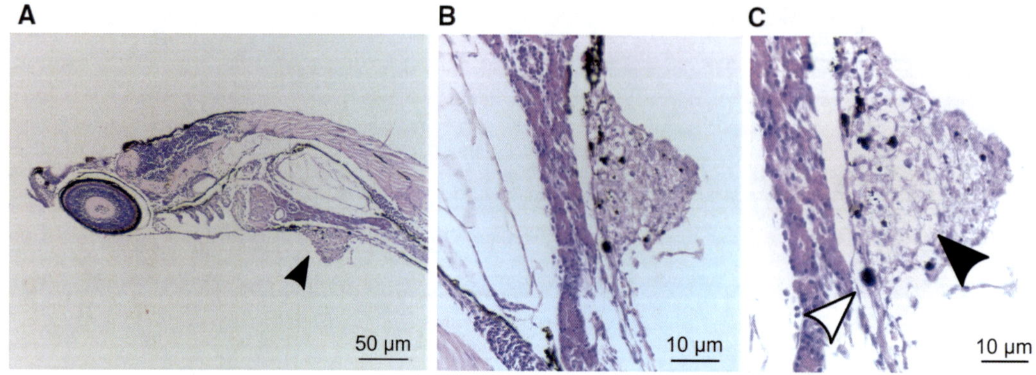

Fig. 3 Light microscopy pictures of xenograft slices after tissue processing and H&E staining. (**a–c**) Human CM-DiI-labeled SKM-1 cells were transplanted into the vessel-free area of the yolk sac and fixed at 3 days posttransplantation. Shown are exemplary slices of sections with tumorlike structures at different resolutions. Arrowheads in (**c**) indicate typical characteristics of human dysplastic myeloid cells like big atypical formed nuclei (white) and high nucleus-cytoplasm ratio (black)

4. Embed the processed embryos in paraffin.
5. Store at −20 °C or directly cut them into 5 μm thick slices with a microtome.
6. Dry the obtained slices overnight on SuperFrost® Plus glass slides, and perform an eosin/hematoxylin (H&E) staining procedure.
7. Acquire representative images using a stereo microscope (Fig. 3) (*see* **Note** 7).

3.8 Analysis of Host Microenvironment Interactions Using Confocal Microscopy

1. Embed embryos in 800 μl of 0.6% low-melting agarose after adding 200 μl 0.4% Tricaine for anesthesia.
2. Screen the injected larvae using a confocal microscope, using the appropriate filters depending on the transgenic zebrafish line and the labeling dye.
3. Define Z-stacks for each larva at the position of interest, and perform automated image acquisition (Fig. 4) (*see* **Note 8**).
4. For longtime observations add egg water into the dish after solidification of the agarose.

4 Notes

1. Several transient dyes are commercially available which allow labeling and follow-up of most cell types [49]. It is however highly recommended to test each dye for each cell type that will be used. Additionally, please consider that dead cells may also contain fluorescence. Make sure to verify that the diameter of

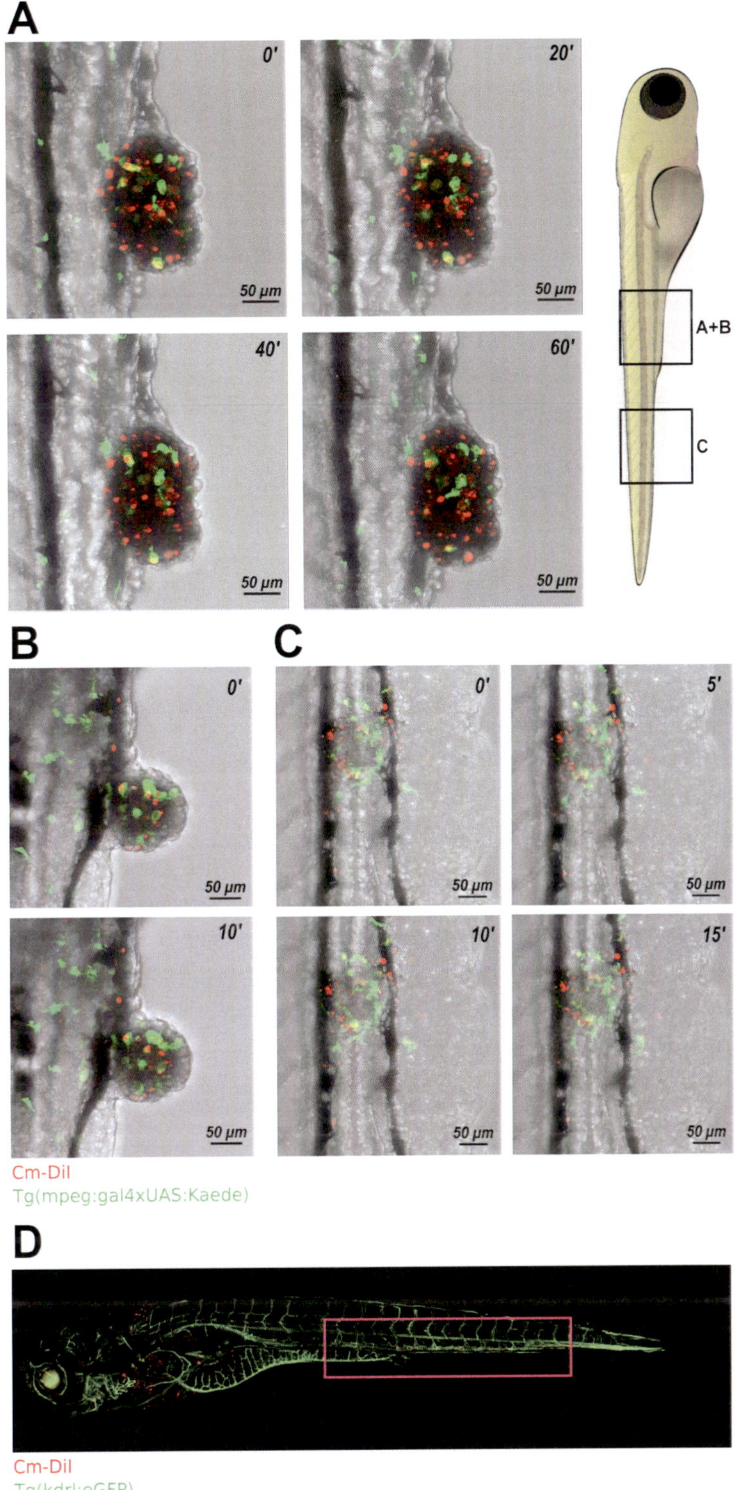

Fig. 4 Analysis of host microenvironment interactions using confocal microscopy. (**a**–**c**) Time series of fluorescent confocal microscopy. Shown are human SKM-1 cells stained with red CM-Dil cell tracker in 5 dpf transgenic *Tg(mpeg:*

what is measured is not smaller than a cell. If possible, use stable transgenic protein expression of any fluorophore of choice.

2. Flick the tube before loading to obtain a homogeneous cell suspension.
3. For some cells it might be necessary to break the tip of the needle. Use forceps and open the needle under the stereomicroscope.
4. Analysis 1 h after transplantation only allows screening for successful transplantation. If you inject for the first time, make sure on the next days if cells are still seen and alive.
5. Screening of single larvae might also be possible.
6. Other live/dead stains might be considered. Please make sure to choose an appropriate fluorochrome that does not interfere with your labeling dye or transgenic line.
7. Depending on the cell line that is used, look for specific features such as human dysplastic myeloid cells which show big atypical formed nuclei and high nucleus-cytoplasm ratio.
8. Different host microenvironment interactions can be analyzed depending on the transgenic zebrafish line that is used.

Acknowledgments

We thank Prof. Dr. Sven Perner for the help with histopathological analyses of tumorlike structures. This work was funded by grants from the Swiss National Science Foundation (164200 and 149735).

Fig. 4 (continued) Gal4;UAS:Kaede) zebrafish in which macrophages are labeled in green. (**a**) Migration of macrophages toward tumorlike structures (left) and schematic overview of the regions shown in (**a**–**c**). (**b**) Different cell morphology of macrophages depending on their location in the fish. (**c**) Red cells and macrophages accumulating in the posterior part of the xenograft tail. (**d**) Human cells detected in the CHT region marked in pink in a transgenic *Tg (kdrl:eGFP)* line after injection into the DoC. Note that the cells extravasated from the vasculature into the in proximity of the circulatory loop between the dorsal aorta and the caudal vein in this region. Fish length at 4 dpf: 3.7 mm

References

1. Mullins MC, Nüsslein-Volhard C (1993) Mutational approaches to studying embryonic pattern formation in the zebrafish. Curr Opin Genet Dev 3:648–654
2. Granato M, Nüsslein-Volhard C (1996) Fishing for genes controlling development. Curr Opin Genet Dev 6:461–468
3. Kari G, Rodeck U, Dicker AP (2007) Zebrafish: an emerging model system for human disease and drug discovery. Clin Pharmacol Ther 82:70–80
4. Feitsma H, Cuppen E (2008) Zebrafish as a cancer model. Mol Cancer Res 6:685–694
5. Amatruda JF, Patton EE (2008) Genetic models of cancer in zebrafish. Int Rev Cell Mol Biol 271:1–34
6. den Hertog J (2005) Chemical genetics: drug screens in zebrafish. Biosci Rep 25:289–297
7. Lam SH, Chua HL, Gong Z, Lam TJ, Sin YM (2004) Development and maturation of the immune system in zebrafish, *Danio rerio*: a gene expression profiling, in situ hybridization and immunological study. Dev Comp Immunol 28:9–28
8. Krauss J, Astrinides P, Frohnhöfer HG, Walderich B, Nüsslein-Volhard C (2013) transparent, a gene affecting stripe formation in Zebrafish, encodes the mitochondrial protein Mpv17 that is required for iridophore survival. Biol Open 2:703–710
9. White RM, Sessa A, Burke C, Bowman T, LeBlanc J, Ceol C, Bourque C, Dovey M, Goessling W, Burns CE, Zon LI (2008) Transparent adult zebrafish as a tool for in vivo transplantation analysis. Cell Stem Cell 2:183–189
10. Theocharides AP, Rongvaux A, Fritsch K, Flavell RA, Manz MG (2016) Humanized hemato-lymphoid system mice. Haematologica 101:5–19
11. Paczulla AM, Dirnhofer S, Konantz M, Medinger M, Salih HR, Rothfelder K, Tsakiris DA, Passweg JR, Lundberg P, Lengerke C (2017) Long-term observation reveals high-frequency engraftment of human acute myeloid leukemia in immunodeficient mice. Haematologica 102:854–864
12. Konantz M, Balci TB, Hartwig UF, Dellaire G, André MC, Berman JN, Lengerke C (2012) Zebrafish xenografts as a tool for in vivo studies on human cancer. Ann N Y Acad Sci 1266:124–137
13. Haldi M, Ton C, Seng WL, McGrath P (2006) Human melanoma cells transplanted into zebrafish proliferate, migrate, produce melanin, form masses and stimulate angiogenesis in zebrafish. Angiogenesis 9:139–151
14. Corkery DP, Dellaire G, Berman JN (2011) Leukaemia xenotransplantation in zebrafish—chemotherapy response assay in vivo. Br J Haematol 153:786–789
15. Pruvot B, Jacquel A, Droin N, Auberger P, Bouscary D, Tamburini J, Muller M, Fontenay M, Chluba J, Solary E (2011) Leukemic cell xenograft in zebrafish embryo for investigating drug efficacy. Haematologica 96:612–616
16. von Mässenhausen A, Sanders C, Brägelmann J, Konantz M, Queisser A, Vogel W, Kristiansen G, Duensing S, Schröck A, Bootz F, Brossart P, Kirfel J, Lengerke C, Perner S (2016) Targeting DDR2 in head and neck squamous cell carcinoma with dasatinib. Int J Cancer 139:2359–2369
17. Queisser A, Hagedorn S, Wang H, Schaefer T, Konantz M, Alavi S, Deng M, Vogel W, von Mässenhausen A, Kristiansen G, Duensing S, Kirfel J, Lengerke C, Perner S (2017) Ecotropic viral integration site 1, a novel oncogene in prostate cancer. Oncogene 36:1573–1584
18. Schaefer T, Wang H, Mir P, Konantz M, Pereboom TC, Paczulla AM, Merz B, Fehm T, Perner S, Rothfuss OC, Kanz L, Schulze-Osthoff K, Lengerke C (2015) Molecular and functional interactions between AKT and SOX2 in breast carcinoma. Oncotarget 6:43540–43556
19. Wang H, Schaefer T, Konantz M, Braun M, Varga Z, Paczulla AM, Reich S, Jacob F, Perner S, Moch H, Fehm TN, Kanz L, Schulze-Osthoff K, Lengerke C (2017) Prominent oncogenic roles of EVI1 in breast carcinoma. Cancer Res 77:2148–2160
20. He S, Lamers GE, Beenakker JW, Cui C, Ghotra VP, Danen EH, Meijer AH, Spaink HP, Snaar-Jagalska BE (2012) Neutrophil-mediated experimental metastasis is enhanced by VEGFR inhibition in a zebrafish xenograft model. J Pathol 227:431–445
21. Nicoli S, Presta M (2007) The zebrafish/tumor xenograft angiogenesis assay. Nat Protoc 2:2918–2923
22. Lee SL, Rouhi P, Dahl Jensen L, Zhang D, Ji H, Hauptmann G, Ingham P, Cao Y (2009) Hypoxia-induced pathological angiogenesis mediates tumor cell dissemination, invasion, and metastasis in a zebrafish tumor model. Proc Natl Acad Sci U S A 106:19485–19490
23. Zhao C, Wang X, Zhao Y, Li Z, Lin S, Wei Y, Yang Y (2011) A novel xenograft model in zebrafish for high-resolution investigating dynamics of neovascularization in tumors. PLoS One 6:e21768

24. Stoletov K, Kato H, Zardouzian E, Kelber J, Yang J, Shattil S, Klemke R (2010) Visualizing extravasation dynamics of metastatic tumor cells. J Cell Sci 123:2332–2341
25. Jacob F, Alam S, Konantz M, Liang CY, Kohler RS, Everest-Dass AV, Huang YL, Rimmer N, Fedier A, Schötzau A, Núñez López M, Packer N, Lengerke C, Heinzelmann-Schwarz V (2018) Transition of mesenchymal and epithelial cancer cells depends on α1-4 galactosyltransferase-mediated glycosphingolipids. Cancer Res 78(11):2952–2965
26. Tobia C, Gariano G, De Sena G, Presta M (2013) Zebrafish embryo as a tool to study tumor/endothelial cell cross-talk. Biochim Biophys Acta 1832:1371–1377
27. Bentley VL, Veinotte CJ, Corkery DP, Pinder JB, LeBlanc MA, Bedard K, Weng AP, Berman JN, Dellaire G (2015) Focused chemical genomics using zebrafish xenotransplantation as a pre-clinical therapeutic platform for T-cell acute lymphoblastic leukemia. Haematologica 100:70–76
28. Liu Y, Asnani A, Zou L, Bentley VL, Yu M, Wang Y, Dellaire G, Sarkar KS, Dai M, Chen HH, Sosnovik DE, Shin JT, Haber DA, Berman JN, Chao W, Peterson RT (2014) Visnagin protects against doxorubicin-induced cardiomyopathy through modulation of mitochondrial malate dehydrogenase. Sci Transl Med 6:266ra170
29. Zhang B, Shimada Y, Hirota T, Ariyoshi M, Kuroyanagi J, Nishimura Y, Tanaka T (2016) Novel immunologic tolerance of human cancer cell xenotransplants in zebrafish. Transl Res 170:89–98.e83
30. Mizgirev IV, Revskoy S (2010) A new zebrafish model for experimental leukemia therapy. Cancer Biol Ther 9:895–902
31. Staal FJ, Spaink HP, Fibbe WE (2016) Visualizing human hematopoietic stem cell trafficking in vivo using a zebrafish xenograft model. Stem Cells Dev 25:360–365
32. Hess I, Iwanami N, Schorpp M, Boehm T (2013) Zebrafish model for allogeneic hematopoietic cell transplantation not requiring preconditioning. Proc Natl Acad Sci U S A 110:4327–4332
33. Shayegi N, Alakel N, Middeke JM, Schetelig J, Mantovani-Löffler L, Bornhäuser M (2015) Allogeneic stem cell transplantation for the treatment of refractory scleromyxedema. Transl Res 165:321–324
34. Moore JC, Tang Q, Yordán NT, Moore FE, Garcia EG, Lobbardi R, Ramakrishnan A, Marvin DL, Anselmo A, Sadreyev RI, Langenau DM (2016) Single-cell imaging of normal and malignant cell engraftment into optically clear prkdc-null SCID zebrafish. J Exp Med 213:2575–2589
35. Iwanami N, Hess I, Schorpp M, Boehm T (2017) Studying the adaptive immune system in zebrafish by transplantation of hematopoietic precursor cells. Methods Cell Biol 138:151–161
36. de Jong JL, Burns CE, Chen AT, Pugach E, Mayhall EA, Smith AC, Feldman HA, Zhou Y, Zon LI (2011) Characterization of immune-matched hematopoietic transplantation in zebrafish. Blood 117:4234–4242
37. Langenau DM, Ferrando AA, Traver D, Kutok JL, Hezel JP, Kanki JP, Zon LI, Look AT, Trede NS (2004) In vivo tracking of T cell development, ablation, and engraftment in transgenic zebrafish. Proc Natl Acad Sci U S A 101:7369–7374
38. Tenente IM, Tang Q, Moore JC, Langenau DM (2014) Normal and malignant muscle cell transplantation into immune compromised adult zebrafish. J Vis Exp. https://doi.org/10.3791/52597
39. Tang Q, Moore JC, Ignatius MS, Tenente IM, Hayes MN, Garcia EG, Torres Yordán N, Bourque C, He S, Blackburn JS, Look AT, Houvras Y, Langenau DM (2016) Imaging tumour cell heterogeneity following cell transplantation into optically clear immune-deficient zebrafish. Nat Commun 7:10358
40. Stachura DL, Svoboda O, Campbell CA, Espín-Palazón R, Lau RP, Zon LI, Bartunek P, Traver D (2013) The zebrafish granulocyte colony-stimulating factors (Gcsfs): 2 paralogous cytokines and their roles in hematopoietic development and maintenance. Blood 122:3918–3928
41. Svoboda O, Stachura DL, Machoňová O, Pajer P, Brynda J, Zon LI, Traver D, Bartůněk P (2014) Dissection of vertebrate hematopoiesis using zebrafish thrombopoietin. Blood 124:220–228
42. Svoboda O, Stachura DL, Machoňová O, Zon LI, Traver D, Bartůněk P (2016) Ex vivo tools for the clonal analysis of zebrafish hematopoiesis. Nat Protoc 11:1007–1020
43. Santos MD, Yasuike M, Hirono I, Aoki T (2006) The granulocyte colony-stimulating factors (CSF3s) of fish and chicken. Immunogenetics 58:422–432
44. Wehmas LC, Tanguay RL, Punnoose A, Greenwood JA (2016) Developing a novel embryo-larval zebrafish xenograft assay to prioritize human glioblastoma therapeutics. Zebrafish 13:317–329

45. Harfouche R, Basu S, Soni S, Hentschel DM, Mashelkar RA, Sengupta S (2009) Nanoparticle-mediated targeting of phosphatidylinositol-3-kinase signaling inhibits angiogenesis. Angiogenesis 12:325–338
46. Cheng J, Gu YJ, Wang Y, Cheng SH, Wong WT (2011) Nanotherapeutics in angiogenesis: synthesis and in vivo assessment of drug efficacy and biocompatibility in zebrafish embryos. Int J Nanomedicine 6:2007–2021
47. Ellett F, Pase L, Hayman JW, Andrianopoulos A, Lieschke GJ (2011) mpeg1 promoter transgenes direct macrophage-lineage expression in zebrafish. Blood 117:e49–e56
48. Choi J, Dong L, Ahn J, Dao D, Hammerschmidt M, Chen JN (2007) FoxH1 negatively modulates flk1 gene expression and vascular formation in zebrafish. Dev Biol 304:735–744
49. Progatzky F, Dallman MJ, Lo Celso C (2013) From seeing to believing: labelling strategies for in vivo cell-tracking experiments. Interface Focus 3:20130001

Chapter 17

Statistical and Mathematical Modeling of Spatiotemporal Dynamics of Stem Cells

Walter de Back, Thomas Zerjatke, and Ingo Roeder

Abstract

Statistical and mathematical modeling are crucial to describe, interpret, compare, and predict the behavior of complex biological systems including the organization of hematopoietic stem and progenitor cells in the bone marrow environment. The current prominence of high-resolution and live-cell imaging data provides an unprecedented opportunity to study the spatiotemporal dynamics of these cells within their stem cell niche and learn more about aberrant, but also unperturbed, normal hematopoiesis. However, this requires careful quantitative statistical analysis of the spatial and temporal behavior of cells and the interaction with their microenvironment. Moreover, such quantification is a prerequisite for the construction of hypothesis-driven mathematical models that can provide mechanistic explanations by generating spatiotemporal dynamics that can be directly compared to experimental observations. Here, we provide a brief overview of statistical methods in analyzing spatial distribution of cells, cell motility, cell shapes, and cellular genealogies. We also describe cell-based modeling formalisms that allow researchers to simulate emergent behavior in a multicellular system based on a set of hypothesized mechanisms. Together, these methods provide a quantitative workflow for the analytic and synthetic study of the spatiotemporal behavior of hematopoietic stem and progenitor cells.

Key words Statistical modeling, Mathematical modeling, Spatial statistics, Point patterns, Cell shape analysis, Cell motility, Cellular genealogies, Cell-based modeling, Cellular Potts model, Center-based model

1 Introduction

Despite major advances in the identification of molecular and genetic components and biomarkers of the local bone marrow environments ("niches"), in which hematopoietic stem and progenitor cells (HSPC) reside, much remains to be learned about the spatiotemporal dynamics of normal and aberrant hematopoiesis [1–3]. There are many remaining open questions concerning, e.g., the localization of HSPC relative to various, potentially different stem cell niches, their chemotactic and migratory behavior within the bone marrow, the heterogeneity of cellular morphologies, the role of niche factors on cell fate decisions, etc.

Gerd Klein and Patrick Wuchter (eds.), *Stem Cell Mobilization: Methods and Protocols*, Methods in Molecular Biology, vol. 2017, https://doi.org/10.1007/978-1-4939-9574-5_17,

Modern microscopy provides image data with increasing spatial and temporal resolution, e.g., high-resolution imaging of deep tissue of the bone marrow [4], live-cell video microscopy of stem cell cultures [5–7], as well as intravital imaging of HSPC within the bone marrow [8–11]. However, in order to interpret such rich image data and use it to test scientific hypothesis, e.g., on the spatiotemporal organization of HSPCs, statistical and mathematical modeling is required. On the one hand, data-driven statistical models provide frameworks to describe and quantify experimentally observed aspects of HSPC behavior, to formulate (null-) hypotheses, and to formally compare and potentially distinguish the observed behavior from a formulated hypothesis, such as random behavior. Hypothesis-driven mathematical models, on the other hand, provide mathematical and computational frameworks to test whether a set of assumptions on the cellular behavior and interactions of cells is, in principle, able to generate the observed spatiotemporal regularities.

In this chapter, we provide introductions to common procedures in statistical modeling to quantify (1) spatial distributions, (2) motility, (3) cell shape, and (4) proliferative behavior of cells. In each of these procedures, we specify the types of questions that can be addressed, how to prepare input data, what steps are required to measure statistical properties, how to formulate the null hypothesis, and how to compare observations to this expectation. Moreover, we describe two cell-based modeling frameworks, i.e., the center-based model and the cellular Potts model, in which hypotheses on cellular behavior can be tested computationally. Together, these statistical and mathematical modeling methods provide a full quantitative workflow for the analytic and synthetic study of the spatiotemporal behavior of hematopoietic stem and progenitor cells.

2 Methods

2.1 Image Segmentation and Tracking

A prerequisite for statistical and mathematical modeling is the presence of quantitative data. Obtaining these from images is a nontrivial process that typically starts with *image segmentation*, in which an image is partitioned into fore- and background to, e.g., separate cells from other structures. If not only static but also dynamic information is of interest, *cell tracking* becomes relevant. Here, moving cells are located and linked between image frames. Subsequently, cellular events, such as cell deaths or divisions, can be detected and related to positions of cells in space and time. Both are active fields of research, and a wide range of segmentation and tracking methods exists, from manual annotation [12] and computer-assisted methods [13–15] to fully automated machine learning algorithms [16, 17]. While the presence of segmented images and/or cell tracks is a precondition for quantification and statistical modeling, image analysis itself is not in the focus of this publication. Therefore, we refer the reader to available review literature and software packages.

2.1.1 Further Reading and Software

A brief historical overview of cell segmentation methods, before the advent of machine learning, is given by Meijering [18], while Kan [19] describes some of the opportunities of machine learning in cell image analysis. Caicedo and co-workers [20] present a series of best practices for image acquisition, image processing, and subsequent data analysis focusing on high-content profiling, while Skylaki et al. [7] provide an overview of the challenges in long-term imaging and quantification.

A number of software packages are available for biological image segmentation, including *Fiji* [21], *Ilastik* [15], *CellProfiler* [22], *CellCognition* [23], and the more recent *fastER* [13]. Some of this software also includes features for cell tracking, such as the *TrackMate* plugin [24] included in *Fiji* and the manual and automatic tracking tools in *Ilastik* [15] and *CellCognition* [23]. Other software packages are designed specifically for tracking live cells such as *tTt* [12]. Wiesmann et al. [25] review a number of additional software tools for quantitative image analysis.

2.2 Spatial Distribution

Distances between cells are often used as a proxy for cellular interactions. Although we would like to emphasize that local proximity and functional interaction is in general not the same, quantitative information on distances between cells (and potentially other structures) allow to describe patterns of cell motility as well as to identify the presence of (stem cell niche-mediated) local functional or regulatory peculiarities. Therefore, aided by high-resolution imaging, many studies focus on measuring distances of stem cells to other cell types and bone marrow structures in order to identify stem cell niches [4, 26, 27]. Understanding the localization of HSPCs can support answering questions such as: Are hematopoietic stem cells randomly distributed within the bone marrow, or does their distribution follow certain structural rules? Are these cells more prominently localized to vascular or endosteum surfaces? Are there signs of attractive or repulsive interactions between different cell types?

Methods from *point pattern analysis*, a statistical framework mostly developed in the context of ecology and geology, can be used to answer these questions. Specifically, this analysis is used to investigate whether a certain distribution of points shows signs of regularities. It typically starts by assuming complete spatial randomness (CSR) as a null hypothesis and investigates whether the observed cellular pattern can be described by such a random pattern-generating process. CSR suggests that (1) the number of points is proportional to the area of a subregion (*homogeneity*) and that (2) the locations of points are independent from each other (*independence*) (*see* **Note 1**). Based on these assumptions, it is possible to describe the number of points in any given region by a Poisson distribution. To statistically test whether the null hypothesis of CSR is plausible (in a probabilistic sense), one can use the *quadrant counting* and *Ripley's K-function* as described below.

- *Input data*: In point pattern analysis, cells are approximated as point particles. These are given by the 2D or 3D coordinates of the center of mass of the segmented cell or, alternatively, by manually tracking. The cell density is described by the *intensity* of the point pattern and can be estimated by $\lambda = \frac{n(\boldsymbol{x})}{|W|}$ where λ is the intensity, assuming the point process is homogeneous, $\boldsymbol{x}$ is the set of point pattern of size $n(\boldsymbol{x})$, and $|W|$ is the size of the observation window.
- *Quantify and testing homogeneity*: To test for homogeneity of the point pattern, one can check whether the regions of equal area contain roughly equal numbers of points (*see* Fig. 1a), i.e., applying *quadrant counting*. To test the null hypothesis of CSR, a typical χ^2 (chi-squared) statistic can be used:

$$\chi^2 = \sum_j \frac{(\text{Observed} - \text{Expected})^2}{\text{Expected}} = \sum_j \frac{\left(n_j - \frac{n}{m}\right)^2}{\frac{n}{m}} \tag{1}$$

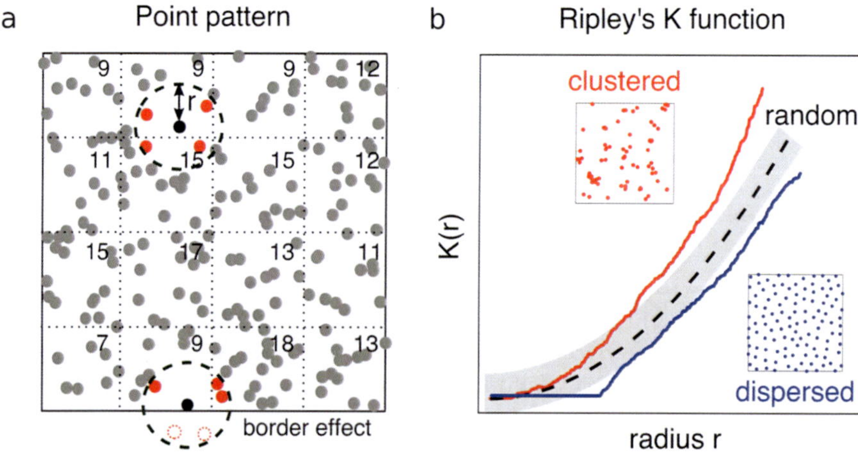

Fig. 1 Spatial distribution. (**a**) By representing cellular populations by their centers of mass, their spatial distribution can be studied as a point pattern within an observation area *A*. To test for homogeneity, the *quadrant counting* method counts the number of points in subregions (dotted lines). To compute Ripley's K-function, one counts for each point the number of neighboring points within an increasing radius, normalized by the maximum number of pairwise distances, and correcting for border effects where neighboring points cannot be counted. (**b**) Ripley's K-function (aka reduced second moment measure) shows an exponentially increasing function $\boldsymbol{K_{\text{Poisson}}(r)} = \boldsymbol{\pi r^2}$ for a 2D completely spatial random (CSR) pattern. Observed K-functions above the CSR expectation indicate clustering, while an observed K-function below the CSR expectation shows that the pattern is regular or dispersed. The gray region indicates the upper and lower global envelopes

where n_j is the number of points observed in subregion j, n is the total number of points in the observation window, and m is the number of (equally sized) subregions. Under the CSR null hypothesis, the test statistic is approximately χ^2-distributed with $m - 1$ degrees of freedom.

- *Quantify spatial correlation*: Independence of the localization of points can be violated either by (1) *clustering* of points, i.e., points are closer together than expected, e.g., due to some attractive interaction, or by (2) *dispersion*, i.e., points are farther apart than expected, e.g., due to a repelling activity. A commonly used method to analyze spatial correlation is the *reduced second moment measure*, better known as *Ripley's K-function* [28]. Ripley's K-function (to be estimated by the empirical K-function given by Eq. 2 counts the average number of neighboring points within a certain radius (*see* Fig. 1a), normalized to the possible number of pairwise distances $n(n - 1)$ within area A:

$$\hat{K}(r) = \frac{A}{n(n-1)} \sum_{i=1}^{n} \sum_{j=1, j \neq i}^{n} k_{ij}(r) e_{ij}(r) \tag{2}$$

where A is the observation area, n is the observed number of points, and k_{ij} indicates whether the pairwise distance d_{ij} is smaller than r:

$$k_{ij}(r) = \begin{cases} 1 & \text{if } d_{ij} \leq r, \\ 0 & \text{otherwise} \end{cases} \tag{3}$$

The edge correction factor $e_{ij}(r)$ compensates for the fact that we cannot count points that lie within the radius r but outside of the observation window (*see* Fig. 1a). Therefore, without correction, the number of points is underestimated for larger radii. Several correction methods exist, ranging from simple border methods that suffice for large data set to more complex methods such as isotropic or translation correction that are advised for smaller data sets [29].

- *Comparison to CSR*: From the homogeneity and independence properties of CSR, it follows that the point pattern can be described as the realization of a *Poisson process*. This can be used to calculate the expected number of points lying within a distance r of a typical random point. In 2D, under the CSR assumption, the theoretical K-function is $K_{\text{Poisson}}(r) = \pi r^2$, and in 3D it is given by $K_{\text{Poisson}}(r) = \frac{4}{3}\pi r^3$. The empirical and theoretical K-functions are typically compared graphically (*see* Fig. 1b). An empirical K-function that is above the theoretical CSR expectation, i.e., having more than expected neighboring points, indicates a clustered pattern. Conversely, a line below the theoretical expectation, i.e., having less than expected neighbors at a certain radius, is a sign of regularity or dispersion (*see* **Notes 2** and **3**).

- *Significance testing of CSR*: To test whether the empirical (i.e., observed) K-function deviates statistically significant from the expected K-function (under CSR), one can perform a *Monte Carlo test* where the K-function under the CSR null hypothesis is repeatedly simulated. For this, the maximum and minimum deviation of the simulated and the theoretical K-function values are used as a global envelop (*see* Fig. 1b and **Note 4**). The observed clustering or dispersion is considered significantly different from CSR when the observed K-function crosses out of the global envelope. The significance level associated with this *Monte Carlo* test is determined by the number of simulations used to calculate the global envelope, i.e., $\alpha = 1/m + 1$, where m is the number of simulations. That is, for $m = 19$ simulations, a test at the typical significance level of $\alpha = 0.05$ would be obtained (*see* Subheading 10.6 in [30]).

2.2.1 Further Reading and Software

In case the above analysis shows signs of inhomogeneous point pattern, more analysis is required, e.g., to reveal dependence of the intensity of the point pattern to other factors. *Spatial covariate analysis* can be used to analyze whether cell density is correlated with covariates such as the concentration of a chemokine or the distance to the vascular surface (Chapter 6.6 in [30]). When localization data on multiple cell types in the same culture or tissue are available, one might ask whether there is an interaction between these cell types. This requires a more elaborate *multitype point pattern* analysis, where one tests the null hypothesis of *complete spatial randomness and independence* (CSRI), e.g., by computing empirical K-functions in a pairwise fashion to reveal attraction or repulsion of cell types around or away from each other (Chapter 14 in [30]). Suggested textbooks on statistics for spatial data are [30–32]. The latter accompanies the key software package for point pattern analysis spatstat (http://spatstat.org) for the statistical software package *R* which includes functions for the methods mentioned above, among many others. For *python*, some functions for, e.g., computing the (edge corrected) K-function are available in astropy (http://www.astropy.org).

2.3 Cell Motility

Cell migration is a fundamental property of cells that changes in response to chemical, mechanical, and genetic perturbations. Disruption of normal cell motility is potentially associated with disease, e.g., due to immunological response or malignant processes. Unraveling such anomalies requires a rigorous analysis of cell migration patterns.

Statistical analysis of cell motility allows answering questions such as: How fast do cells migrate? Do cells migrate persistently, or are the cells constrained, e.g., by a crowded environment? Is there evidence of directionality in the migration patterns?

The most basic statistics about a cell's trajectory are its length and its curviness or tortuosity. However, to characterize the type of motility a cell exhibits, more elaborate analysis is required. Analogously to the CSR in analyzing spatial distributions of cells, here, we assume randomness as a null hypothesis. Under this assumption, the amount of space that a particle "explores" is proportional to the time interval, which can be measured by calculating the mean squared displacement (MSD). The MSD relates the mean displacement of a particle to different time intervals.

The most common model to describe deviations from a complete-random cell motility is the *persistent random walk* (PRW) model, which accounts for a degree of persistence of motion. While this model has been shown to accurately describe cell migration on 2D surfaces [33], recent studies have found that it fails to describe cell motility in more biologically relevant 3D environments [34]. Although the PRW model can account for faster-than-diffusive (superdiffusive) persistent motion, it cannot describe slower-than-diffusive (subdiffusive) motility, which is commonly found in cells in confined environments including porous media such as trabecular bone [35]. In this case, the general *anomalous diffusion* (AD) model, which relates the cell displacement to time interval with a simple power law, $MSD_{AD}(\tau) = D\tau^{\alpha}$, is more appropriate to describe both population average and individual cell migration paths.

- *Input data*: Cell trajectories should be recorded in the form of a table with columns: cell ID, time, and x, y, and z coordinates of the cell centroid in consistent units (typically minutes and microns) and equally spaced time intervals (*see* **Notes 5** and **6**).
- *Quantify length and tortuosity of trajectory*: The most basic statistic about the trajectory is its length. The total length of a trajectory is simply the sum of the trajectory segments $L = \sum_{t=0}^{t_{\max}} \sqrt{(x_t - x_{t+1})^2 + (y_t - y_{t+1})^2}$. The curviness or tortuosity of the trajectory can also be calculated in a straightforward fashion as the *arc over chord ratio*: the ratio between total length of the trajectory L and the length of a straight line between start and end points: $T_{AOC} = {}^{L}/_{L_X}$ where $L_X = \sqrt{(x_{t_{\max}} - x_0)^2 + (y_{t_{\max}} - y_0)^2}$. However, this measure fails to capture differences between long curved trajectories and short twisted ones because it quantifies the global deviation of the trajectory from a straight line and does not account for local measures such as curvature. A more robust measure of tortuosity calculates the *mean of the local curvatures* along to the trajectory:

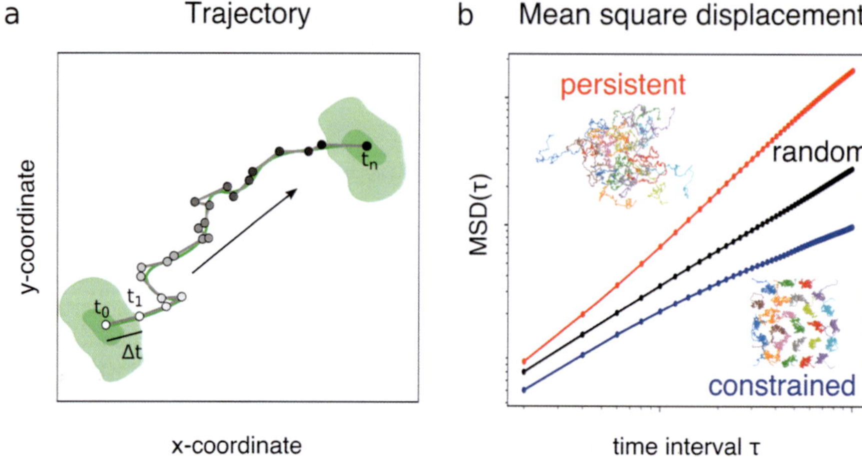

Fig. 2 Cell motility. (**a**) The trajectories of cells measured over regular time intervals can be studied by investigating how the displacement of the cell relates to the time interval, assuming the process is time-invariant. This can be quantified by the mean square displacement, calculated for each individual cell or averaged over the population. (**b**) The mean square displacement is shown on a log-log plot and can be described by the anomalous diffusion model **MSD** $= \boldsymbol{D\tau^{\alpha}}$. For a normal diffusion (random) process, the displacement is expected to be proportional to the time interval and appears as a straight line and has anomality parameter $\boldsymbol{\alpha} = 1$. Cell trajectories above this expectation with $\boldsymbol{\alpha} > 1$ move faster-than-diffusion (superdiffusive), indicating persistent movement. Cell below the expectation with $\boldsymbol{\alpha} < 1$ move closer-than-diffusion (subdiffusive) which indicates constrained movement, often encountered in crowded environments

$$T_C = \frac{1}{N} \sum_{t=0}^{t_{\max}} \frac{x'(t)y''(t) - y'(t)x''(t)}{\left[x'(t)^2 + y'(t)^2\right]^{3/2}} \tag{4}$$

where $x'(t)$ and $x''(t)$ denote the first and second derivative of the spline approximating the trajectory (*see* Fig. 2a). Alternatively, if interested mainly in the cell's persistence, one can compute the *direction autocorrelation* (*see* Subheading 2.3.1).

- *Test for time invariance*: Quantities such as the MSD assume time invariance of the trajectories, i.e., the velocities should be approximately constant over the course of the experiments. To test this, one first calculates the displacements for each cell and for every time point $dx = x(t + \tau) - x(t)$ and $dy = y(t + \tau) - y(t)$, where τ is the time interval (typically simply 1). Then take the mean and standard deviation of the velocity over all cells $v_t = \left\langle \sqrt{(dx^2 - dy^2)}/\tau \right\rangle$. As a graphical test, one can plot the mean velocities and their standard deviations over time (*see* **Note 7**). If both remain approximately constant over time, they can be considered as being time-invariant (*see* **Note 8**). As a formal test, one can perform a linear regression on the mean velocity (*see* **Note 9**). A time dependency is indicated by a slope deviating from zero, which can be statistically confirmed by testing the corresponding null hypothesis (i.e., slope = 0) and observing a p-value smaller than a predetermined significance level (e.g., $\alpha = 0.05$).

- *Compute mean square displacement*: The mean square displacement (MSD) of individual trajectories is computed by calculating the mean displacement for different time intervals (here for 2D):

$$\mathrm{MSD}_{\mathrm{cell}}(\tau) = \frac{\Delta t}{t_{\max} - \tau} \times \sum_{t=0}^{t_{\max}-\tau} \left((x(t+\tau) - x(t))^2 + (y(t+\tau) - y(t))^2 \right) \quad (5)$$

where with $\tau = n\Delta t$ with $n = 1, 2, \ldots, n_{\max}$ and Δt is the time between individual frames (*see* **Note 10** and Fig. 2a). The MSD of the whole population of cell trajectories, called aggregate or ensemble MSD, is calculated by averaging over the MSD for all cells at each particular τ:

$$\mathrm{MSD}_{\mathrm{ensemble}}(\tau) = \frac{1}{n_{\max}} \sum_{n=1}^{n_{\max}} \mathrm{MSD}_{\mathrm{cell}\ n}(\tau) \quad (6)$$

- *Fit anomalous diffusion model*: To estimate the parameters of the anomalous diffusion (AD) model, i.e., to estimate the diffusion coefficient D and the anomality parameter α, we fit the MSD of a cell or ensemble using a nonlinear regression (*see* **Note 11**). Importantly, only the first 20–30% part of the data (containing the smallest time intervals) should be used in the fitting procedure (*see* **Note 12**). For a normal diffusion (random) process, the displacement is expected to be proportional to the time interval and appears as a straight line and has anomality parameter $\alpha = 1$. An MSD with $\alpha > 1$ exhibits faster-than-diffusion motion (superdiffusion) which indicates persistence. In contrast, an MSD with an estimated $\alpha < 1$ move slower-than-diffusion (subdiffusion) which indicates constrained movement that is often associated with crowded environments.
- *Significance testing*: To test whether the fitted AD model is significantly different from the random expectation, one can apply *Monte Carlo testing* as described in Subheading 2.2 by fitting the AD model to the MSD of a number of simulated random walk trajectories (see **Note 13**).

2.3.1 Further Reading and Software

Further analysis can consist of spatial autocorrelation metrics, such as the velocity and direction autocorrelation, which measure how a quantity correlates with itself over different time scales. For instance, to compare the persistence of cells between conditions, the *direction autocorrelation* can be calculated, e.g., with the Excel-based *DiPer* software tool [36]. With such analysis, one is able to better distinguish between the common models for cell migration, the PRW, anisotropic PRW [37], anomalous diffusion [38], and fractional diffusion models [39]. An excellent review on computational methods for measuring cell migration is given in [40].

Several software packages offer tools to analyze cell trajectories, including *CellProfiler Tracer* (http://cellprofiler.org/tracer); *MotilityLab*, a website as well as an *R* package (http://www.motilitylab.net/); the *python* package *TrackPy* (https://soft-matter.github.io/trackpy); and *TrackMate*, an *ImageJ* plugin (https://imagej.net/TrackMate).

2.4 Cell Shape Analysis

Cell shape is one of the most common properties used to characterize cellular phenotypes, in particular in high-content imaging and drug screening. This is due to the fact that cell morphology can be used as a proxy for a range of cellular processes including cell death, division, polarity, and motility. Moreover, the shape of a cell is a relatively easy accessible property given appropriate cytoplasmic staining. Statistical analysis of cell shape aims to answer questions such as: What is the source of heterogeneity in cell shape within a population? Can one identify subpopulations with similar cell shapes? Are cell shapes correlated with cell lineages?

Here, we present a procedure to explore the heterogeneity of cell shapes in which shapes, represented by regions in binary masks, are quantified using a number of 2D shape descriptors. Subsequently, differences within the population and the potential presence of subpopulations are analyzed and visualized by creating a "shape space" using principal component analysis.

- *Input data*: The input for cell shape analysis consists of a set of binary images in which cells are separated and segmented from the background (*see* Subheading 2.1). If multiple cells are present in an image, each cell must be uniquely identified using *connected-component labeling*, which detects and uniquely labels connected regions in an image, after which a separate binary image can be generated for each cell.
- *Region-based statistics*: Each 2D labeled region can be described using a wide range of statistics. These statistics are either based on the region itself or on one of several geometrical approximations of the region. Some of the most commonly used shape statistics are listed below:
  - The region itself can be quantified in terms of the *area* A, i.e., the number of pixels that form the area A, and the *perimeter* P, i.e., the length of a line through the centers of the border pixels, or the *maximum Feret* (aka *caliper*) diameter as the longest distance between two points in the region.

- Based on a *circular approximation*, one can quantify the *equivalent diameter* or a circle with the same area $= \sqrt{4A/\pi}$. The *acircularity* is defined as the ratio between the observed perimeter and the expected perimeter of a circle, $a = P^2/4A\pi$.
- By calculating the *elliptic approximation* (using the moments of inertia), one can measure the length of the major (long) and minor (short) axes and define the *elongation* as the ratio between the two, quantify the shape's *orientation* as the angle between the x-axis and the major axis, and measure the *eccentricity*, defined as the ratio between the two focal points over the major axis length.
- Properties related to the *bounding box*, the minimal rectangle containing all the points of the region, are the *extent* (aka *rectangularity*), the ratio between the area of the region and its bounding box and the *aspect ratio*, and the ratio between the width and height of the bounding box, but these should be handled with care (*see* **Note 14**).
- The *convex hull* of a region is the smallest convex polygon (nonintersecting polygon with all interior angles less than 180°) that contains all points of the region and can be used to calculate the area A_{hull} and perimeter P_{hull} of the convex hull. Based on those, one can calculate the *convexity* as the ratio between the perimeters of the convex hull and the original region $c = P_{\text{hull}}/P$ and the *solidity*, the ratio between the areas $s = A/A_{hull}$, where the area of the convex hull is equal or larger than the regions area.

• *Feature selection and dimensionality reduction*: After quantifying each cell shape using the descriptors above, a matrix is obtained with rows containing samples (shapes) and columns corresponding to a specific shape descriptor, often containing more than 20 features per cell shape. However, many of these shape descriptors quantify similar aspects of the shape, i.e., there may be strong correlations between descriptors. Moreover, depending on the most relevant shape properties in a particular study, some descriptors may be redundant. To reduce the dimensionality of the data and find the most relevant features, one can apply, e.g., a *principal component analysis* (PCA). PCA allows to represent the high-dimensional data in a lower-dimensional (typically 2D or 3D) "shape space" that still captures most of the variance in the data (*see* Fig. 3b and **Note 15**). To interpret the principal components in terms of the descriptors, one can look at the *component loadings*, which describe the correlation between descriptors and the principal components.

a Cell shapes

region circle

ellipse convex hull

b Principle component analysis

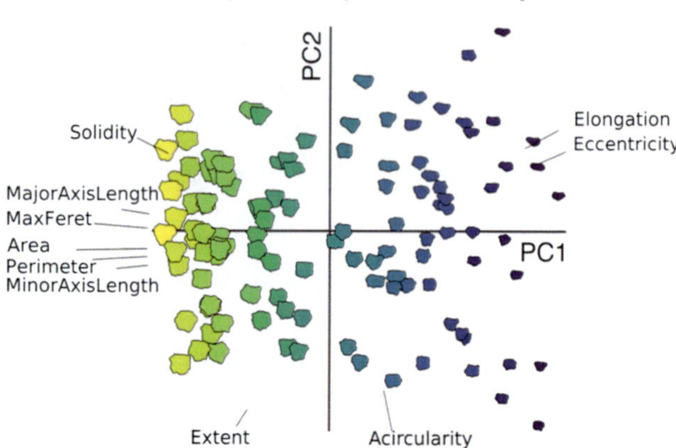

Fig. 3 Cell shape analysis. (**a**) After segmentation, cell shapes can be quantified according to a large number of shape descriptors called region properties, ranging from simple quantities such as area and perimeter to more complex ones based on various geometric models such as the circular, elliptic approximations, as well as the convex hull. (**b**) To explore the variance in a population of cell shapes described by shape descriptor, one can apply a dimensionality reduction method such as principal component analysis (PCA) to explore the distribution of cell shape and investigate the most distinctive descriptors (feature selection)

2.4.1 Further Reading and Software

Apart from binary masks, cell shapes can also be represented by distance maps or polygonal outlines and subjected to different encodings such as Fourier, elliptic Fourier [41], and Zernike decompositions. Pincus and Theriot [42] present an extensive review and quantitative comparison of these methods. A number of studies focus on the dynamics of cell shape rather than static shapes, e.g., by using time series of shape descriptors [43] or modeling trajectories in shape space [44].

A number of software packages exist to extract region-based statistics from binary image masks, including regionprops in the image processing toolbox in *MATLAB*, and a function by the same name in the skimage.measure *python* package. Implementations for PCA are offered as *pca* in the statistics and machine learning toolbox in *MATLAB* and as in the sklearn.decomposition.PCA package in *python*.

2.5 Cell Lineage Tree Analysis

Cellular genealogies, also denoted as cell lineage trees, are pedigree-like structures that represent the complete divisional history of a cell (Fig. 4a). Analyzing these structures can help answering questions like: Are certain cellular characteristics inherited upon division, and if so, how far-reaching are these correlations? Do certain events, e.g., cell death, occur preferentially in the genealogical vicinity of other such events? When does the decision to differentiate take place? Is a differentiation process regulated in an instructive (i.e., by differential differentiation propensities) or selective manner (i.e., by differential death rates)?

a Cellular genealogy

i j

$r_{ij} = 4$

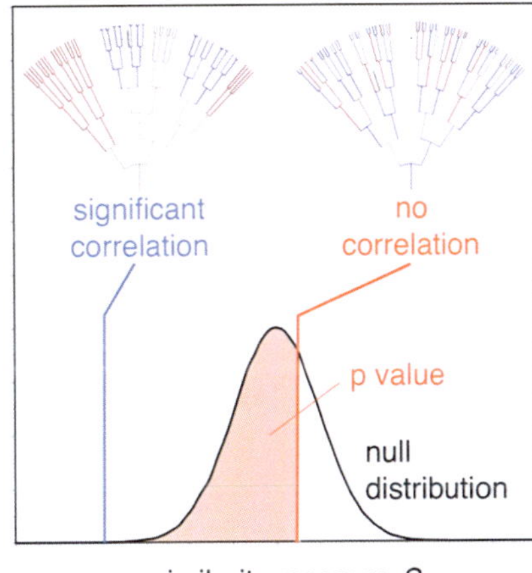

Fig. 4 Lineage tree analysis. (**a**) Cellular genealogies (also denoted as cell lineage trees) represent the complete divisional history of a founder cell (at the bottom). Cells are depicted as straight lines with the length corresponding to the cell's lifetime. Upon division, a mother cell is connected to its two daughter cells. The topological distance r_{ij} between two cells i and j is defined as the number of divisions that separate the cells within the tree. (**b**) In order to detect genealogical correlations of a feature of interest, e.g., the length of cell cycle (as color-coded here within the genealogies), a similarity measure is calculated and compared to a null distribution of randomly expected values. A small p-value (defined as the fraction of values generated under the null hypothesis that are less or equal to the actually observed value) indicates a clustering of the feature of interest (left genealogy), while a large p-value indicates a random distribution of the feature of interest (right genealogy)

Genealogical information can be used to detect correlations of discrete events, e.g., cell death or onset of differentiation marker, or continuous (i.e., metric) features, e.g., cell cycle length or cell motility [45, 46] (*see* **Note 16**). Four steps are necessary to detect potential correlation structures: (1) represent the genealogical information in an efficient way, (2) define a similarity measure that reflects genealogical correlations, (3) find an appropriate null model that reflects the correlation measure's expected distribution without any correlation, and finally (4) compare the actual value with this null distribution.

- *Input data*: An efficient representation of the genealogical structure is a cell numbering according to the following rule: For a mother cell i, label its two daughter cells as $2i$ and $2i + 1$. Using this numbering, it is easy to calculate a cell's generation as well as the topological distance r_{ij} between a pair of cells i and j, i.e., the number of divisions that separate those events within the genealogy (Fig. 4a):

$$\text{generation}(i) = \lfloor \log_2 i \rfloor$$
$$r_{ij} = \lfloor \log_2 i \rfloor + \lfloor \log_2 j \rfloor - 2\lfloor \log_2 c(i,j) \rfloor + \begin{cases} 1 & \textit{if } c(i,j) = i \text{ or } j \\ 0 & \text{otherwise} \end{cases}, \tag{7}$$

with $c(i,j)$ being the last common ancestor of cells i and j, which is calculated by iteratively determining the ancestors of both cells. $\lfloor \cdot \rfloor$ here denotes rounding to the nearest lower integer.

- *Quantify genealogical correlations*: Depending on the outcome of interest, i.e., a discrete event or a continuous feature, two different similarity measures can be defined. Both are based on the topological distance r_{ij} between a pair of cells i and j.
  - *Discrete events*, e.g., cell death or onset of markers: For every event i in the genealogy, the minimal distance to the closest other event is determined. This is averaged over the set of all events E:

$$S = \frac{1}{|E|} \sum_{i \in E} \min r_{ij} \tag{8}$$

  - *Continuous quantities*, e.g., motility or cell cycle length: For every pair of cells, the absolute difference of the feature of interest m is determined and weighted according to their topological distance. This is averaged over the number of pairs:

$$S = \frac{1}{n^2} \sum_{i,j} \frac{|m_i - m_j|}{2^{r_{ij}}} \tag{9}$$

To especially estimate the range of correlation structures, the measure can be restricted to pairs of cells that have the same topological distance k:

$$S^{(k)} = \frac{1}{n^2} \sum_{r_{ij}=k} |m_i - m_j| \tag{10}$$

- *Null model for randomly expected distribution*: The defined similarity measures are only meaningful relative to a "neutral" control, i.e., considering what would be expected without any correlation. To determine this so-called null distribution, a permutation procedure is suggested that randomizes the correlation structure. There are two general ways of randomization: either (a) by reassigning the attribution of mother and daughter cells or (b) by reassigning the feature of interest. The randomization procedure is repeated a sufficiently large number of times (e.g., 100,000 times), each time calculating the chosen correlation measure. This permutation procedure then gives the (empirical) null distribution (Fig. 4b).
  - *Randomize mother-daughter attribution*: For each generation, the assignment of cells to mother cells of the previous generation is randomized. This randomization changes the genealogical topology and thus makes it necessary to

recalculate topological distances of cell pairs for each permutation step (*see* **Note 17**).

- *Randomize feature of interest*: Using this randomization procedure, the topology of the genealogy is not changed, but the feature of interest, e.g., cell motility, is reassigned to the cells. This reassignment should be restricted to cell of the same generation to avoid intermingling with general temporal effects, i.e., changes of the feature over time that are independent of genealogical correlations.

- *Compare similarity measure to null distribution*: The actual value of the similarity measure S is now compared to this null distribution. An empirical p-value can be computed to decide whether there is a genealogical correlation of the feature of interest. This p-value is defined as the fraction of values of the similarity measure generated under the null hypothesis that are less or equal to the actual value (Fig. 4b). A large p-value would indicate that the actual value is within the range of values that are expected (i.e., rather likely) without any correlation (Fig. 4b, right genealogy), while a small p-value indicates a low probability to observe such a genealogy without any correlation, i.e., suggesting a clustering of the feature of interest that is beyond the randomly expected (Fig. 4b, left genealogy). Formal testing can be done by comparing the empirical p-value with a predefined significance level.

  Restricting the measure to cell pairs that have the same topological distance k (*see* Eq. 9) and repeating the analysis for a range of values of k will provide information about how far-reaching correlations are.

2.5.1 Further Reading and Software

The statistical approach described above had already been used to disentangle differentiation process in a hematopoietic cell line [46]. An approach to cluster cell lineage trees according to common patterns and defining centroid trees that represent the characteristic patterns can be found at [47]. Another statistical method to discriminate distinct groups of genealogies is described in [48]. Branching process models can be used to estimate differentiation rates from genealogical data [49, 50]. A method to infer and discriminate different spatiotemporal effects on cell state transitions is described in [51]. In [52] the authors provide a way to fit stochastic models to lineage trees. To our knowledge, there is no publicly available software implementing methods to analyze cellular genealogies, unfortunately.

2.6 Cell-Based Modeling

The above (data-driven) statistical modeling provides a framework to analyze (image-based) experimental data in order to reveal patterns, regularities, and interactions in distribution, motility, shapes, and/or genealogies that indicate the presence of regulatory mechanisms. However, statistical modeling does not necessarily provide enough information to pinpoint what mechanism may be responsible for the observed regularities. At this point, hypothesis-driven mathematical modeling can be very useful. Such approaches provide formal mathematical or computational frameworks in which one can formulate hypotheses on the suspected responsible mechanisms in mathematical expressions that can be simulated in a computer. This process forces one (1) to make all assumptions explicit, (2) to formulate the proposed mechanisms in an unambiguous fashion, and (3) to provide a complete framework including, e.g., the presence of unknown factors (as free parameters) and stochastic processes (as random variables). Such models are *generative* in the sense that they generate artificial data, based on a set of hypotheses and corresponding assumptions. A consistent model will include parameters that are biologically meaningful, interpretable, and, under appropriate parameter choices, be able to generate the same patterns or regularities as observed in the experimental system, under normal conditions as well as perturbations. However, even in such an ideal case, i.e., having identified a model that is consistent with observed data, one must interpret results with care. In principle, from such a consistency, one can only conclude that the model's assumptions on regulatory mechanisms are *sufficient* to generate the observed pattern, but one cannot necessarily exclude other mechanisms of regulation. In contrast, if a model description does not allow to consistently describe/explain the data under consideration, one can exclude this set of hypothesis/assumptions as potential explanations (i.e., *falsification* strategy).

One can distinguish two general approaches to mathematical and computational modeling of biological populations: cellular populations are either described as continuous or as discrete quantities. *Continuous models*, implemented, e.g., by ordinary or partial differential equations, have the advantage that they, when formulated in simple terms, allow rigorous mathematical analysis that can reveal general relationships between processes. However, continuous models are generally not well-suited to capture heterogeneity in populations or to predict the behavior of small populations where stochastic effects might play a dominant role. Therefore, many modeling studies, in particular those concerning stem cell behavior (e.g., in the intestinal crypt or in the bone marrow niches), have used *discrete models* instead. In a discrete approach, cells are modeled as individual objects with certain properties and a set of (local) interaction rules, using so-called *agent-based* or, more specifically, *cell-based models* [53]. Since these models are often too complex to allow rigorous mathematical analysis, they are typically simulated by a computer and are hence often denoted as *computational models*. A number of such models have been proposed in the area of hematopoietic stem cell organization [54, 55], but these did not explicitly account for spatial aspects including spatial distribution or cell migration.

For spatial cell-based modeling that allows these aspects to be explicitly represented and studied, several well-established theoretical frameworks exist. The most common methods include *cellular automata* (CA) models [56], *cellular Potts models* (CPM) [57], *center-based models* (CBM) [58], *vertex models* (VM) [59, 60], and *subcellular element models* (SEM) [61], ordered by increasing complexity. Within each of these frameworks, it is possible to simulate self-organization in tissues and heterogeneous populations with cellular processes such as motility, interactions, division, and death [62]. However, there are also notable differences. One key difference is in how cells are represented spatially. Whereas some (*on-lattice*) methods use a discretized space (CA, CPM), others (CBM, VM, SEM) use continuous space (*off-lattice*). Some methods treat cells as point particles without explicit volume (CA, CBM), while others model cells with explicit shapes (CPM, VM, SEM). Another key difference is in how *model dynamics* are generated. While CA models are rule-based, CPM and some VM models formulate an energy function that is minimized, and in other methods, dynamics are generated by explicit calculation of forces. Here, we describe two of these methods that represent a cross-section of these difference: the center-based model (off-lattice, particle-like, force-based dynamics; Fig. 5a) and the cellular Potts model (on-lattice, explicit cell shape, energy-based dynamics; Fig. 5b).

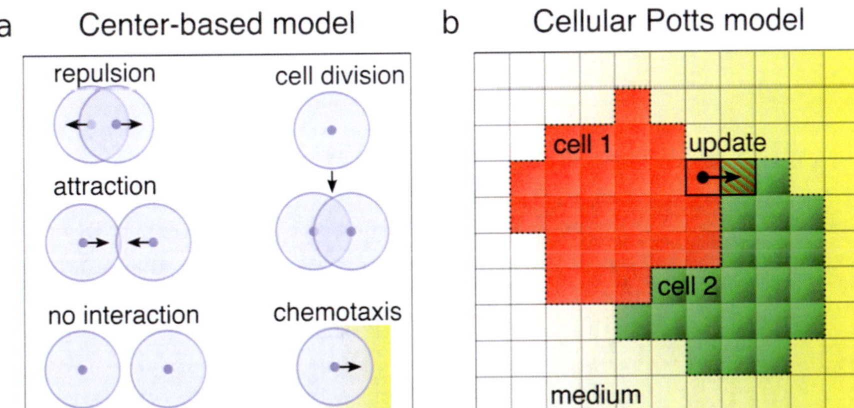

Fig. 5 Cell-based modeling. (**a**) In the center-based model, cells are modeled by their centers in continuous space, and forces are applied according to the distances from their neighbors: a repelling force when cells are too close (volume exclusion) and an attractive force when cells are in intermediate distance (cell-cell adhesion). Cells can also divide by placing daughter cells according to a random unit vector and can respond to external fields of chemokine gradients. (**b**) In the cellular Potts model, cells are modeled on as domains on a lattice and are governed by an energy function with terms to constraint volume and perimeter as well as cell-cell surface lengths. By randomly attempting to copy states between neighboring lattice sites, the cells change shape and move in order to minimize the energy

- *Center-based model (CBM)*: In center-based models (following the notation in [62]), cells are represented as particle-like objects defined by their centers (*see* Fig. 5a), modeled as a set of points $\{\mathbf{x}_1, \ldots, \mathbf{x}_N\}$, that are free to move in space. Each cell has a radius R, and two cells are assumed to be neighbors if their centers are close $\|\mathbf{x}_i - \mathbf{x}_j\| = \|\mathbf{r}_{ij}(t)\| < r_{\max}$ where $r_{\max}$ is the interaction radius. We can then specify a *repelling force* between cells if they are closer than a natural separation distance $\|\mathbf{r}_{ij}(t)\| < s_{ij}$ to model volume exclusion and an *attractive force* if the distance between cells is between the natural separation and interaction radius $s_{ij} \leq \|\mathbf{r}_{ij}(t)\| \leq r_{\max}$ to model cell-cell adhesion and *no interaction force* in case the distance between cells is larger than the interaction radius $\|\mathbf{r}_{ij}(t)\| > r_{\max}$:

$$\mathbf{F}_{ij}(t) = \begin{cases} \mu_{ij} s_{ij} \hat{\mathbf{r}}_{ij}(t) \log\left(1 + \frac{\|\mathbf{r}_{ij}(t)\| - s_{ij}}{s_{ij}}\right), & \text{for } \|\mathbf{r}_{ij}(t)\| < s_{ij} \\ \mu_{ij}\left(\|\mathbf{r}_{ij}(t)\| - s_{ij}\right)\hat{\mathbf{r}}_{ij}(t) e^{-k_c \frac{\|\mathbf{r}_{ij}(t)\| - s_{ij}}{s_{ij}}}, & \text{for } s_{ij} \leq \|\mathbf{r}_{ij}(t)\| \leq r_{\max} \\ 0, & \text{for } \|\mathbf{r}_{ij}(t)\| > r_{\max} \end{cases} \tag{11}$$

where μ_{ij} is a spring constant controlling the size of the force, depending on the interacting cell types, $\hat{\mathbf{r}}_{ij}$ is the unit vector giving the force direction, and k_c is a parameter controlling how the attractive force decays with distance between centers.

To compute the dynamics for this model, for simplicity, we assume all cells have identical mechanical properties and use force balance to calculate the model equation. The new position of each cell can then be calculated by the sum of the forces acting on the cell:

$$\mathbf{x}_i(t + \Delta t) = \mathbf{x}_i + \frac{\Delta t}{\eta} \sum_{j \in \mathcal{N}_i(t)} \mathbf{F}_{ij}(t) \tag{12}$$

where $\mathbf{F}_{ij}(t)$ is the current force between cell i and j, $\mathcal{N}_i(t)$ is the set of neighboring cells of cell i, η is the damping constant, and Δt is the time step which must be selected appropriately small to guarantee numerical stability.

- *Cellular Potts model (CPM)*: In the cellular Potts model, each cell is spatially represented by a domain on a lattice (Fig. 5b). Instead of calculating forces directly, forces are encoded implicitly by an energy function that provides constraints for cell shape and motility. In particular, the energy function consists of multiple terms that represent the interaction energy, accounting for adhesive interactions between cell types τ_i and τ_j, and the area and perimeter constraints, which penalize deviations of the actual cell area a and perimeter p from a target area A_T and target perimeter constraint P_T, respectively:

$$E = \sum_{\text{interfaces } i,j} J_{\tau(\sigma_i),\tau(\sigma_j)} + \sum_{\text{cells},\sigma>0} \lambda_A (a - A_T)^2 + \sum_{\text{cells},\sigma>0} \lambda_P (p - P_T)^2 \quad (13)$$

where the λ terms act as weights on the different terms.

Motility is modeled by random sampling of the lattice (modified Metropolis algorithm), whereby, for every sampled lattice point $\boldsymbol{x}_S$, we evaluate the change in energy ΔE *if* we would copy the state of the lattice to a randomly sampled neighboring site $\boldsymbol{x}_T$. The probability of actually performing the update depends on the change of energy:

$$P(\Delta E_{\mathbf{x}_S \rightarrow \mathbf{x}_T}) = \begin{cases} 1 & \text{for } \Delta E \leq 0 \\ e^{-\frac{\Delta E}{T}} & \text{otherwise} \end{cases} \quad (14)$$

where proposed updates that decrease energy E are always accepted, while updates that increase energy are only accepted with a (Boltzmann) probability that decreases with the energy difference. The parameter T can be used to modulate this probability decrease and models the amount of allowed membrane fluctuations. Effects of external signals such as chemoattractant can be incorporated into the CPM by adding a chemotaxis energy:

$$\Delta E_{\mathbf{x}_S \rightarrow \mathbf{x}_T} = \lambda_C (c_{\mathbf{x}_T} - c_{\mathbf{x}_S}) \quad (15)$$

Both cell-based models described above are generic frameworks that can be used to simulate the dynamics of multicellular systems such as the behavior of hematopoietic stem cells in vitro or in vivo. Basic cellular behavioral mechanisms such as cell motility, chemotaxis, and cell division can be simulated in isolation or in combination to investigate the type and variety of emergent patterns these give rise to. In particular, the results of the statistical modeling described above, formulated as a set of assumptions and parameters on cellular behavior and interactions, can be formalized and simulated under different sets of conditions, ceteris paribus. Moreover, by recording the spatial locations, cell trajectories, cell shape parameters, and divisional history of the simulated cells, one can perform the same statistical analysis on the simulated data and on the experimental data, thus allowing for a direct quantitative comparison and, therefore, checking for consistency.

There are several subtle but key differences in the manner in which basic cellular mechanisms such as motility, adhesion, chemotaxis, shape changes, and cell division are represented in the CBM and the CPM. For instance, some strengths of the CBM are the ability to simulate large populations and the modeling long-range mechanical effects. However, this comes with the cost of implicit cell shape and less sensitive cell-cell adhesion which renders it less suitable to study the effects of, e.g., cell sorting [62]. Conversely, the computational costs of the CPM make the simulation of large-scale cellular populations expensive. However, the explicit representation of cell shapes and the cell-cell contacts in CPM makes it highly suitable to model tissue dynamics through cell surface mechanics [63].

2.6.1 Further Reading and Software

A detailed review of the full range of cell-based modeling formalisms is presented in [64]; *see* [65] for a more general overview. An interesting comparison between various cell-based modeling formalisms, according to a number of common use cases, is presented in [62].

There are a growing number of dedicated software tools available for cell-based modeling, including several that provide implementations for the cell-based models described in this section. Some platforms such as *PhysiCell* [66] focus on scalability to simulate large-scale populations. Other software tools, most notably *Chaste* [67], are designed for flexibility by providing implementations of various modeling framework within a unified framework, including the ones described in this section. Whereas these software tools typically require substantial programming expertise, the modeling environment *Morpheus* [68] is designed for usability and allows also non-experts to construct multicellular models using a graphical user interface.

3 Notes

1. The locations of cells are, strictly speaking, not independent because cells have a finite volume and cannot overlap (volume exclusion). Therefore, cells will always be more dispersed at short range than expected by CSR.
2. Due to volume exclusion, cellular point patterns will always show dispersion for small radii.
3. In more complex situations, it is possible that the empirical function crosses the theoretical one, e.g., when there is short-range dispersion and long-range clustering.
4. Local (i.e., pointwise) envelopes are calculated for every radius separately, by considering the maximum and minimum value of the simulated K-function. The global envelop is calculated by taking the maximum deviation of all local envelopes, i.e., $L(r) = K_{\mathrm{CSR}} - \max_r |K(r) - K_{\mathrm{CSR}}|$ and $R(r) = K_{\mathrm{CSR}} + \max_r |K(r) - K_{\mathrm{CSR}}|$where $K_{\mathrm{CSR}} = \pi r^2$ in 2D and $K_{\mathrm{CSR}} = \frac{4}{3}\pi r^3$ in 3D situations.

5. The time intervals, the time between individual frames, should be identical over the whole trajectory and between trajectories that are being compared.
6. The length of the trajectories, in terms of the number of time intervals, should be as long as possible. Short trajectories may be ignored altogether.
7. Time can be defined in two different ways: When defining time relatively to each trajectory's start, one can detect potential dependencies of the velocity on the cell cycle progress. Conversely, when taking absolute time, one can detect potential temporal dependencies that act globally on all cells in the culture.
8. In case an initial transient is observed followed by constant behavior, one can consider discarding the initial transient time points and limit the analysis for the experiment, excluding the initial transient.
9. Functions for linear regression include lm in *R* and scipy.stats.linregress in *python*.
10. Typically, cellular trajectories are relatively short. To augment data, one can use overlapping time intervals. For example, $\tau = 3$ one takes the average over overlapping times $\{t_0, t_2\}$, $\{t_1, t_3\}$, $\{t_2, t_4\}$, etc. instead of the average over $\{t_0, t_2\}$, $\{t_3, t_5\}$, $\{t_6, t_8\}$, etc.
11. Functions for nonlinear regression include nls in *R* and scipy.optimize.curve_fit in *python*.
12. For large time intervals, only limited number of samples are averaged over and are therefore unreliable.
13. Analytically derived confidence intervals based on student's *t*-value are based on the assumption that the data is normal-distributed, which might not be guaranteed. Determining confidence regions by repeated simulation avoids this issue.
14. Most implementations provide axis-aligned bounding boxes, where the bounding box is oriented according to the *x*- and *y*-axes of the image instead of the object itself. This implies the statistics are not rotation invariant (diagonal objects have larger bounding box), and results depend on the arbitrary orientation in which they have been imaged. Thus, these properties should be handled with care.
15. PCA is sensitive to the scales of the feature. Therefore, a preprocessing step is necessary to standardize the features. This can be done by subtracting the mean (zero-centering) and dividing by the standard deviation for each column.
16. Inferring correlation structures from cellular genealogies requires highly accurate cell tracking over several generations

that so far cannot be achieved by automated tracking methods and thus needs manually created or corrected tracking data.

17. Due to recalculating topological distances, this randomization procedure is computationally expensive. It has to be used, e.g., when studying cell death events, since a pure reassigning of death status would lead to genealogies with dead cells having progeny.

Acknowledgments

The work presented in this paper is supported by Deutsche Krebshilfe (SyTASC grant number 70111969) and the German Ministry of Education and Research (BMBF) (HaematoOPT grant number 031A424).

References

1. Krause DS, Scadden DT (2015) A hostel for the hostile: the bone marrow niche in hematologic neoplasms. Haematologica 100 (11):1376–1387
2. Krinner A, Roeder I (2014) Quantification and modeling of stem cell–niche interaction. In: A systems biology approach to blood. Springer, pp 11–36
3. Nombela-Arrieta C, Manz MG (2017) Quantification and three-dimensional microanatomical organization of the bone marrow. Blood Adv 1(6):407–416
4. Acar M, Kocherlakota KS, Murphy MM, Peyer JG, Oguro H, Inra CN, Jaiyeola C, Zhao Z, Luby-Phelps K, Morrison SJ (2015) Deep imaging of bone marrow shows non-dividing stem cells are mainly perisinusoidal. Nature 526(7571):126–130
5. Etzrodt M, Endele M, Schroeder T (2014) Quantitative single-cell approaches to stem cell research. Cell Stem Cell 15(5):546–558
6. Schroeder T (2011) Long-term single-cell imaging of mammalian stem cells. Nat Methods 8(4s):S30
7. Skylaki S, Hilsenbeck O, Schroeder T (2016) Challenges in long-term imaging and quantification of single-cell dynamics. Nat Biotechnol 34(11):1137–1144
8. Foster K, Lassailly F, Anjos-Afonso F, Currie E, Rouault-Pierre K, Bonnet D (2015) Different motile behaviors of human hematopoietic stem versus progenitor cells at the osteoblastic niche. Stem Cell Rep 5(5):690–701
9. Kim S, Lin L, Brown GA, Hosaka K, Scott EW (2017) Extended time-lapse in vivo imaging of tibia bone marrow to visualize dynamic hematopoietic stem cell engraftment. Leukemia 31 (7):1582–1592
10. Lo Celso C, Lin CP, Scadden DT (2011) In vivo imaging of transplanted hematopoietic stem and progenitor cells in mouse calvarium bone marrow. Nat Protoc 6(1):1–14
11. MacLean AL, Smith MA, Liepe J, Sim A, Khorshed R, Rashidi NM, Scherf N, Krinner A, Roeder I, Lo Celso C (2017) Single Cell Phenotyping Reveals Heterogeneity Among Hematopoietic Stem Cells Following Infection. Stem Cells 35(11):2292–2304
12. Hilsenbeck O, Schwarzfischer M, Skylaki S, Schauberger B, Hoppe PS, Loeffler D, Kokkaliaris KD, Hastreiter S, Skylaki E, Filipczyk A, Strasser M, Buggenthin F, Feigelman JS, Krumsiek J, van den Berg AJ, Endele M, Etzrodt M, Marr C, Theis FJ, Schroeder T (2016) Software tools for single-cell tracking and quantification of cellular and molecular properties. Nat Biotechnol 34(7):703–706
13. Hilsenbeck O, Schwarzfischer M, Loeffler D, Dimopoulos S, Hastreiter S, Marr C, Theis FJ, Schroeder T (2017) fastER: a user-friendly tool for ultrafast and robust cell segmentation in large-scale microscopy. Bioinformatics 33 (13):2020–2028
14. Molnar C, Jermyn IH, Kato Z, Rahkama V, Östling P, Mikkonen P, Pietiäinen V, Horvath P (2016) Accurate morphology preserving segmentation of overlapping cells based on active contours. Sci Rep 6:32412
15. Sommer C, Straehle C, Koethe U, Hamprecht FA (2011) Ilastik: interactive learning and

segmentation toolkit. In: Biomedical imaging: from nano to macro, 2011 IEEE International Symposium on, 2011. IEEE, pp 230–233

16. Pelt DM, Sethian JA (2018) A mixed-scale dense convolutional neural network for image analysis. Proc Natl Acad Sci U S A 115 (2):254–259
17. Ronneberger O, Fischer P, Brox T (2015) U-net: convolutional networks for biomedical image segmentation. In: International Conference on Medical image computing and computer-assisted intervention. Springer, pp 234–241
18. Meijering E (2012) Cell segmentation: 50 years down the road [life sciences]. IEEE Signal Process Mag 29(5):140–145
19. Kan A (2017) Machine learning applications in cell image analysis. Immunol Cell Biol 95 (6):525–530
20. Caicedo JC, Cooper S, Heigwer F, Warchal S, Qiu P, Molnar C, Vasilevich AS, Barry JD, Bansal HS, Kraus O (2017) Data-analysis strategies for image-based cell profiling. Nat Methods 14(9):849–863
21. Schindelin J, Arganda-Carreras I, Frise E, Kaynig V, Longair M, Pietzsch T, Preibisch S, Rueden C, Saalfeld S, Schmid B (2012) Fiji: an open-source platform for biological-image analysis. Nat Methods 9(7):676–682
22. Carpenter AE, Jones TR, Lamprecht MR, Clarke C, Kang IH, Friman O, Guertin DA, Chang JH, Lindquist RA, Moffat J (2006) CellProfiler: image analysis software for identifying and quantifying cell phenotypes. Genome Biol 7(10):R100
23. Held M, Schmitz MH, Fischer B, Walter T, Neumann B, Olma MH, Peter M, Ellenberg J, Gerlich DW (2010) CellCognition: time-resolved phenotype annotation in high-throughput live cell imaging. Nat Methods 7(9):747–754
24. Tinevez J-Y, Perry N, Schindelin J, Hoopes GM, Reynolds GD, Laplantine E, Bednarek SY, Shorte SL, Eliceiri KW (2017) TrackMate: an open and extensible platform for single-particle tracking. Methods 115:80–90
25. Wiesmann V, Franz D, Held C, Münzenmayer C, Palmisano R, Wittenberg T (2015) Review of free software tools for image analysis of fluorescence cell micrographs. J Microsc 257(1):39–53
26. Nilsson SK, Johnston HM, Coverdale JA (2001) Spatial localization of transplanted hemopoietic stem cells: inferences for the localization of stem cell niches. Blood 97 (8):2293–2299
27. Gomariz A, Helbling PM, Isringhausen S, Suessbier U, Becker A, Boss A, Nagasawa T, Paul G, Goksel O, Székely G, Stoma S (2018) Quantitative spatial analysis of haematopoiesis-regulating stromal cells in the bone marrow microenvironment by 3D microscopy. Nature communications 9(1):2532.
28. Ripley BD (1976) The second-order analysis of stationary point processes. J Appl Probab 13 (2):255–266
29. Baddeley A (1999) Spatial sampling and censoring. In: Barndorff-Nielsen O, Kendall W, van Lieshout H (eds) Stochastic geometry: likelihood and computation. Chapman and Hall, London, pp 37–78
30. Baddeley A, Rubak E, Turner R (2015) Spatial point patterns: methodology and applications with R. CRC Press, Boca Raton
31. Cressie N (2015) Statistics for spatial data. Wiley, New York
32. Gelfand AE, Diggle P, Guttorp P, Fuentes M (2010) Handbook of spatial statistics. CRC Press, Boca Raton
33. Tranquillo RT, Lauffenburger DA, Zigmond S (1988) A stochastic model for leukocyte random motility and chemotaxis based on receptor binding fluctuations. J Cell Biol 106 (2):303–309
34. Wu P-H, Giri A, Sun SX, Wirtz D (2014) Three-dimensional cell migration does not follow a random walk. Proc Natl Acad Sci 111 (11):3949–3954
35. Luzhanskey ID, MacMunn JP, Cohen JD, Barney LE, Jansen LE, Schwartz AD, Peyton S (2017) Anomalous diffusion as a descriptive model of cell migration. bioRxiv:236356
36. Gorelik R, Gautreau A (2014) Quantitative and unbiased analysis of directional persistence in cell migration. Nat Protoc 9(8):1931–1943
37. Wu PH, Giri A, Wirtz D (2015) Statistical analysis of cell migration in 3D using the anisotropic persistent random walk model. Nat Protoc 10(3):517–527
38. Dieterich P, Klages R, Preuss R, Schwab A (2008) Anomalous dynamics of cell migration. Proc Natl Acad Sci 105(2):459–463
39. Makarava N, Menz S, Theves M, Huisinga W, Beta C, Holschneider M (2014) Quantifying the degree of persistence in random amoeboid motion based on the Hurst exponent of fractional Brownian motion. Phys Rev E 90 (4):042703
40. Masuzzo P, Van Troys M, Ampe C, Martens L (2016) Taking aim at moving targets in

computational cell migration. Trends Cell Biol 26(2):88–110

41. Sánchez-Corrales YE, Hartley M, van Rooij J, Marée AF, Grieneisen VA (2018) Morphometrics of complex cell shapes: lobe contribution elliptic Fourier analysis (LOCO-EFA). Development. pii: dev156778
42. Pincus Z, Theriot J (2007) Comparison of quantitative methods for cell-shape analysis. J Microsc 227(2):140–156
43. Driscoll MK, McCann C, Kopace R, Homan T, Fourkas JT, Parent C, Losert W (2012) Cell shape dynamics: from waves to migration. PLoS Comput Biol 8(3):e1002392
44. Gordonov S, Hwang MK, Wells A, Gertler FB, Lauffenburger DA, Bathe M (2016) Time series modeling of live-cell shape dynamics for image-based phenotypic profiling. Integr Biol 8(1):73–90
45. Glauche I, Lorenz R, Hasenclever D, Roeder I (2009) A novel view on stem cell development: analysing the shape of cellular genealogies. Cell Prolif 42(2):248–263
46. Bach E, Zerjatke T, Herklotz M, Scherf N, Niederwieser D, Roeder I, Pompe T, Cross M, Glauche I (2014) Elucidating functional heterogeneity in hematopoietic progenitor cells: a combined experimental and modeling approach. Exp Hematol 42 (9):826–837 e821–817
47. Khakhutskyy V, Schwarzfischer M, Hubig N, Plant C, Marr C, Rieger MA, Schroeder T, Theis FJ (2014) Centroid clustering of cellular lineage trees. In: International conference on information technology in bio-and medical informatics. Springer, pp 15–29
48. Stadler T, Skylaki S, DK K, Schroeder T (2018) On the statistical analysis of single cell lineage trees. J Theor Biol 439:160–165
49. Marr C, Strasser M, Schwarzfischer M, Schroeder T, Theis FJ (2012) Multi-scale modeling of GMP differentiation based on single-cell genealogies. FEBS J 279(18):3488–3500
50. Nordon RE, Ko K-H, Odell R, Schroeder T (2011) Multi-type branching models to describe cell differentiation programs. J Theor Biol 277(1):7–18
51. Strasser MK, Feigelman J, Theis FJ, Marr C (2015) Inference of spatiotemporal effects on cellular state transitions from time-lapse microscopy. BMC Syst Biol 9(1):61
52. Feigelman J, Ganscha S, Hastreiter S, Schwarzfischer M, Filipczyk A, Schroeder T, Theis FJ, Marr C, Claassen M (2016) Analysis of cell lineage trees by exact Bayesian inference identifies negative autoregulation of Nanog in mouse embryonic stem cells. Cell Sys 3 (5):480–490.e413
53. d'Inverno M, Luck M, Luck MM (2004) Understanding agent systems. Springer, Berlin
54. Krinner A, Roeder I, Loeffler M, Scholz M (2013) Merging concepts-coupling an agent-based model of hematopoietic stem cells with an ODE model of granulopoiesis. BMC Syst Biol 7(1):117
55. Roeder I, Horn M, Glauche I, Hochhaus A, Mueller MC, Loeffler M (2006) Dynamic modeling of imatinib-treated chronic myeloid leukemia: functional insights and clinical implications. Nat Med 12(10):1181–1184
56. Deutsch A, Dormann S (2007) Cellular automaton modeling of biological pattern formation: characterization, applications, and analysis. Springer, Berlin
57. Graner F, Glazier JA (1992) Simulation of biological cell sorting using a two-dimensional extended Potts model. Phys Rev Lett 69(13):2013–2016
58. Drasdo D (2007) Center-based single-cell models: an approach to multi-cellular organization based on a conceptual analogy to colloidal particles. In: Single-cell-based models in biology and medicine. Springer, pp 171–196
59. Alt S, Ganguly P, Salbreux G (2017) Vertex models: from cell mechanics to tissue morphogenesis. Phil Trans R Soc B 372 (1720):20150520
60. Fletcher AG, Osterfield M, Baker RE, Shvartsman SY (2014) Vertex models of epithelial morphogenesis. Biophys J 106 (11):2291–2304
61. Sandersius SA, Newman TJ (2008) Modeling cell rheology with the subcellular element model. Phys Biol 5(1):015002
62. Osborne JM, Fletcher AG, Pitt-Francis JM, Maini PK, Gavaghan DJ (2017) Comparing individual-based approaches to modelling the self-organization of multicellular tissues. PLoS Comput Biol 13(2):e1005387
63. Magno R, Grieneisen VA, Marée AF (2015) The biophysical nature of cells: potential cell behaviours revealed by analytical and computational studies of cell surface mechanics. BMC Biophys 8(1):8
64. Van Liedekerke P, Palm M, Jagiella N, Drasdo D (2015) Simulating tissue mechanics with agent-based models: concepts, perspectives and some novel results. Comput Part Mech 2 (4):401–444
65. Tanaka S (2015) Simulation frameworks for morphogenetic problems. Computation 3 (2):197–221

66. Ghaffarizadeh A, Heiland R, Friedman SH, Mumenthaler SM, Macklin P (2018) PhysiCell: an open source physics-based cell simulator for 3-D multicellular systems. PLoS Comput Biol 14(2):e1005991
67. Mirams GR, Arthurs CJ, Bernabeu MO, Bordas R, Cooper J, Corrias A, Davit Y, Dunn S-J, Fletcher AG, Harvey DG (2013) Chaste: an open source C++ library for computational physiology and biology. PLoS Comput Biol 9(3):e1002970
68. Starruß J, de Back W, Brusch L, Deutsch A (2014) Morpheus: a user-friendly modeling environment for multiscale and multicellular systems biology. Bioinformatics 30 (9):1331–1332

INDEX

Gerd Klein and Patrick Wuchter (eds.), *Stem Cell Mobilization: Methods and Protocols*, Methods in Molecular Biology, vol. 2017, https://doi.org/10.1007/978-1-4939-9574-5, © Springer Science+Business Media, LLC, part of Springer Nature 2019

D

E

F

G

H

I

L

M

N

O

P

R

S

Zeitfracht Medien GmbH
Ferdinand-Jühlke-Straße 7
99095 Erfurt, Deutschland
produktsicherheit@kolibri360.de